SpringerBriefs in Physics

Series Editors

Balasubramanian Ananthanarayan, Centre for High Energy Physics, Indian Institute of Science, Bangalore, India

Egor Babaev, Department of Physics, Royal Institute of Technology, Stockholm, Sweden

Malcolm Bremer, H. H. Wills Physics Laboratory, University of Bristol, Bristol, UK

Xavier Calmet, Department of Physics and Astronomy, University of Sussex, Brighton, UK

Francesca Di Lodovico, Department of Physics, Queen Mary University of London, London, UK

Pablo D. Esquinazi, Institute for Experimental Physics II, University of Leipzig, Leipzig, Germany

Maarten Hoogerland, University of Auckland, Auckland, New Zealand

Eric Le Ru, School of Chemical and Physical Sciences, Victoria University of Wellington, Kelburn, New Zealand

Dario Narducci, University of Milano-Bicocca, Milan, Italy

James Overduin, Towson University, Towson, USA

Vesselin Petkov, Montreal, Canada

Stefan Theisen, Max-Planck-Institut für Gravitationsphysik, Golm, Germany

Charles H. T. Wang, Department of Physics, University of Aberdeen, Aberdeen, UK

James D. Wells, Department of Physics, University of Michigan, Ann Arbor, MI, USA

Andrew Whitaker, Department of Physics and Astronomy, Queen's University Belfast, Belfast, UK

SpringerBriefs in Physics are a series of slim high-quality publications encompassing the entire spectrum of physics. Manuscripts for SpringerBriefs in Physics will be evaluated by Springer and by members of the Editorial Board. Proposals and other communication should be sent to your Publishing Editors at Springer.

Featuring compact volumes of 50 to 125 pages (approximately 20,000–45,000 words), Briefs are shorter than a conventional book but longer than a journal article. Thus, Briefs serve as timely, concise tools for students, researchers, and professionals.

Typical texts for publication might include:

- A snapshot review of the current state of a hot or emerging field
- A concise introduction to core concepts that students must understand in order to make independent contributions
- An extended research report giving more details and discussion than is possible in a conventional journal article
- A manual describing underlying principles and best practices for an experimental technique
- An essay exploring new ideas within physics, related philosophical issues, or broader topics such as science and society

Briefs allow authors to present their ideas and readers to absorb them with minimal time investment. Briefs will be published as part of Springer's eBook collection, with millions of users worldwide. In addition, they will be available, just like other books, for individual print and electronic purchase. Briefs are characterized by fast, global electronic dissemination, straightforward publishing agreements, easy-to-use manuscript preparation and formatting guidelines, and expedited production schedules. We aim for publication 8–12 weeks after acceptance.

Vittorio Canuto

Langevin Stochastic Equations: Treatment of Ocean, Planetary Boundary Layer, and Stellar Turbulence

Vittorio Canuto
NASA Goddard Institute for Space Studies
Columbia University
New York, NY, USA

ISSN 2191-5423 ISSN 2191-5431 (electronic)
SpringerBriefs in Physics
ISBN 978-3-031-86543-5 ISBN 978-3-031-86544-2 (eBook)
https://doi.org/10.1007/978-3-031-86544-2

© The Editor(s) (if applicable) and The Author(s), under exclusive license to Springer Nature Switzerland AG 2025

This work is subject to copyright. All rights are solely and exclusively licensed by the Publisher, whether the whole or part of the material is concerned, specifically the rights of translation, reprinting, reuse of illustrations, recitation, broadcasting, reproduction on microfilms or in any other physical way, and transmission or information storage and retrieval, electronic adaptation, computer software, or by similar or dissimilar methodology now known or hereafter developed.
The use of general descriptive names, registered names, trademarks, service marks, etc. in this publication does not imply, even in the absence of a specific statement, that such names are exempt from the relevant protective laws and regulations and therefore free for general use.
The publisher, the authors and the editors are safe to assume that the advice and information in this book are believed to be true and accurate at the date of publication. Neither the publisher nor the authors or the editors give a warranty, expressed or implied, with respect to the material contained herein or for any errors or omissions that may have been made. The publisher remains neutral with regard to jurisdictional claims in published maps and institutional affiliations.

This Springer imprint is published by the registered company Springer Nature Switzerland AG
The registered company address is: Gewerbestrasse 11, 6330 Cham, Switzerland

If disposing of this product, please recycle the paper.

This book is dedicated to my granddaughter Aura Sophia Canuto

Preface

This book derives, solves, and assesses *the Langevin stochastic equations (LSE) as a new tool to treat turbulent mixing in the ocean, planetary boundary layer (PBL), and stars.* Previous work demonstrated the LSE's ability to successfully describe non-geophysical turbulent flows. In contrast, this work concentrates on geophysical flows, specifically Chap. 1 treats modeling of oceanic mesoscales (M) and sub-mesoscales (SM), and Chap. 2 treats vertical mixing in ocean, PBL, and stars. This book is meant for advanced students and teachers with interest in future climate change and the role played by the ocean. The key difficulty in describing oceanic M and SM is that they are governed by *nonlinear interactions* for which no satisfactory model exists. Since the use of the time-honored Navier–Stokes equations (NSE) turned out to be unsuccessful in describing non-linearity, the only remaining tools were heuristic models, many of which are still in use today. Their use presents a puzzle: to be reliable, *projections of future climate must be predictive, but their heuristic treatments of M and SM have no predictive power at all.* The goal of this book is to show that the transition from NSE to LSE helps that problem, opening the way to a predictive treatment of M and SM, endowing future climate predictions with the credibility they need.

New York, NY, USA Vittorio M. Canuto
September 2024

Acknowledgements

Climate modeling at GISS was supported by the NASA Modeling, Analysis, and Predictions program, and resources supporting this work were provided by the NASA High-End Computing (HEC) Program through the NASA Center for Climate Simulations (NCCS) at the Goddard Space Flight Center (NASA 509496.02.08.04.24).

Contents

1 **Mesoscales and Sub-Mesoscales**........................... 1
   1.1   From Navier-Stokes to Langevin Equations................ 1
   1.2   Langevin Stochastic Equations, LSE..................... 3
   1.3   Modelling Turbulent Forces........................... 5
   1.4   Predicting Basic Turbulence Laws...................... 8
   1.5   Turbulent Thermal Convection......................... 13
   1.6   Mesoscales. Adiabatic Regime.......................... 17
   1.7   Some Results of the New Model........................ 20
   1.8   Mesoscales. Diabatic-D Regime......................... 24
   1.9   Diffusivity Tensor, Kinetic Energy, Mesoscale Diffusivity....... 27
   1.10 OGCM Results...................................... 35
   1.11 Subduction by Mesoscales............................. 40
   1.12 Upwelling by Mesoscales.............................. 43
   1.13 Sub-Mesoscales, SM.................................. 45
   1.14 SM Contribution to Subduction, Re-Stratification and Eddy
         Compensation....................................... 48
   1.15 SM Contribution to Re-Stratification and Eddy Compensation... 50
   1.16 PV Destruction by SM................................ 52
   1.17 SM Kinetic Energy................................... 55
   1.18 Vertical Velocities................................... 59
   1.19 Sub-Mesoscales and Mesoscale Interaction................. 63
   1.20 Conclusions.. 64
   Appendices.. 65
        Appendix A...................................... 65
        Appendix B: The $p(z)$ Functions in (1.138)............... 66
        Appendix C: Mesosale Vertical Buoyanc Flux. Extent of the
        D-Regime (Figs. 1.36 and 1.37)......................... 67
        Appendix D: Volume and Ensemble Averages............... 68
   References... 70

2	**Vertical Mixing in Oceans, PBL and Stars**	77
2.1	LSE Equations for Buoyancy	77
2.2	Derivation of the RSM Equations from the LSE	80
2.3	Heat and Momentum Turbulent Diffusivities	82
2.4	Shear, Vorticity, Heat and Salt Fluxes	87
2.5	Solutions for the Salt-Fingers Case	91
2.6	Reynolds Stresses	91
2.7	The EKE Equation	92
2.8	Results for the SF Case	92
2.9	Diffusive Convective DC	93
2.10	Passive Tracers	99
2.11	Solar Convection, Rotation, Overshooting and Angular Momentum	105
2.12	Overflows. Gravity Currents. New Turbulence Models	121
2.13	Symmetric Instabilities, SI. The Role of Ro	130
2.14	Deep Convection, DC. New Plume Model	136
2.15	Derivation of the ε Equation from a Two-Point Closure	147
2.16	Planetary Boundary Layer, PBL	151
	Appendices	152
	Appendix A: Derivation of the Structure Function $S_b(Ri)$	152
	Appendix B: SM Momentum, Buoyancy Fluxes and Shear Production	156
	References	158

Acronyms

BBL	Bottom boundary layer
BI	Baroclinic instability
DC	Deep convection
DD	Double diffusion
DIA	Direct interaction approximation
EDQNM	Eddy-damped quasi-normal approximation
LSE	Langevin stochastic equations
MLT	Mixing length theory
NSE	Navier–Stokes equations
OGCM	Ocean global circulation models
PBL	Planetary boundary layer

Chapter 1
Mesoscales and Sub-Mesoscales

Abstract The first goal of this chapter is to address the suggestion to move from NSE to LSE. The second goal is to derive the LSE specifically, modelling of the turbulent forcing acting on an eddy of arbitrary size. The LSE are then shown to predict basic turbulence relations, data from measurements and numerical simulations and thermal convection. Mesoscales in the adiabatic and diabatic regime are discussed together with several results. Mesoscale diffusivity and kinetic energy are derived. Results from an OGCM are presented and compared with data of other groups. Subduction and upwelling by mesoscales are discussed. Next, the text treats Sub-mesoscales, their contributions to subduction, re-stratification and eddy compensation, destruction of potential vorticity, SM eddy kinetic energy, induced vertical velocity and coexistence of meso and sub-mesoscales.

1.1 From Navier-Stokes to Langevin Equations

Turbulence is a ubiquitous phenomenon e.g., in stars, earth's atmosphere and oceans, because is a dynamical transport process several orders of magnitude more efficient than kinematic processes. A reliable description of turbulent mixing is therefore both a necessity and a challenge. The challenge is because turbulence is a macroscopic manifestation of non-linearity. We recall that in a model without non-linearity, instabilities make the kinetic energy grow in time without limit which is unphysical, while in the presence of non-linear interactions, energy is distributed (cascaded is the technical term) among a large number of eddies ranging from the size of the system to the much smaller ones affected by kinematic viscosity. A stationary state can thus be achieved. It is important to stress that turbulence is not a property of a *fluid* but a property of a *flow* that requires energy to be sustained. Without a source of energy, turbulence dies out and the flow becomes laminar. The most well-known attempt to construct a turbulence model using the Navier-Stokes Equations NSE, was DIA (directed interaction approximation) by Kraichnan (1965) who however could not reproduce the well-known $k^{-5/3}$ Kolmogorov law. In the one-point closure models, the difficulty discussed in sec. 3 of Leith (1971) is that the second-order velocity correlation entails third-order moments which in turn entail fourth-order moments giving rise to more unknowns than equations (the closure problem), a sequence that must be terminated at some level. As an example, Ogura (1963) showed that terminating the series at the fourth order assumed to be quasi-normal (Gaussian), led to negative energy spectra, a result attributed to an excessive build-up of third-order moments. Orszag (1970) suggested a pragmatic solution known as the *eddy-damped quasi-normal approximation* EDQNM that introduced eddy-damping terms intended to limit the growth of third-order moments. Phenomenology was thus introduced leading to the conclusion (Leith 1971) that" *unfortunately, a completely satisfactory fundamental theory of turbulence does not yet exist"*. For an instructive account of Leith Atmospheric Model, LAM, see Hamilton (2020). This 50+ years old conclusion is still valid today and heuristic models such as EDQNM are difficult to generalize to the types of turbulence generated by shear instabilities, buoyancy instabilities, etc. that occur in the ocean. Modeling of turbulence processes has reached an impasse that represents a significant impediment considering the large number of mixing ocean processes generated by mesoscales, sub-mesoscales, vertical shear, internal gravity waves, gravity currents etc. To be able to describe all these processes, one must first construct a reliable treatment of non-linearity. Given the known difficulties of past NSE-based approaches, in this work we no longer attempt to describe turbulence using the NSE instead, we use the latter as a template to construct a representation known as the *Langevin Stochastic Equations* LSE:

$$\text{Brownian motion equations} + \text{NSE} \rightarrow \text{LSE} \qquad (1.1)$$

whose most significant features are:

(a) LSE preserve the key feature of turbulence, non-linearity,
(b) while the NSE are non-linear in the velocity and linear in the viscosity terms, the LSE have the opposite structure, they are linear in the velocity and temperature fields and non-linear in the viscosity-diffusivity terms,
(c) they can describe different types of turbulent flows within the same framework,
(d) property b) avoids the *closure problem* and allows analytic solutions, for example, the analytic form of the mesoscale horizontal-vertical tracer fluxes that are then used to study analytically processes such as stratification, subduction and upwelling.

It must be pointed out that transition (1.1) was first suggested by Kraichnan (1971, K71) who wrote *"a model equation of Langevin type for the turbulent velocity is constructed in which the non-linear terms in the NSE are replaced by a dynamical damping term and a random forcing term"*. The key differences with the present model are as follows: (1) K71 employed the DIA version of the NSE whereas we use the original NSE as rearranged by Wyld (1961) to highlight the role of non-linearity, (2) while K71 did not provide a model of the random forcing, we parameterize the work done on an eddy with wavenumber k by both larger and smaller eddies making the model complete. The new approach reproduces key turbulence relations, e.g.,

Kolmogorov and Obukhov spectral laws, Kolmogorov and Batchelor constants, data on shear and buoyancy driven turbulence for a total of 80+ turbulence statistics from numerical simulations and laboratory measurements. On that basis, the LSE are then used to describe *mesoscales M and sub-mesoscales SM*. In the M case, we present a comparison with data from seven ocean global circulation models (OGCMs) with encouraging results. In the SM case, we derive new fluxes for arbitrary tracers needed in biogeochemical studies, new subduction rates, de-stratification effect in the ACC, sizeable reduction of the PV destruction by down-front winds and the first derivation of the algebraic equation for the SM kinetic energy in terms of large-scale variables. The above results represent a concrete example of a unified treatment of turbulent mixing processes. In Part II, we treat 1D vertical mixing models for buoyancy, shear and buoyancy plus shear.

1.2 Langevin Stochastic Equations, LSE

In the spirit of relation (1.1), we begin by considering the classical Brownian motion of a particle under the effect of an external force and a fluctuating one. Both Einstein and Langevin studied the problem using two different approaches reaching the same result, the root-mean-squared displacement of a Brownian particle increases with the square root of time. Einstein solved the Fokker-Planck equation while Langevin who, in his own words, used an *"infinitively simpler approach"*, applied Newton second law to Brownian motion (Langevin, 1908; Lemons 1997) thus creating an analytic approach to random processes that we extend to turbulent flows with the resulting equations known as the *Langevin Stochastic Equations*, LSE. The Newton law describing Brownian motion has the form (sec. 1.3 of Coffey and Kalmykov 2017):

$$mx'' = -\xi x' + F(t) \qquad (1.2)$$

where primes represent time derivatives, the first term on the right-hand side is a friction taken to be the Stokes law and F(t) is a *fluctuating force* with zero average. The extension of (1.2) to turbulence yields the stochastic equation for the velocity $u_i(k)$ of an eddy of wavenumber k:

$$\frac{\partial}{\partial t} u_i(k) = f_i(ext) + f_i^<(turb) + f_i^>(turb) \qquad (1.3)$$

where the Stokes's forcing in (1.2) is now generalized to $f_i(ext)$ representing shear, buoyancy etc., while the fluctuating force F(t) is represented by two turbulent forces, one acting on an eddy of wavenumber k by all eddies *smaller* than k and the other acting on an eddy of size k by all eddies *larger* than it. Combining (1.3) for u_i with the one for u_j^*, using $\partial_t u_i = i\omega u_i$ and introducing the following correlations φ, $\widetilde{\varphi}$:

$$<f_i^{ext}f_j^{ext}> = P_{ij}\phi, \qquad <f_i^t f_j^t> = P_{ij}\widetilde{\phi}, \qquad <f_i^{ext}f_j^t> = 0 \qquad (1.4)$$

where $P_{ij} = \delta_{ij} - k_i k_j/k^2$, together with Wyld (1961) result:

$$f_i^>(\text{turb}) = -k^2 \nu_d(k) u_i(k) \qquad (1.5)$$

one obtains that the velocity correlation $Q_{ij} = <u_i u_j^*> = P_{ij}Q$ has the form:

$$Q(k) = \frac{\phi(k) + \widetilde{\phi}(k)}{\omega^2 + k^4 \nu_d^2(k)} \qquad (1.6)$$

In the next section, we show how the NSE yield a relation with the same structure as in (1.6).

Wyld treatment of the Navier-Stokes Equations. The form of the NSE to be used in (1.2) is a choice. In the previous section, we pointed out that K71 used the DIA formulation of the NSE. We have chosen a different form of the NSE derived by Wyld (1961) who used a Feynman diagram technique to obtain an interesting representation of the NSE that highlighted the effects of non-linearity. Contrary to the choice of DIA, we employ a form of the NSE that contains no approximations, it is not a model/theory of turbulence but a physically useful regrouping of the non-linear terms in the NSE that highlight the main effect of non-linearity. We begin with the case without non-linear terms. The form of the velocity correlation $Q_{ij} = <u_i u_j^*> = QP_{ij}$ is derived using the Fourier representation of the NSE, $\partial \mathbf{u}(k)/\partial t = i\omega \mathbf{u}(k)$ and the kinematic viscosity term $\nu k^2 \mathbf{u}(k)$. In this case, we have (see Appendix A):

$$Q^0 = \frac{\phi(k)}{\omega^2 + \nu^2 k^4} \qquad (1.7)$$

where $\phi(k)$ is the correlation function of an arbitrary external force. Wyld's regrouping of the NSE non-linear terms showed that the correlation function (1.7) is changed to the new form:

$$Q^0(\kappa) \rightarrow Q(\kappa) = G(\kappa)\left[\phi(k) + \widetilde{\phi}(k)\right]G(\kappa)^*, \quad G(\kappa)^{-1} = -i\omega + \nu k^2 - Z(k) \qquad (1.8)$$

Comparing (1.159) with (1.8), one sees that non-linearity has changed both kinematic viscosity and the correlation $\varphi(k)$ as follows:

$$\nu \rightarrow \nu_d = \nu - k^{-2} Z(\kappa), \qquad \phi(k) \rightarrow \phi(k) + \widetilde{\phi}(k) \qquad (1.9)$$

and the final form of the correlation then reads as follows:

1.3 Modelling Turbulent Forces

$$Q(k) = \frac{\phi(k) + \tilde{\phi}(k)}{\omega^2 + k^4 \nu_d^2(k)}, \quad \nu_d(k) = \nu + \nu_t(k) \qquad (1.10)$$

Since relation (1.10) has the same structure as (1.6), the inference is that the LSE can be viewed as a proxy of the NSE themselves. Some comments about (1.10) may be useful. The result shows that *non-linearity has changed the size-independent kinematic viscosity to a size-dependent turbulent viscosity* $\nu_t(k)$ *whose sum with ν yields the dynamical viscosity* $\nu_d(k)$. Since Fig.1 of C1 shows that in Wyld's analysis both $\nu_t(k)$ and $\tilde{\phi}$ are given by infinite series of terms, approximations were made in the literature which we cite for completeness though none of them is used in this work. In sec. IV, Wyld showed the approximation needed to recover Eq. (2) of Chandrasekhar (1955); using what is known as one-loop approximation, one recovers Kraichnan's DIA model; using the renormalization group, Yakhot and Orszag (1986) assumed $\tilde{\phi} = 0$ and concentrated on the turbulent viscosity, but the choice precluded the possibility of applying their formalism to realistic turbulent flows which is the main goal of this work.

1.3 Modelling Turbulent Forces

Wyld (1961) derived (1.5) but not the form of the turbulent viscosity $\nu_t(k)$. The first derivation is due to Yakhot and Orszag (1986) but their Eq. (3.2) did not encompass large and small eddies in a smooth fashion, a relation that was derived 10 years later in C1, Eq. (24). Since the latter derivation is somewhat complex, we next present a heuristic procedure leading to essentially the same result. Consider Prandtl (1925) *mixing length theory* that gives the following expression for the turbulent viscosity felt by a large eddy k_0^{-1}:

$$\nu_t(k_0) \propto k_0^{-1} K^{1/2}, \quad K = \int E(k) dk \qquad (1.11)$$

which we heuristically generalize to an expression of the turbulent viscosity felt by an eddy of arbitrary k:

$$k_0^{-1} K^{1/2} \rightarrow \left[\int_k^\infty q^{-2} dq E(q) \right]^{1/2} \qquad (1.12)$$

$$\nu_t(k) = \left[\int_k^\infty q^{-2} dq E(q) \right]^{1/2} \qquad (1.13)$$

which we further extend to include the kinematic viscosity ν:

$$\nu_d(k) = \left\{ \nu^2 + \int_k^\infty q^{-2} dq E(q) \right\}^{1/2}, \nu_d(k) = \nu + \nu_t(k) \tag{1.14}$$

which is quite close to the one in Eq. (24) of C1 based on the RNG (renormalization group):

$$\nu_d(k) = \left\{ \nu^2 + \frac{2}{5} \int_k^\infty q^{-2} E(q) dq \right\}^{1/2} \tag{1.15}$$

This relation exhibits some features worth noticing.

(a) since $\nu_d(k)$ depends on E(k), Eq. (1.15) is highly non-linear,
(b) for a small eddy (large k), the integral is small, $\nu_t(k) \to 0$, as expected for small eddies,
(c) for large eddies (small k), the integral is the largest and ν becomes irrelevant,
(d) thus, (1.15) encompasses both large and small eddies in a smooth fashion,
(e) if data and/or numerical simulations provided E(k) (e.g., Wang et al. 2019), the dynamical viscosity could be computed using (1.15).

Modelling $f_i^<$(turb). Since neither Yakhot and Orszag (1986) nor subsequent authors, provided an expression for this forcing, no application to real turbulence flows was possible and RNG remained an interesting but incomplete framework. A key contribution of our work is the determination of $f_i^<$(turb) which we discuss next. We begin with the known, model independent equation for the time evolution of the energy spectrum E(k) (Batchelor 1971; Monin and Yaglom 1971; Lesieur 1991; McComb 1992):

$$\partial_t E(k) = T(k) - 2\nu k^2 E(k) + A_{ext} \tag{1.16}$$

where T(k) is *the transfer function* that was the subject of many studies over many years as documented in the above textbooks. Because of the conservation law:

$$\int T(k) dk = 0 \tag{1.17}$$

the transfer function T(k) can be expressed in terms of an *energy flux* $\Pi(k)$:

$$T(k) = -\frac{\partial}{\partial k} \Pi(k), \quad \Pi(0) = \Pi(\infty) = 0 \tag{1.18}$$

To determine $\Pi(k)$, we adopt the same procedure used by Kolmogorov in his work, namely *the local nature of the transfer in k-space* which translates into:

1.3 Modelling Turbulent Forces

$$\Pi(k) = r(k)E(k) \tag{1.19}$$

where r(k) is the velocity or rapidity of the transfer. If (1.19) is viewed as the analog of a local mass current in a pipe:

$$j(x) = \rho(x)v(x) \tag{1.20}$$

using the mapping:

$$j \to \Pi, \quad \rho \to E, \quad v \to r \tag{1.21}$$

relation (1.18) then gives:

$$T(k) = -r(k)\partial_k E(k) - E(k)\partial_k r(k) \tag{1.22}$$

To identify the two terms, consider the relation:

$$A_t(k) = <f_i^< u_i>, \quad T(k) = A_t(k) - 2k^2\nu_t(k)E(k) \tag{1.23}$$

where A_t represents the action of large eddies on a given eddy k while the second term represents the action on the eddy k by smaller eddies. Comparison with (1.22), suggests the identifications:

$$A_t(k) = -r(k)\partial_k E(k), \quad \partial_k r(k) = 2k^2\nu_t(k) \tag{1.24}$$

Integrating the second relation, yields the rapidity in terms of the turbulent viscosity:

$$\frac{1}{2}r(k) = \int_0^k q^2\nu_t(q)dq \tag{1.25}$$

and the final form of Eq.(1.16) becomes:

$$\partial_t E(k) = -r(k)\partial_k E(k) - 2\nu_d k^2 E(k) + A_{ext} \tag{1.26}$$

or, more compactly:

$$\partial_t E(k) = A_t(k) - 2\nu_d k^2 E(k) + A_{ext} \tag{1.27}$$

Eqs. (1.27) and (1.15) are a close set of equations that can be solved for any A_{ext}. This kinetic energy equation includes the contributions of both small and large eddies.

Temperature field spectrum. The temperature variance is defined as follows:

$$E_\theta = \frac{1}{2}\overline{\theta^2} = \int E_\theta(k)dk \qquad (1.28)$$

and the analog of Eq. (1.26) reads as follows:

$$\partial_t E_\theta(k) = -r_\theta(k)\partial_k E_\theta(k) - 2k^2\chi_d(k)E_\theta(k) + A_{ext} \qquad (1.29)$$

where:

$$\frac{1}{2}r_\theta(k) = \int_0^k q^2\chi_t(q)dq, \qquad \chi_d(k) = \chi + \chi_t(k) \qquad (1.30)$$

where the *dynamical conductivity* $\chi_d(k)$ is related to the *dynamical viscosity* as follows:

$$\frac{d\chi_d}{d\nu_d} = \frac{10}{3}\frac{\nu_d}{\nu_d + \chi_d} \qquad (1.31)$$

with the initial condition $\chi_d(\nu) = \nu$, where ν/χ is the kinamatic Prandtl number. The analytic solution of (1.31) is presented in Eq. C18 of C1.

1.4 Predicting Basic Turbulence Laws

To be credible, a turbulence model must reproduce well-known regimes whose spectra have been assessed in detail (Batchelor 1971; McComb 1992).

Inertial regime. If we take $\partial_t E = 0$, $\Pi(k) = \varepsilon$, $A_{ext} = 0$ and seek a solution of the form $E(k) = Ak^{-m}$, $\nu_d(k) = Bk^{-n}$, the result is:

$$E(k) = Ko\,\varepsilon^{2/3}k^{-5/3}, \qquad Ko = \frac{5}{3} \qquad (1.32)$$

Therefore, the model predicts the well-known Kolmogorov spectrum (1.32) and a Kolmogorov constant Ko in agreement with measured data (Lesieur 1991).

Inertial-convective regime. In an analogous manner, one derives the relations:

$$E_\theta(k) = Ba\,\varepsilon^{-1/3}\varepsilon_\theta k^{-5/3}, \qquad Ba = \sigma_t Ko, \qquad \sigma_t = \frac{\nu_d}{\chi_d} \qquad (1.33)$$

1.4 Predicting Basic Turbulence Laws

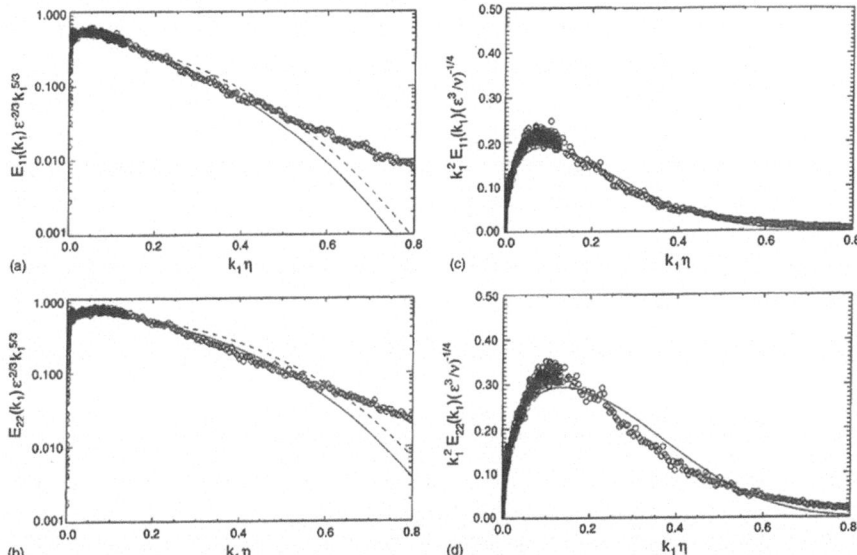

Fig. 1.1 (a)–(b) shows the 1D compensated spectra $k_1^{5/3}E_{11}(k_1)$, $k_1^{5/3}E_{22}$ vs k_1 compared with experimental results with/without back-scatter; (c)–(d) show the one-dimensional spectra $k_1^2 E_{11}(k_1)$, $k_1^2 E_{22}(k_1)$ compared with experimental data; $\eta = (\nu^3/\varepsilon)^{1/4}$. (Reprinted with permission from Canuto et al. (1996c). © NASA all rights reserved b)

where ε_θ is the rate of dissipation of temperature variance and Ba is the Batchelor constant. The model predicts the Corrsin-Obukhov spectrum (1.33) and the Batchelor constant Ba in accordance with the data (Lesieur 1991).

1D energy spectra. Once the 3D spectrum E(k) is known, one can compute the 1D spectra (Fig. 1.1):

$$E_{11}(k_1) = \int_{k_1}^{\infty} E(k) k^{-1}(1-k_1^2/k^2) dk, \qquad E_{22}(k) = \frac{1}{2}\left(E_{11}(k) - k\frac{dE_{11}(k)}{dk}\right) \tag{1.34}$$

2D turbulence. In 2D turbulence, there is an *inverse energy cascade* which is viewed as the most important result in turbulence studies after Kolmogorov *direct energy cascade* in 3D turbulence. The 2D case distinguishing feature is that it is characterized by two types of cascades, *energy and enstrophy* H(k) represented by the two conservation laws:

$$\int dk T(k) = 0, \qquad \int dk k^2 T(k) = 0 \qquad (1.35)$$

Previous models of 2D turbulence suggested abandoning Kolmogorov's basic assumption that the energy transfer is a local process. In the present model, we retain the locality assumption in the inertial range meaning the constancy of the enstrophy flux $\Pi_H(k)$ = constant, where $\partial_k \Pi_H = -k^2 T(k)$. Thus, to satisfy both the latter and the conservation law (1.35), one must extend the relationship between the flux $\Pi(k)$ and the enstrophy spectrum $H(k) = k^2 E(k)$. Using Eq. (13) of C6, the model yielded the following results:

$$\text{first regime } k < k_* : \quad E(k) \propto \varepsilon^{2/3} k^{-5/3} \qquad (1.36)$$

$$\text{second regime } k_* < k : \quad E(k) \propto h^{2/3} k^{-3}, \quad H(k) \propto h^{2/3} k^{-1} \qquad (1.37)$$

where k_* is the location of the external forcing. The above relations reproduce the results of numerical simulations result (Lesieur 1991).

Reynolds stresses. Shear driven flows. We begin with the definition of the Reynolds stresses:

$$< u_i(\mathbf{k}) u_j(\mathbf{k'}) >_s = R_{ij}(k) \delta(\mathbf{k} + \mathbf{k'}) \qquad (1.38)$$

Equation (40a) of C1 gives the time derivative of (1.38):

$$\frac{1}{2} \partial_t R_{ij}(\mathbf{k}) = A_{ij}^t(\mathbf{k}) - k^2 \nu_d(k) R_{ij}(\mathbf{k}) + A_{ij}^{ext}(\mathbf{k}), \quad A_{ij}^t(\mathbf{k}) = (8\pi k^2)^{-1} A_t(k) P_{ij}(k) \qquad (1.39)$$

where $A_t(k)$ is given by the first of (1.24). The k-space form of the first term on the right-hand side of Eq. (1.39) was discussed in sec. II of C2 where the following result was derived (Eq.9):

$$f_i(k) = -i k_j u_i U_j - \widetilde{G}_{ij} u_j + k_j G_{jm} P_{i\ell} \partial u_\ell / \partial k_m, \quad \widetilde{G}_{ij} \equiv P_{ik} U_{k,m} P_{mj} \qquad (1.40)$$

In sec. III of C2, it was shown how to transform (1.39) to three differential equations for three scalar functions, Eq. (25)–(27). Figure 1.2 shows the model prediction of the Reynolds stress spectrum with experimental data. The three differential

1.4 Predicting Basic Turbulence Laws

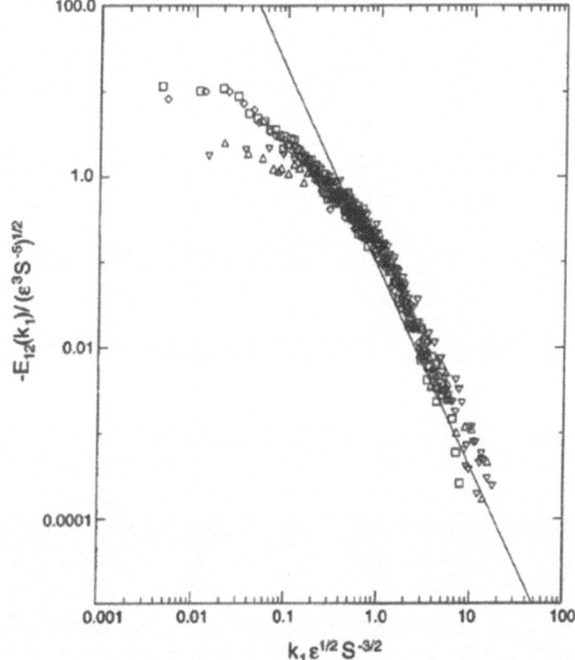

Fig. 1.2 Model prediction of the Reynolds-stress spectrum $E_{12}(k_1)$ vs. k_1, with experimental data, (Reprinted with permission from Canuto and Dubovikov (1996a, b). © NASA all rights reserved)

equations were solved for the following cases *plane strain, axisymmetric contraction, homogeneous shear*. Figure 1.3 shows the model prediction for the case of plane strain and Fig. 1.4 shows the 1D spectrum compared with experimental data in the inertial and dissipative regions.

K-ε model. Turbulence studies based on second-order closure models traditionally employ two equations, one for K representing large eddies and the other for ε representing the dissipation of K. Pope (2000) provided an extensive discussion of this model. The equation for K contains no adjustable parameters whereas the one for ε is heavily parameterized. Despite such a freedom, the K- ε model does not reproduce the data better than in Fig. 1.3.

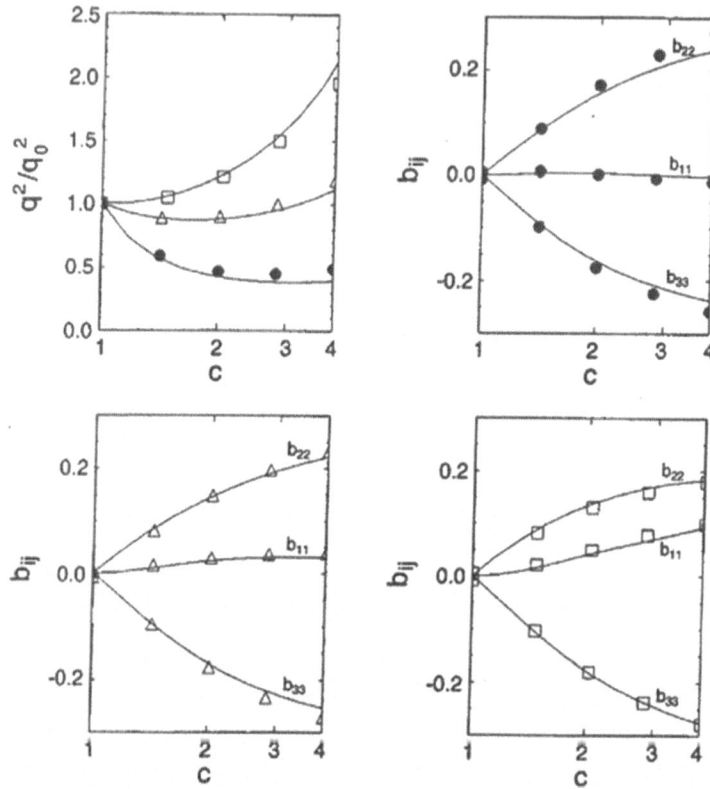

Fig. 1.3 Plane strain. Time evolution of the ratio of the kinetic energies $q^2(t)/q_o^2$ and the dimensionless Reynolds-stress tensor $b_{ij}(t)$ versus total strain. DNS data are given by dots (skewness $S = 0.65$), triangles ($S = 2.6$), and squares ($S = 25$). Model predictions are given by solid lines. (Reprinted with permission from Canuto et al. (1996c). © NASA all rights reserved)

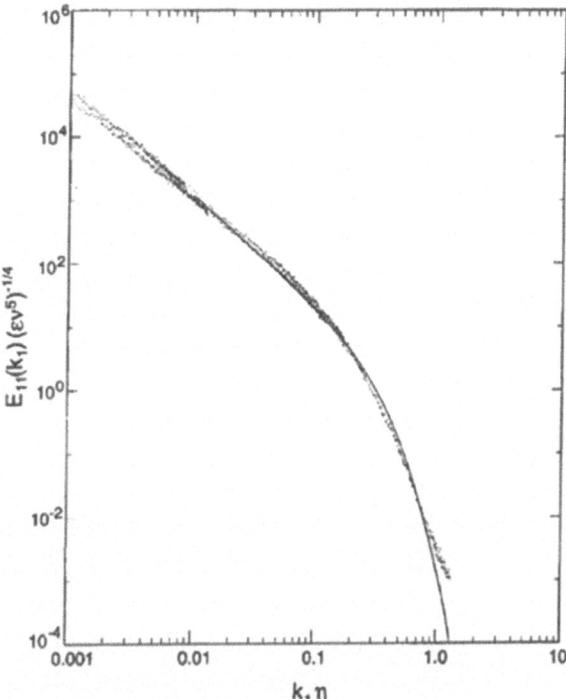

Fig. 1.4 1D spectrum compared with experimental data in the inertial and dissipative regions. (Reprinted with permission from Canuto et al. (1996c). © NASA all rights reserved)

1.5 Turbulent Thermal Convection

A time-honored test of a turbulence model is the prediction of the Nusselt number Nu (ratio of the convective to the thermal-conductive fluxes) vs. Rayleigh number Ra:

$$\mathrm{Nu} \propto \mathrm{Ra}^n, \quad \mathrm{Ra} = \frac{g\alpha d^3 \Delta T}{\nu \chi} \tag{1.41}$$

where d is the distance of the two plates between which there is a temperature difference ΔT and α is the thermal expansion coefficient (Malkus 1954). A long-standing suggestion was that n = 1/3, but in 1969 high accuracy measurements at the University of Chicago (Castaing et al. 1989) rekindled the interest in this basic process since the result was not n = 1/3 but:

$$\mathrm{Nu} \propto \mathrm{Ra}^{2/7} \tag{1.42}$$

Direct numerical simulation results by Kerr (1996) were confirmed by the Chicago data and further provided the scaling of the thermal boundary layer vs. Ra, z-profile of the mean temperature, temperature variance, turbulent kinetic energy, convective flux and vertical kinetic energy. Other than the numerical simulations by Shraiman and Siggia (1990) and Wu and Libchaber (1990), we know of no model that reproduced such data. Next, we present the scaling laws derived from the LSE. We employ Eq. (1.3) and the analog for the temperature field:

$$\frac{\partial \theta(k)}{\partial t} = f_\theta(\text{ext}) + f_\theta^< (\text{turb}) - \chi_d(k)k^2\theta(k) \tag{1.43}$$

Next, we multiply the velocity equation and (1.43) by $u_i(\mathbf{k})$, w and θ and introduce the spectra:

$$E(k), \ E_\theta(k) = \int k^2 d\Omega_k [e(\mathbf{k}), e_\theta(\mathbf{k})], \quad J(k), \ E_z(k) = \int k^2 d\Omega_k [j(\mathbf{k}), e_z(\mathbf{k})] \tag{1.44}$$

It is a matter of algebra to derive the following equations for the kinetic energy e, the buoyancy variance e_θ, the z-component of the kinetic energy e_z and the heat flux $j = \overline{w\theta}$ ::

$$\partial_t e = a^t + g\alpha j - 2k^2 \nu_d e, \quad \partial_t e_\theta = a^t_\theta + \beta j - 2k^2 \chi_d e_\theta$$

$$\partial_t e_z = \frac{1}{2} P_{zz} a^t + g\alpha P_{zz} j - 2k^2 \nu_d e_z, \quad \partial_t j = 2g\alpha P_{zz} e_\theta + 2\beta e_z - 2k^2(\nu_d + \chi_d)j \tag{1.45}$$

where $\beta = -dT/dz$ is the vertical temperature gradient. The Reynolds Stress Model (Canuto et al. 2010) yields analogous equations with adjustable parameters while in (1.45) there are none (Fig. 1.5).

1.5 Turbulent Thermal Convection

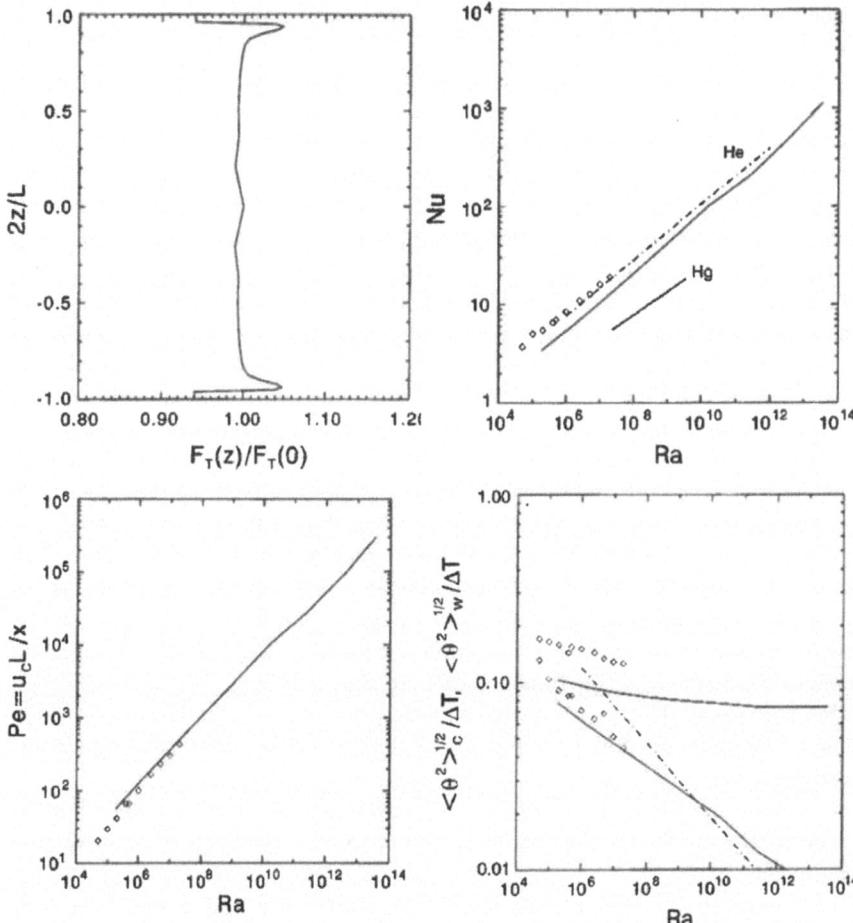

Fig. 1.5 Vertical profile of the normalized total flux; Nusselt number vs. Ra (dash-dotted line, Chicago data, squares, DNS data (Kerr 1996); Peclet number vs. Ra (solid line, model results; square, DNS data); temperature fluctuation, notation as in the Nu case. (Reprinted with permission from Canuto et al. (1997a, b). © NASA all rights reserved)

Summary of the LSE results in the 1996–1999 *Physics of Fluids* papers. It is useful to summarize the results presented in the above papers.

In Canuto and Dubovikov (1996a) we presented the LSE formalism, the derivation of Kolmogorov and Batchelor constants, the equation for the turbulent diffusivity in terms of the turbulent viscosity, the equations for the Reynolds stresses and the equation for the fluctuating temperature.

In Canuto and Dubovikov (1996b), we studied the case of shear driven flows, specifically axisymmetric contraction, plane strain and homogeneous shear for which the predicted Reynolds stress spectral function scaling like $k^{-7/3}$ is in good agreement with the data.

In Canuto et al. (1996c), we presented the numerical solutions of the equations for turbulence developed in C1. Among them, the predicted energy spectrum in the dissipation region was shown to be in agreement with laboratory measurements; in the inertial-convective region, the temperature variance spectrum was closer to the spectrum $-11/3$ obtained by LES when the velocity field was rapidly stirred at all scales rather than $-17/3$ when the velocity field was frozen in time and has a Gaussian statistics; in the case of freely decaying turbulence, the power law spectra for energy and temperature variance, as well as the velocity and temperature integral scales, agreed with LES data; after a few evolutionary times, the skewness S reached $S = 0.5$, in accordance with a variety of data; for shear-driven flows, the Reynolds stress spectrum exhibited an inertial regime with a power $-7/3$ in accordance with the data; for two shear-driven flows, plane strain and axisymmetric contraction, turbulent kinetic energy, Reynolds stress tensor, and dissipation rate ε_{ij} versus time compare very well with DNS data; the slow and rapid parts of the pressure–strain correlation tensor compared with DNS data better than with the three widely used heuristic models. The rapid parts were also in good agreement with the DNS data and for homogeneous shear, turbulent kinetic energy and the Reynolds stress tensor vs. time reproduced LES data. We suggest consulting fig.1–3 and for the case of decaying turbulence, Fig.4–6, while for the skewness behavior in time, see Fig.8. Comparisons with experimental data are shown in Fig.9, 11, 12, 14, 15.

In Canuto et al. (1997a), we studied buoyancy driven flows and the comparison to laboratory data on the Nusselt number vs. Raleigh number relation and we recommend Fig.1–6.

In Canuto and Dubovikov (1997b) we considered the effect of rotation on the kinetic energy spectra and the retarding effect of rotation on the energy cascade was quantified.

In Canuto et al. (1997c), we studied 2D turbulence and comparison with DNS data.

In Canuto and Dubovikov (1999a), we constructed the five independent, orthogonal tensors necessary to represent shear driven flows.

In Canuto et al. (1999b), the Reynolds stresses for shear driven flows were constructed and several laboratory and DNS data were reproduced, together with the Smagorinsky-Lilly constant that had been widely used to represent unresolved small-scale turbulence.

In Canuto et al. (1999c) we constructed a one-point closure form of the Reynolds stresses and compared the results with laboratory data and DNS data of channel and homogeneous flows. The model equations contain no adjustable parameters.

1.6 Mesoscales. Adiabatic Regime

Mesoscales (10–100 km) are energetic, long lived (several months), coherent oceanic structures that are not generally numerically resolved by coarse resolution ocean global circulation models OGCMs and must therefore be parameterized. Canuto and Dubovikov (2005) showed that in the high frequency regime $\omega > f$, non-linearity is unimportant but not in the low frequency regime $\omega < f$, leading to the suggestion that such a turbulent flow could be viewed as a system of "broken" low frequency Rossby waves. Furthermore, since the ocean contains two physically distinct regimes, *adiabatic-A* (deep ocean) and *diabatic-D* (upper layers), one needs two different parameterizations to account for the different conservation laws satisfied in the two regimes. In the A-regime, water parcels move along surfaces of constant density (isopycnals) with a small diapycnal flux. In the D-regime, water parcels no longer move along density surfaces but along equipotential z-surfaces and exhibit strong diapycnal fluxes. Since the bulk of the ocean (a few km depth) is adiabatic, the D-regime, though important because of its contact with the atmosphere, has received less attention over the years. In fact, there is no dynamically based D-regime parameterization making the one suggested in sec. 1.8 the only available one. OGCMs have been run that included a D-regime which, in the absence of a physically based model, was treated with tapering functions to join the A-D regimes and/or as a boundary-value model (Ferrari et al. 2010) whose shortcomings were analyzed in Canuto and Dubovikov (2011). In what follows, we show that such approaches are no longer needed since a dynamically based model of the D-regime is now available. In this section we treat the A-regime and show that altimetry T-P data point out the existence of a mesoscale *drift velocity* that was not part of the GM model (Gent and McWilliams 1990) but which is now part of the new model and plays a crucial role in solving a long-standing problem of the original GM model.

Traditionally, mesoscales were modeled as a *diffusion process* reminiscent of the Fickian models of small-scale mixing processes. The procedure was not successful because it was incomplete since it omitted a second mechanism *advection*. Consider an arbitrary scalar field φ whose turbulent flux $\Phi_i = \overline{u_i'\varphi'}$ can be expressed in the general form:

$$\Phi_i = -K_{ij}\frac{\partial \overline{\varphi}}{\partial x_j} \quad , \quad K_{ij} = K_{ij}(s) + K_{ij}(a) \tag{1.46}$$

where the diffusivity tensor K_{ij} has a symmetric part $K_{ij}(s)$ representing *diffusion* and an anti-symmetric part $K_{ij}(a)$ representing *advection*. Since the dynamic equation of the mean field $D_t\overline{\varphi} + \partial_i\Phi_i = Q$ contains the divergence of the flux (1.46), the antisymmetric component $K_{ij}(a)$ yields two terms, $K_{ij}(a)\partial_i\partial_j\overline{\varphi}$ which is identically zero being the product of symmetric and anti-symmetric tensors, while the second term can be written as:

$$\frac{\partial K_{ij}(a)}{\partial x_i}\frac{\partial \overline{\varphi}}{\partial x_j} = -u_j^+\frac{\partial \overline{\varphi}}{\partial x_j} \tag{1.47}$$

where u_j^+ is an *advective* or *bolus velocity*. It is important to stress that it is only because of the anti-symmetric nature of $K_{ij}(a)$ that the bolus velocity satisfies the required zero divergence condition, $\partial_i u_i^+ = 0$. As for the symmetric part $K_{ij}(s)$, one could in principle define a term $\partial_i K_{ij}(s)$ with the dimension of velocity but which is not a physical velocity since it is not divergent free. GM were the first to suggest the following heuristic form of the bolus velocity $(\mathbf{u}^+, w^+)$:

$$\mathbf{u}^+ = -\kappa_M \frac{\partial \mathbf{s}}{\partial z}, \qquad w^+(z) = -\int_{-H}^{z} \nabla_h \cdot \mathbf{u}^+(z') dz' \qquad (1.48)$$

where κ_M is the mesoscale diffusivity, $\mathbf{s} = -N^{-2} \nabla_H \bar{b}$ is the slope of the isopycnals and the second relation follows from the zero-divergence relation, $\nabla_H \cdot \mathbf{u}^+ + \partial_z w^+ = 0$. The three velocities $\bar{\mathbf{u}}$, $\mathbf{u}^+$, $\bar{\mathbf{u}} + \mathbf{u}^+$ are called *mean, eddy and residual*. Liang et al. (2017) employed the results of the ECCO program and their Fig.6 shows vertical transports that exhibit maxima of $\pm 100 Sv$, where $1Sv = 10^6 m^3 s^{-1}$. The physical basis of the down-gradient in the first term of (1.48) suggested by GM is that mesoscales act as a sink of mean potential energy which is the key novelty of the model. Though the mesoscale diffusivity κ_M was not provided by GM, further studies showed it to be surface enhanced (Danasaboglu and Marshall 2007). This feature is relevant since w^+ must vanish $w^+(0) = 0$ at the surface where $\kappa_M(0) \neq 0$. Killworth (1997) suggested that $w^+(0) = 0$ is satisfied if:

$$\int_{-H}^{0} dz\, \mathbf{u}^+(z) = 0 \qquad (1.49)$$

which is not satisfied by the GM model. The derivation of the bolus velocity based on the LSE yielded the following result (Canuto and Dubovikov 2005):

$$\mathbf{u}^+ = -\kappa_M \frac{\partial \mathbf{s}}{\partial z} - \frac{\kappa_M}{\mathrm{fr}_d^2} \mathbf{e}_z \times (\bar{\mathbf{u}} - \mathbf{u}_d) \equiv -\frac{\partial \kappa_M \boldsymbol{\xi}}{\partial z} \qquad (1.50)$$

The first term in (1.50) is the GM model (Gent, 2011). In (1.50), f is the Coriolis parameter, r_d is the first Rossby deformation radius and $\mathbf{u}_d$ is the *barotropic mesoscale drift velocity* for which the LSE yields the form:

$$\mathbf{u}_d = <\bar{\mathbf{u}}> -\mathrm{fr}_d^2 \mathbf{e}_z \times < \frac{\partial \mathbf{s}}{\partial z} >, \qquad <\varphi> = \frac{\int_{-H}^{0} \varphi(z) \kappa_M(z) dz}{\int_{-H}^{0} \kappa_M(z) dz} \qquad (1.51)$$

for an adiabatic ocean of dept. H. Using (1.51) into the first of (1.50) yields the symmetric form:

1.6 Mesoscales. Adiabatic Regime

$$\mathbf{u}^+ = -\kappa_M \left(\frac{\partial \mathbf{s}}{\partial z} - \left< \frac{\partial \mathbf{s}}{\partial z} \right> \right) - \frac{\kappa_M}{\mathrm{fr}_d^2} \mathbf{e}_z \times (\bar{\mathbf{u}} - \left< \bar{\mathbf{u}} \right>) \tag{1.52}$$

which in an *adiabatic ocean*, satisfies relation (1.49) identically for any κ_M. It must be stressed that is the presence of the drift velocity, not tapering functions, that allowed condition (1.49) to be satisfied. At the depth where $\bar{\mathbf{u}} = \mathbf{u}_d$ (*steering level*), (1.50) reduces to the GM model. Figure 1.6 shows that the drift velocity (1.51) reproduces satisfactorily the T/P data (Fu 2009; Chelton et al. 2011). The presence of κ_M in (1.51) differs from the straight vertical average used by Klocker and Marshall (2014) since κ_M is surface enhanced.

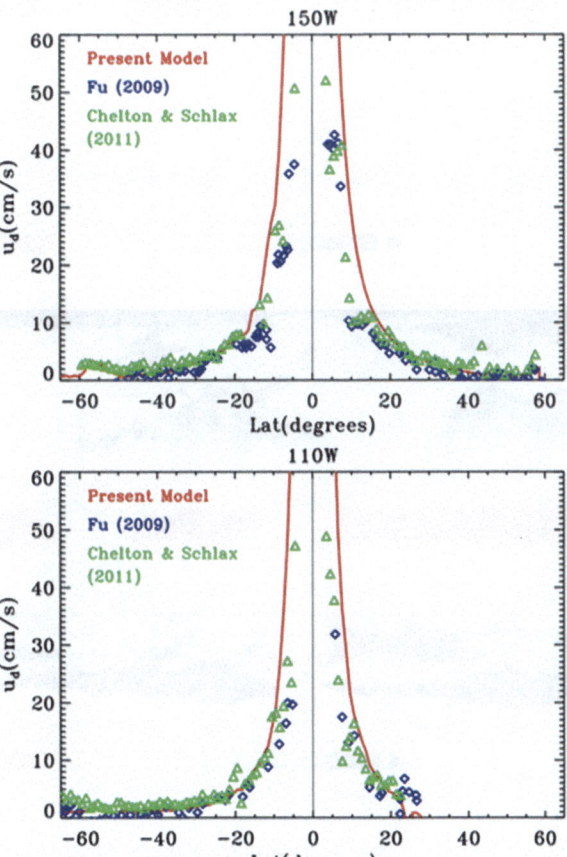

Fig. 1.6 Comparison of $|\mathbf{u}_d|$ Eq. (1.51) with the data of Fu (2009) and Chelton et al. (2011) at 150°W and 110°W. The data are reproduced satisfactorily. The model results were obtained from an average of the last 3 years of a simulation of the GISS OGCM that was run for 300 years. (Reprinted with permission of Canuto et al. (2018a, b). ©NASA all rights reserved)

1.7 Some Results of the New Model

In Fig. 1.7 we present some results from the new model.

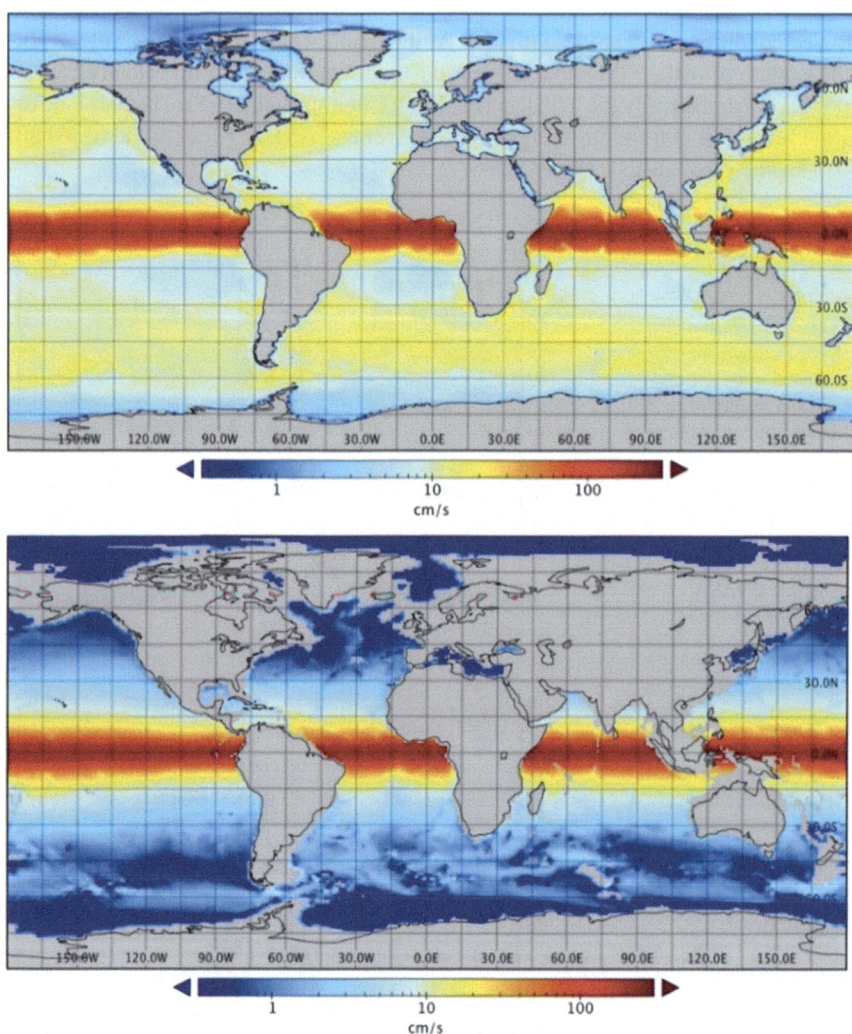

Fig. 1.7 Shows $u_D = \bar{u} - u_d$ at two different depths. Since the drift velocity is barotropic while the mean velocity $\bar{u}$ depends on z, near the depth where they coincide, the difference $u_D = \bar{u} - u_d$ decreases significantly, as shown in Fig. 1.8 for the vertical profile of the ACC zonal-meridional components of $\bar{u} - u_d$. (Reprinted with permission from Canuto et al. (2018a, b). © NASA all rights reserved)

1.7 Some Results of the New Model

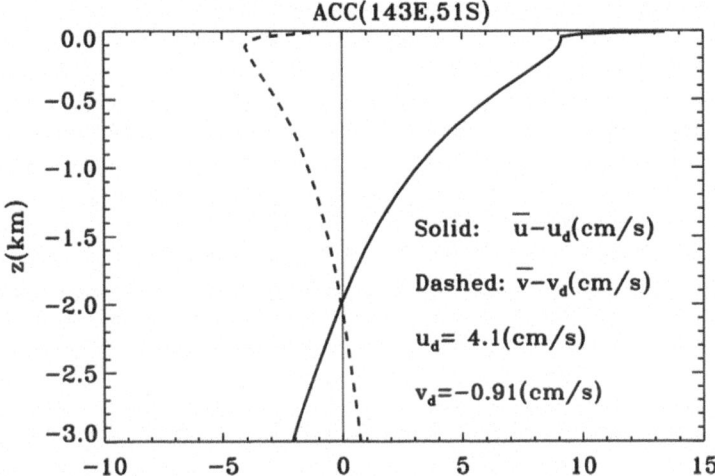

Fig. 1.8 Vertical profiles of the x,y components of the relative velocity $\bar{\mathbf{u}} - \mathbf{u}_d$. (Reprinted with permission from Canuto et al. (2018a, b). © NASA all rights reserved)

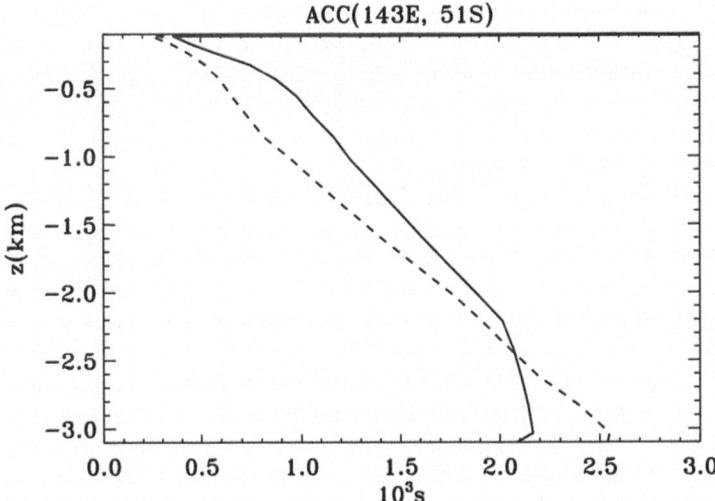

Fig. 1.9 Vertical profiles of the isopycnal slopes with (solid) and without (dashed) the new term in the eddy-induced velocity Eq. (1.50). (Reprinted with permission from Canuto et al. (2018a, b). © NASA all rights reserved)

$$|z| < 2\text{km} : u^+, v^+ > u_{GM}^+, v_{GM}^+, \quad |z| > 2\text{km} : u^+, v^+ < u_{GM}^+, v_{GM}^+ \quad (1.53)$$

and thus, in the ocean upper layers, the GM model underestimates the bolus velocity. As for the effect of the new term in (1.52) on the isopycnal slopes, Fig. 1.9 shows the vertical profiles of the isopycnal slopes with and without the new term. The larger slopes in the present model implies that the "flattening of the isopycnals" is not as large as in the GM model.

This is further reflected in the fact that the energy drawn from the mean potential energy MPE is about 20% lower in the present model (1TW= 10^{12}Watts):

$$\int \rho \kappa_M N^2 (s^2, \ s\cdot\boldsymbol{\xi}) dV = (0.73, \ 0.5) TW \tag{1.54}$$

In *summary*, in the A-regime, the horizontal and vertical buoyancy fluxes are given by:

$$A-regime \quad \mathbf{F}_H = -N^2 \int_{-H}^{z} \mathbf{u}^+(z')dz' = \kappa_M N^2 \boldsymbol{\xi}, \quad F_v = \mathbf{s}\cdot\mathbf{F}_H = \kappa_M N^2 \mathbf{s}\cdot\boldsymbol{\xi}$$

$$\tag{1.55}$$

Stream functions. The following form was derived in Canuto and Dubovikov (2011):

$$\boldsymbol{\Psi} = -\kappa_M T(z) \mathbf{s} \times \mathbf{e}_z, \quad T(z) = \frac{N^2 F_v(\overline{b})}{\kappa_M |\nabla_H \overline{b}|^2}, \quad \mathbf{U}^+ = \nabla \times \boldsymbol{\Psi} \tag{1.56}$$

which in the A-regime becomes:

$$A-regime: \quad T(z) = \frac{\mathbf{s}\cdot\boldsymbol{\xi}}{|\mathbf{s}^2|} \tag{1.57}$$

Summary of the results. (a) analytic solution of the non-linear mesoscale dynamic equations, (b) non-linearity was treated in the spirit of the mixing length theory, (c) the form of the bolus velocity was presented in both isopycnal and z-coordinates, (d) it was predicted that, in addition to GM, the bolus velocity contains a new term, the *drift velocity* $\mathbf{u}_d$ representing the velocity with which mesoscales travel in the ocean, (e) such a velocity is barotropic (z independent), (f) both existence and magnitude of $\mathbf{u}_d$ could not be assessed at the time the CD5,6 models were published, (g) the confirmation had to wait 6 years until 2011 when Chelton et al. (C11) published the results of a census of mesoscales based on altimetry data, (h) form and magnitude of $\mathbf{u}_d$ predicted by CD5,6 reproduced well the C11 altimetry data, as shown in Fig. 1.6, (i) C11 considered $\mathbf{u}_d$ the *"most germane of all non-linear metrics"*, (j) the form of the bolus velocity derived in CD5,6 is as follows:

1.7 Some Results of the New Model

$$\mathbf{u}^+ = \mathbf{u}_{GM}^+ - \frac{\kappa_M}{fr_d^2}\mathbf{e}_z \times (\bar{\mathbf{u}} - \mathbf{u}_d) \tag{1.58}$$

where κ_M is the mesoscale diffusivity, f is the Coriolis parameter, r_d is the Rossby radius, $\bar{\mathbf{u}}$ is the 2D mean velocity and $\mathbf{e}_z = (0, 0, 1)$. At the depth where $\bar{\mathbf{u}} = \mathbf{u}_d$ (*steering level*), the new bolus velocity reduces to the GM model:

$$\mathbf{u}_{GM}^+ = -\kappa_M \frac{\partial \mathbf{s}}{\partial z} \tag{1.59}$$

where $\mathbf{s} = -N^{-2}\nabla_H \bar{b}$ is the slope of the isopycnals, k) CD6 predicted the following form of the barotropic drift velocity:

$$\mathbf{u}_d = \mathbf{c}_R + <\bar{\mathbf{u}}> - fr_d^2 \mathbf{e}_z \times <\partial \mathbf{s}/\partial z> \tag{1.60}$$

where the first term is the Rossby phase velocity and, for a fully adiabatic ocean, the average $<..>$ is defined as follows (H is the ocean depth):

$$<A(z)> = \frac{\int_{-H}^{0} dz A(z)\kappa_M(z)}{\int_{-H}^{0} dz \kappa_M(z)} \tag{1.61}$$

The presence of κ_M differs substantially from the straight vertical average used by Klocker and Marshall (2014) since κ_M is surface enhanced (Danasaboglu and Marshall 2007). ℓ) The CD prediction of a bolus velocity was not only confirmed by the T-P data, but has beneficial effects as well. Substituting $\mathbf{u}_d$, the bolus velocity acquires the symmetric form:

$$\mathbf{u}^+(z) = -\kappa_M \left(\frac{\partial \mathbf{s}}{\partial z} - <\frac{\partial \mathbf{s}}{\partial z}>\right) - \frac{\kappa_M}{fr_d^2}\mathbf{e}_z \times (\bar{\mathbf{u}} - <\bar{\mathbf{u}}>) \tag{1.62}$$

Killworth (1997) suggested that the condition to satisfy $w^+(0) = 0$ is:

$$\int_{-H}^{0} dz\, \mathbf{u}^+(z) = 0 \tag{1.63}$$

which is identically satisfied by (1.62) for any diffusivity, which is not the case with the GM model since $\kappa_M(0)$ is surface enhanced (Danasaboglu and Marshall 2007) and thus $\kappa_M(0) \neq 0$, m) *PV vs. thickness down-gradient.* Treguier (1999) suggested that the bolus velocity be parameterized as a down-gradient of potential vorticity $q = h^{-1}(f + \zeta)$ rather than a down-gradient of thickness layer $\partial s/\partial z = \bar{h}^{-1}\nabla_\rho \bar{h}$. However, since the substitution $\bar{h}^{-1}\nabla_\rho \bar{h} = -\bar{q}^{-1}\nabla_\rho \bar{q} + \beta f^{-1}$ introduces the second term that diverges at the Equator, the suggestion was not adopted in OGCMs. However, since β does not depend on z, it cancels out in the $<..>$ since $\bar{h}^{-1}\nabla_\rho \bar{h}$ - $<\bar{h}^{-1}\nabla_\rho \bar{h}> = -\bar{q}^{-1}\nabla_\rho \bar{q} + <\bar{q}^{-1}\nabla_\rho \bar{q}>$ showing that the two representations are identical, a conclusion consisted with the results of Robert and Marshall (2000) who found no significant differences in the results of the two representations, n) EKE. CD6 suggested a model for the vertical profile of the mesoscale EKE which reproduces the WOCE data satisfactorily, p) EPE. CD6 derived a model of the eddy potential energy EPE whose vertical profile is shown in Fig.3 of CD6 to be close to the result of the eddy-resolving data of Boning and Budich.

Summary. CD5,6 is presently the only existing derivation of the GM bolus velocity, it predicted a drift velocity 6 years before it was confirmed by altimetry data, baroclinicity (1.63) is identically satisfied and thickness and PV down gradient parameterizations are equivalent.

1.8 Mesoscales. Diabatic-D Regime

The heuristic parameterizations by Aiki et al. (2004) and Ferrari et al. (2008, 2010; F8,10) were studied analytically by Canuto and Dubovikov (2011, C11) with the following results:

a) F8's bolus velocity does not induce re-stratification of ML. This motivated F10 to generalize A4 boundary-value model to include stratification as follows $(\mathbf{u}^+ = \partial_z \mathbf{Y}, \ w^+ = -\nabla \cdot \mathbf{Y})$:

$$A4: \ C^{-1}\partial_{zz}^2 \mathbf{Y} = -\nabla_H \bar{b} \ , \quad F10: \ \left(c^2 \partial_{zz}^2 - N^2\right)\mathbf{Y} = -\kappa_M \nabla_H \bar{b} \quad (1.64)$$

b) to the first order in the smallness parameter h/H (h is the depth of the mixed layer and H is the depth of the ocean), the stream functions of both models do not depend on stratification N^2, c) such dependence is present in F10 but only at a higher order in h/H,

c) A4-F10 differ only in locations when h≃H, in the rest of the ocean the two are quite similar,

d) the main argument that affects both A4-F10 is based on the behavior of the *residual flux* that was not considered in either model that were thus unaware of the following difficulty. We begin by recalling the definitions of the buoyancy flux in terms of *stream function* Ψ and *residual flux*:

1.8 Mesoscales. Diabatic-D Regime

$$\mathbf{F}(\overline{b}) = \overline{\mathbf{U}'b'} = \mathbf{\Psi} \times \nabla \overline{b} + \mathbf{F}_r(\overline{b}), \quad \mathbf{F}_r(\overline{b}) = \frac{\mathbf{F}(\overline{b}) \cdot \nabla \overline{b}}{|\nabla \overline{b}|^2} \nabla \overline{b} \qquad (1.65)$$

e) CD11 showed that in F10, the residual flux has the form:

$$\mathbf{F}_r = -\kappa_{res} \nabla_H \overline{b}, \quad \kappa_{res} = A\kappa_M \frac{h}{H}\left(1 + \frac{z}{h}\frac{N^2}{N_1^2}\right) \qquad (1.66)$$

where A is a constant. The flux vanishes at the bottom of the D-regime and:

$$\kappa_{res} < < \kappa_M \qquad (1.67)$$

a result that is not compatible with the numerical simulations of Gille and Davis (1999) and Radko and Marshall (2004) showing that $\mathbf{F}_r$ is very small in the A-regime but large in the D-regime, which is the opposite of what is predicted by the above models.

In contrast with the above treatments, C11 employed the LSE to derive the D-regime parameterization. We begin with the horizontal buoyancy flux given by:

$$\mathbf{F}_H = -\kappa_M N^2 \mathbf{s} \qquad (1.68)$$

In C11 it was shown that the vertical buoyancy flux is given by:

$$F_v(\overline{b}) = -\kappa_M \boldsymbol{\gamma} \cdot \nabla_H \overline{b}, \quad \boldsymbol{\gamma} = \boldsymbol{\omega} \times \mathbf{e}_z, \quad fr_d^2 \boldsymbol{\omega}(z) = z\mathbf{u}_D - \int_z^0 dz' \mathbf{u}_D(z'), \quad \mathbf{u}_D \equiv \overline{\mathbf{u}} - \mathbf{u}_d \qquad (1.69)$$

Using a high-resolution numerical simulation, Luneva et al. (2015) showed that (1.69) works well within but not near the bottom of the D-regime. It is therefore necessary to divide the D-regime into two parts, a mixed layer proper where mixing is strong and a transition-T regime where mixing is no longer as strong. Regrettably, we do not have a parameterization for the T-regime based on mesoscale dynamic equations as in the A-regime; however, since such a T-regime is of limited extent, a heuristic parameterization was suggested that matches the D-A-regimes parameterizations in the appropriate limits. To do so, we adopted the following criteria: a) in the ML, it must recover (1.69), b) the diapycnal flux $F_d = F_v - \mathbf{s} \cdot \mathbf{F}_H$ must vanish at the bottom of the D-regime and c) going from the bottom of the D-regime across the intermediate regime, the terms in the new flux, must decrease fast enough to satisfy a). To do so, the vertical flux is written as:

$$F_v(\overline{b}) = -\kappa_M \boldsymbol{\Omega} \cdot \nabla_H \overline{b} \qquad (1.70)$$

As discussed in detail in Canuto et al. (2018a, b), the final result was:

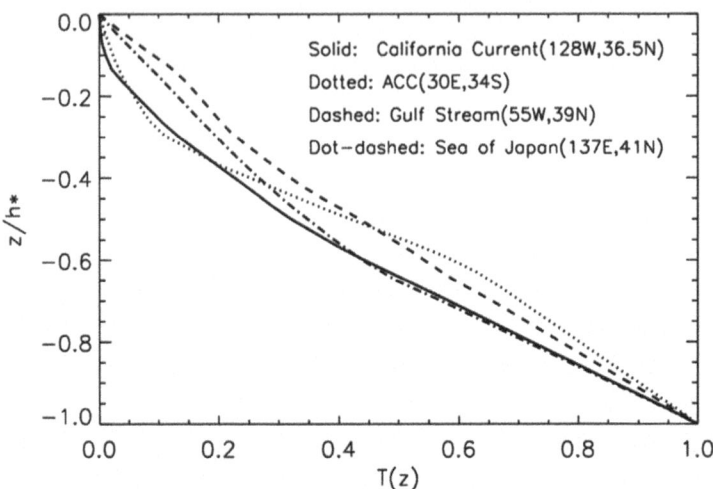

Fig. 1.10 Tapering function T(z) for the California Current, ACC, Gulf Stream and Sea of Japan. All the results yield values lower than the usually assumed straight line. (Reprinted with permission from Canuto et al. (2018a, b). © NASA all rights reserved)

$$\mathbf{\Omega} = [\omega(z) - \Phi(z)\omega_*] \times \mathbf{e}_z + \Phi(z)\frac{N^2(z)}{N_*^2}\mathbf{s}(z), \qquad \Phi(z) = \frac{z^2}{h_*^2}\frac{N^2(z)}{N_*^2} \qquad (1.71)$$

where s_* is the slope at the depth h_* where the A-D regimes match. It then follows that the tapering function is given by (Fig. 1.10):

$$T(z) = \frac{\mathbf{\Omega}\cdot\mathbf{s}}{|\mathbf{s}^2|} \qquad (1.72)$$

In the D-regime, bolus and drift velocities are given by Eqs. (2.8) and (2.5) of C18:

$$\mathbf{u}^+ = -\kappa_M\left(\frac{\partial \mathbf{s}}{\partial z} - <\frac{\partial \mathbf{s}}{\partial z}>\right) - \frac{\kappa_M}{\mathrm{fr}_d^2}\mathbf{e}_z \times (\overline{\mathbf{u}} - <\overline{\mathbf{u}}>) - \kappa_M\frac{\mathbf{s}_*}{H_*}$$

$$\mathbf{u}_d = <\overline{\mathbf{u}}> - \mathrm{fr}_d^2\mathbf{e}_z \times \left(<\frac{\partial \mathbf{s}}{\partial z}> - \frac{\mathbf{s}_*}{H_*}\right) \qquad (1.73)$$

where we have defined the following length scale:

$$<H_*> = \frac{\int_{-H}^{-h_*} \kappa_M(z) dz}{\kappa_M(h_*)} \qquad (1.74)$$

1.9 Diffusivity Tensor, Kinetic Energy, Mesoscale Diffusivity

OGCMs usually express the 3D tracer flux in terms of a diffusivity tensor **K**:

$$\mathbf{F}(\bar{c}) = - \kappa_M \mathbf{K} \cdot \nabla \bar{c} \qquad (1.75)$$

which in the present model is given by:

$$\textbf{A-regime}: \quad \mathbf{K} = \begin{pmatrix} 1 & 0 & s_x \\ 0 & 1 & s_y \\ s_x & s_y & s^2 \end{pmatrix} + \begin{pmatrix} 0 & 0 & -\xi_x \\ 0 & 0 & -\xi_y \\ \xi_x & \xi_y & 0 \end{pmatrix} \qquad (1.76)$$

The first tensor represents the symmetric part given by the Redi (1982) term, while the second tensor represents the anti-symmetric part that accounts for the difference between the slopes **s**, **ξ**.

Furthermore, one has:

$$\textbf{D-regime}: \quad \mathbf{K} = \begin{pmatrix} 1 & 0 & 0 \\ 0 & 1 & 0 \\ K_{31} & K_{32} & K_{33} \end{pmatrix} \qquad (1.77)$$

where:

$$K_{31} = (1-\Phi)\Omega_x + 2\Phi\Omega_x^{\|}, \quad K_{32} = (1-\Phi)\Omega_y + 2\Phi\Omega_y^{\|}, \quad K_{33}$$
$$= \mathbf{\Omega} \cdot \mathbf{s}, \quad \mathbf{\Omega}^{\|} = |\mathbf{s}|^{-2}(\mathbf{s} \cdot \mathbf{\Omega})\mathbf{s} \qquad (1.78)$$

At the A-D interface, (1.76) and (1.77) match since $\xi(-h_*) = s(-h_*)$. Figure 1.11 shows the z profiles of the three components (1.78). The normalization was chosen so that at the bottom of the D-regime, the diffusivities match those of the A-regime.

Mesoscale diffusivity. As discussed in Canuto et al. (2019, C19), this variable has received a great deal of attention though most of the formulations were either heuristic or employed linear stability analysis which in a turbulence context is of doubtful validity since, as cited by Lumley and Yaglom (2001), Corrsin and

Fig. 1.11 z-profiles of the three components of D-regime diffusivities K_{31}, K_{32} and K_{33} (1.78) in the ACC and Gulf Stream. The normalization was chosen so that at the bottom of the D-regime, the diffusivities match those of the A-regime. (Reprinted with permission from Canuto et al. (2018a, b). © NASA all rights reserved)

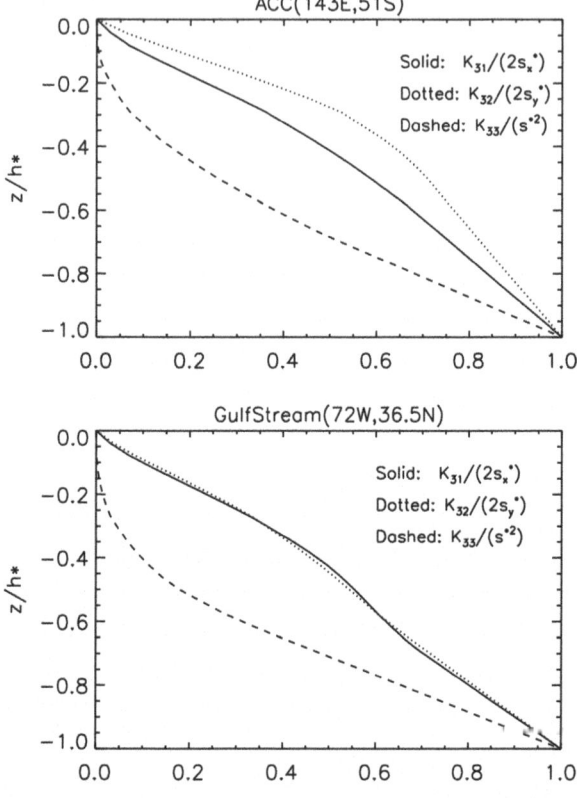

Liepmann concluded that " *stability theory has nothing to do with turbulence*". The LSE framework yielded the following result (CD5):

$$\kappa_M = \frac{1}{2} r_d K^{1/2} \varpi(\mathbf{u}_D, K), \quad \varpi = \left[1 + \frac{1}{K}|\mathbf{u}_D|^2\right]^{-1/2}, \quad \mathbf{u}_D \equiv \overline{\mathbf{u}} - \mathbf{u}_d \quad (1.79)$$

where $\varpi(\mathbf{u}_D, K)$ represents a retardation effect due to the interaction of mesoscales with the mean flow. If one assumes that $K^{-1}|\mathbf{u}_D|^2$ is smaller than unity, $\varpi(\mathbf{u}_D, K)$ can be written in the form:

$$\varpi \cong 1 - \frac{|\mathbf{u}_D|^2}{2K} \quad (1.80)$$

that was employed by previous authors (Ferrari and Nikurashin 2010; Meredith et al. 2012; Bates et al. 2014; Roach et al. 2018; Klocker and Abernathey 2014; Busecke and Abernathey 2019, BA). Both forms of $\varpi(\mathbf{u}_D, K)$ still require a parameterization of drift velocity $\mathbf{u}_d$ and kinetic energy which was found to be:

1.9 Diffusivity Tensor, Kinetic Energy, Mesoscale Diffusivity

$$K(x,y,z) = K_s(x,y)\, \Gamma(x,y,z) \tag{1.81}$$

where altimetry T/P (Schaffenberg and Stammer 2010) provide data on $K_s(x,y)$ and WOCE (2002) provide data on the vertical profile. None of the studies cited above parameterized $\Gamma(x,y,z)$ however, e.g., BA assumed a constant value. By contrast, we proceed as follows: the drift velocity is given by (1.51) and Canuto and Dubovikov (2006, CD6) derived the following vertical profile:

$$\Gamma(x,y,z) = |1 + a_0|^{-2}|a_0 + B_1(z)|^2 \tag{1.82}$$

where $B_1(z)$ is the first baroclinic mode solution of the eigenvalue problem $\partial_{zz}\varphi + (N/fr_d)^2\varphi = 0$, $\varphi = N^{-2}\partial_z B_1$ with $\partial_z B_1 = 0$ at $z = -H, 0$ and $B_1(0) = 1$, where H is the depth of the ocean. Relation (1.82) represents a partition between baroclinic and barotropic components, $B_1(z)$ and a_0. Wunsch (1997) concluded that the first baroclinic mode $B_1(z)$ dominates near $z = 0$; $a_0^2 = |B_1(-H)|$ was a good representation of a CD5 more complete form (Figs. 1.12 and 1.13).

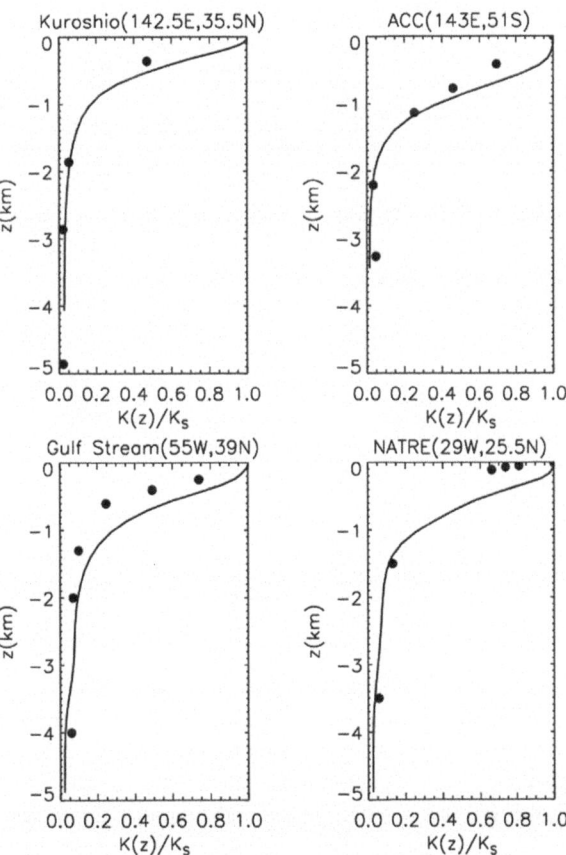

Fig. 1.12 Vertical profiles of the eddy kinetic energy Eq. (1.82) solid line vs. WOCE (2002) data, black dots. The form of the barotropic contribution a_0 is the same in all locations but the specific value depends on the depth of the ocean at that location. The results are from the GISS- OGCM averaged over the last 3 years of a 500 year simulation. The data are reproduced satisfactorily. (Reprinted with permission from Canuto et al. (2019). © NASA all rights reserved)

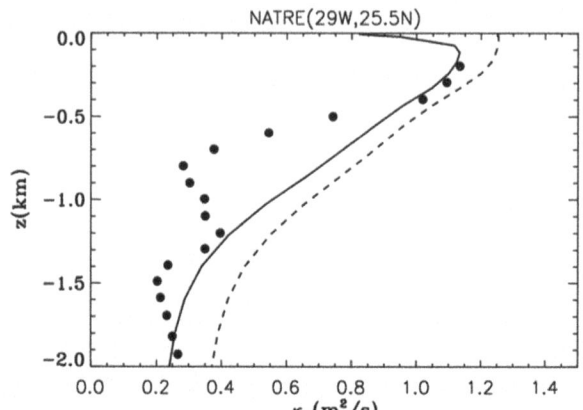

Fig. 1.13 Vertical profiles of the mesoscale diffusivity (1.79), represented by the solid line vs. the data at NATRE (dots, Ferrari and Polzin 2005). The dashed line denotes the model result with the suppression factor $\varpi=1$. The data are from Fig.10a of Bates et al. (2014). (Reprinted with permission from Canuto et al. (2019). © NASA all rights reserved)

Surface kinetic energy K_s. We determined $K_s(x, y)$ as the solution of the volume integrated, stationary eddy kinetic energy equation which becomes the production = dissipation balance:

$$\underbrace{\int_{-H}^{-h_*} F_v(\overline{b})dz}_{A-regime} + \underbrace{\int_{-h_*}^{0} F_v(\overline{b})dz}_{D-regime} = \int_{-H}^{0} \varepsilon(z)dz \quad (1.83)$$

where $F_v(\overline{b})$ is the vertical buoyancy flux due to *baroclinic instabilities* and $\varepsilon(z)$ is the rate of dissipation of $K(x,y,z)$. Since both the A-D fluxes depend linearly on κ_M, one has the relations:

$$\int_{-H}^{-h_*} F_v(\overline{b})dz = K_s^{1/2} K_A, \quad \int_{-h_*}^{0} F_v(\overline{b})dz = K_s^{1/2} K_D \quad (1.84)$$

where:

$$K_A = \frac{1}{2} r_d \int_{-H_b}^{-h_*} \varpi \Gamma^{1/2}(z) N^2 s \cdot \xi dz, \quad K_D = \frac{1}{2} r_d \int_{-h_*}^{0} \varpi \Gamma^{1/2}(z) N^2 s \cdot \Omega dz \quad (1.85)$$

which show that both the A-D productions scale like $K_s^{1/2}$.

Dissipation. In C19, it was shown that use of the Kolmogorov kinetic energy spectrum yielded a dissipation proportional to $K_s^{3/2}$ in which case (1.83) becomes a linear algebraic equation in K_s. The solution exhibited a global pattern similar to that of the T/P data (Schaffenberg and Stammer 2010, see Fig.16 lower panel) but the

1.9 Diffusivity Tensor, Kinetic Energy, Mesoscale Diffusivity

values were too large in the equatorial region and too small in the ACC. We then adopted the formulation of Cessi (2008) whose relations (13) and (19) yielded the following expression for the surface eddy kinetic energy:

$$\alpha_K K_s = (1 + BD)^{-1}(K_A + K_D) \quad , \quad BD \equiv \left[\int_{-H}^{0} \Gamma(z)dz\right]^{-1} \int_{-H}^{-h_*} \gamma(z)\Gamma(z)dz \tag{1.86}$$

where:

$$\alpha_K = (C_K r_d)^{-1} \int_{-H}^{0} \Gamma^{1/2}(z)dz, \, C_K = \left(\frac{3}{2}Ko\right)^{3/2} \tag{1.87}$$

The Kolmogorov constant is $4 < Ko < 8$, BD stands for bottom dissipation and $\gamma(z) = (2/\pi)^{1/2}(H/\delta_b)\exp(-\xi^2/2), \xi \equiv (z+H)\delta_b^{-1}$ with $\delta_b = 40$m. Figures 1.14, 1.15, 1.16, 1.17, 1.18, 1.19, 1.20, 1.21, 1.22 and 1.23, compare several model results with a variety of data.

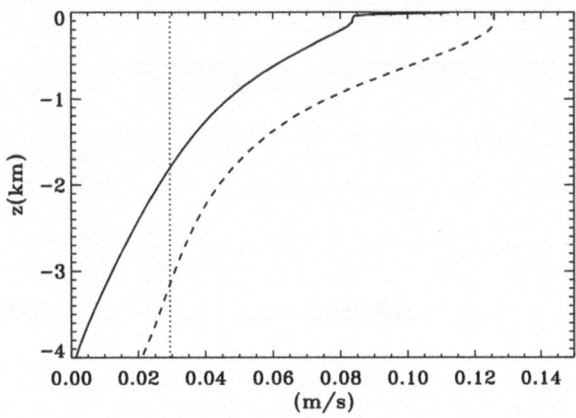

Fig. 1.14 z-profiles averaged between 61S and 56S and 110 W and 80 W of mean velocity- solid line; eddy drift velocity-dotted line; square root of eddy kinetic energy-dashed line. (Reprinted with permission from Canuto et al. (2019). © NASA all rights reserved)

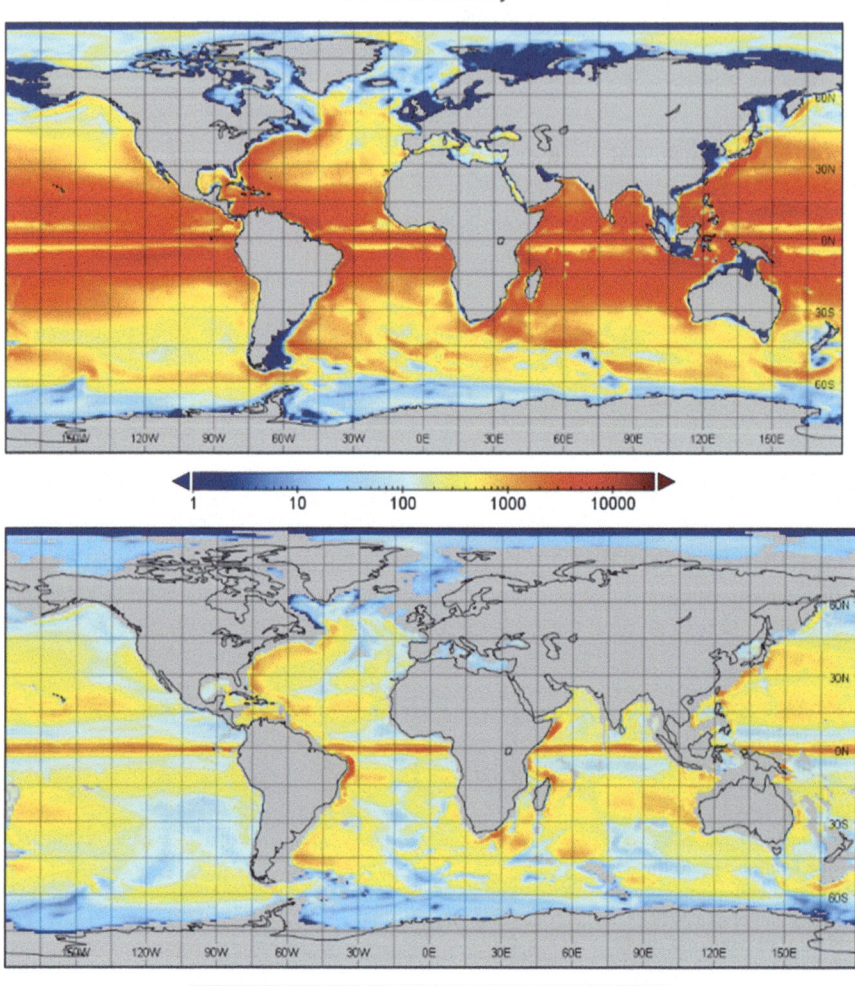

Fig. 1.15 Global maps of the mesoscale diffusivity averaged over the 498th–500th years of the simulation: top panel 6 m depth, bottom panel 2003 m depth. (Reprinted with permission from Canuto et al. (2019). C NASA all rights reserved)

1.9 Diffusivity Tensor, Kinetic Energy, Mesoscale Diffusivity

Surface mesoscale eddy kinetic energy

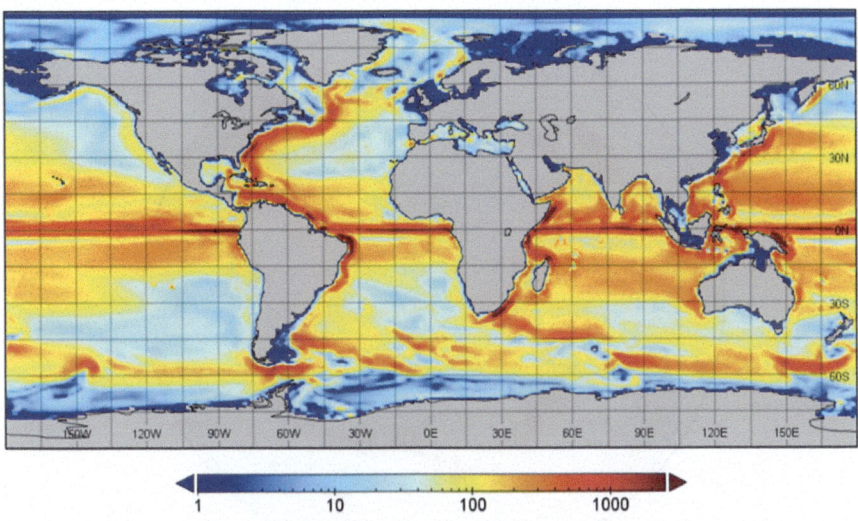

Surface mesoscale eddy kinetic energy (Scharffenberg and Stammer, 2010)

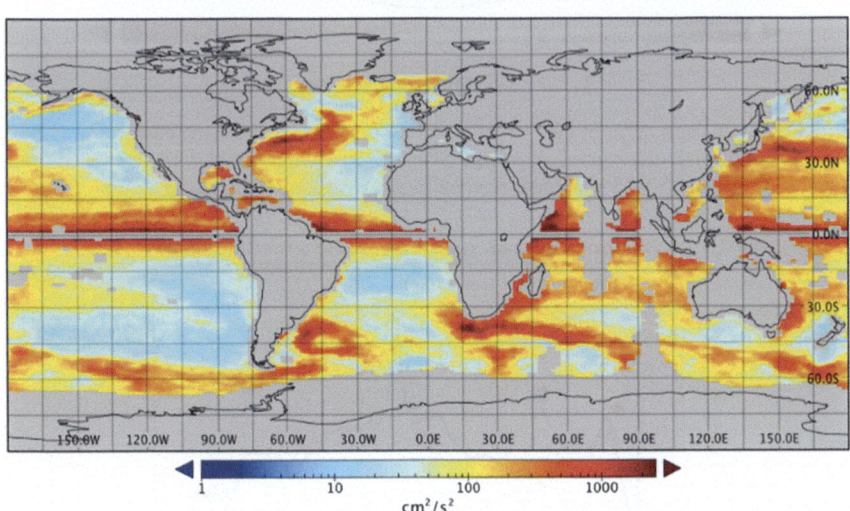

cm²/s²

Fig. 1.16 Upper panel: map of the present model surface eddy kinetic energy K_s, Eq. (1.86). Lower panel: surface eddy kinetic from the T/P data (Schaffenberg and Stammer 2010). (Reprinted with permission from Canuto et al. (2019). © NASA all rights reserved)

Surface mesoscale kinetic energy (Scharffenberg and Stammer, 2010)

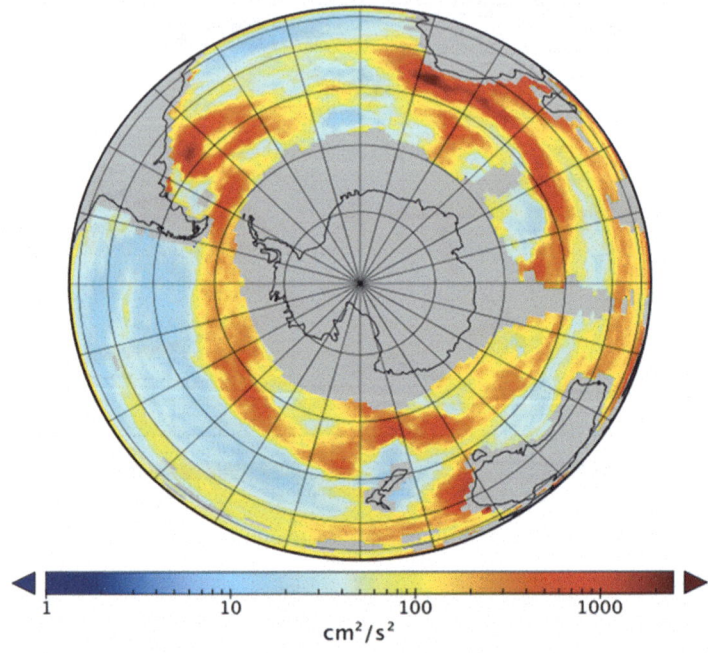

Fig. 1.17 Upper panel. Map of the present model ACC surface eddy kinetic energy, Eq. (1.86). Lower panel, the T/P data (Schaffenberg and Stammer 2010). (Reprinted with permission from Canuto et al. (2019). © NASA all rights reserved)

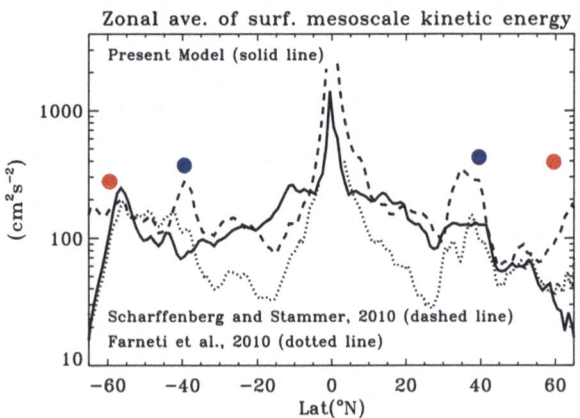

Fig. 1.18 Zonal average of the present model surface eddy kinetic energy Eq. (1.86) represented by the solid line vs. T/P data represented by the dashed line. The red-blue dots include suggested contributions of shear. The dotted line represents the zonal average of the surface eddy kinetic energy from the numerical simulations of Farneti et al. (2010) using the CM2.4 version of the GFDL code. (Reprinted with permission from Canuto et al. (2019). © NASA all rights reserved)

Fig. 1.19 Map of the ACC surface diffusivity (1.79). The results compare well with Fig.3f of Sallée et al. (2008) and Fig.3 of Le Sommer et al. (2011). (Reprinted with permission from Canuto et al. (2019). © NASA all rights reserved)

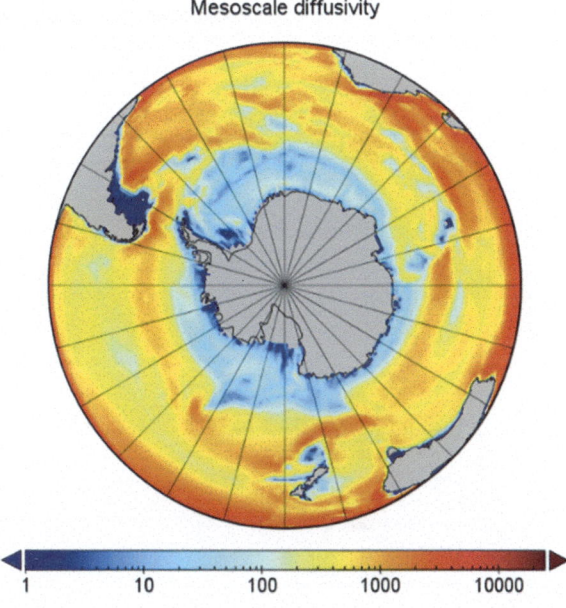

1.10 OGCM Results

We employed the ocean component of the coupled NASA-GISS OGCM model-E (Russell et al. 1995; Russell et al. 2000; Liu et al. 2003), the non-local version of the KPP vertical mixing scheme (Large et al. 1994), the 3D diffusivity tensor for an arbitrary tracer given in sec. 1.9, the mesoscale diffusivity (1.79) and the surface kinetic energy (1.86). The Code employs a mass coordinate approximately proportional to pressure with 32 vertical layers with thickness from ≈ 12 m near the surface to ≈ 200 m at the bottom. The horizontal resolution is $1.25°$ (longitude) by $1°$ (latitude). It is a fully dynamic, non-Boussinesq, mass-conserving free-surface ocean model using a quadratic upstream scheme for the horizontal advection of tracers and a centered difference scheme in the vertical. A 1800s time step is used for tracer evolution. Sea-ice dynamics, thermodynamics and ocean–sea-ice coupling are represented as in the CMIP5 model-E configuration (Schmidt et al. 2014), save that here ice is on the ocean model grid. To force the model, we used the CORE-I Protocol (Griffies et al. 2009) with fluxes obtained from bulk formulae the inputs to which are the ocean model surface state and atmospheric conditions derived from a synthesis of observations that repeat the seasonal cycle of a "normal year". The results presented below correspond to the output of the final 3 years of a 300-year run.

Comments (Frenger et al., 2015). Figure 1.14 shows the depth profiles averaged between (61–56)S and (110–80)W of the mean velocity (solid line), eddy drift velocity (dotted line) and square root of eddy kinetic energy (dashed line). If one

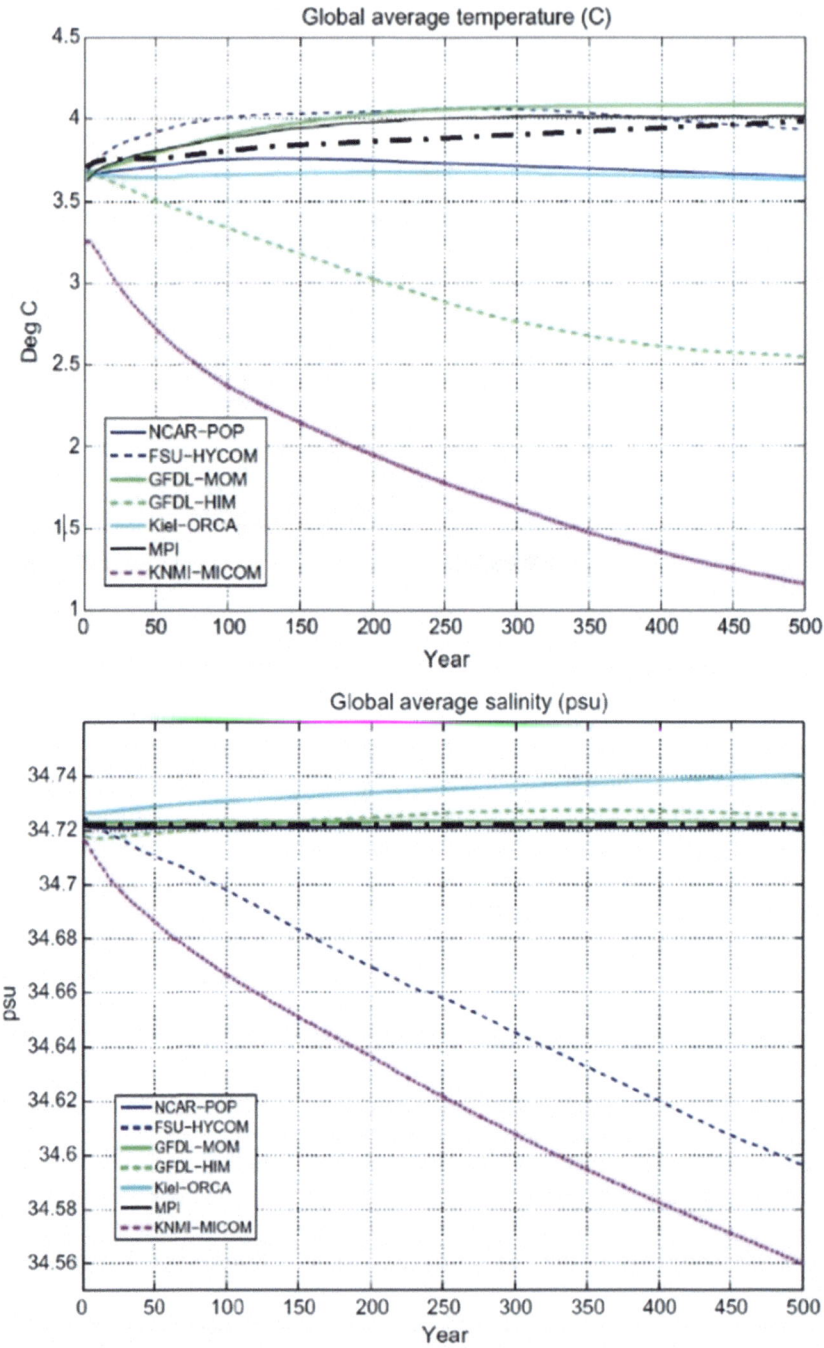

Fig. 1.20 Time series of globally and annually averaged ocean potential temperature (upper panel) and salinity (lower panel) (dash dotted thick black curves) compared with the results of seven OGCMs shown in Figs.3–4 of Griffies et al. (2009). (Reprinted with permission from Canuto et al. (2019). © NASA all rights reserved)

1.10 OGCM Results

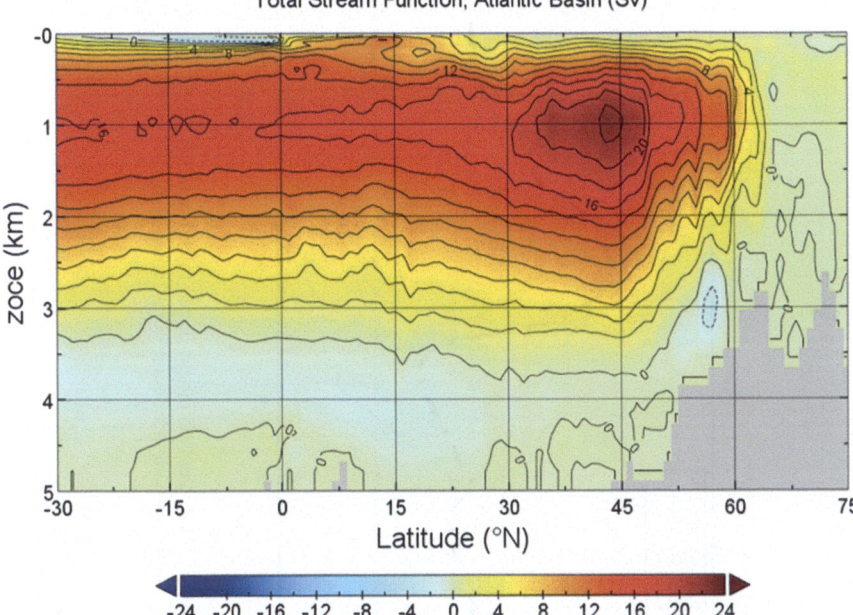

Fig. 1.21 Atlantic overturning circulation from seven OGCMs shown in Fig.23 of Griffies et al. (2009). (Reprinted with permission from Canuto et al. (2019). © NASA all rights reserved)

compares these results with those in the lower panel of Fig.10b of Tulloch et al. (2014), one notices several differences, the present eddy kinetic energy is steeper in the upper two hundred meters, in Tulloch et al. (2014) the drift velocities $\mathbf{u}_d$ was computed using altimetry data while in the present work is given by (1.51), the steering level is at 1.6 km while in the present case is at about 2 km. Finally, in Tulloch et al. (2014), K was also computed using data while in the present case is parameterized and later assessed as discussed above. As expected, the square root of the eddy kinetic energy is larger than that of the mean flow. Figure 1.15 shows global maps of the 3D mesoscale diffusivity (1.79) averaged over the 298th–300th years of the simulation: top panel, 6 m depth, bottom panel, 2 km depth. The decrease with depth is an indication of the surface enhanced eddy kinetic energy shown in Fig. 1.12. The results can be compared with those of Fig.8b of Klocker and Abernathey (2014). Figure 1.16 upper panel shows a map of $K_s(x, y)$ from Eq. (1.86) and the lower panel shows the T/P data (Schaffenberg and Stammer 2010). Considering that (1.86) is analytic, the fact that it captures the main features of $K_s(x, y)$ is a confirmation that baroclinic instabilities are a major contributor to the surface eddy kinetic energy. Figure 1.17 (upper panel) shows a map of the ACC surface eddy kinetic energy (1.86) and the lower panel shows the corresponding T/P data (Schaffenberg and Stammer 2010). Figure 1.18 shows the zonal average of Eq. (1.86), the T/P data and the results of a high-resolution numerical simulation by Farneti et al. (2010); CM2.4 version of the GFDL code, $1/4^0$ resolution, 27.75 km at

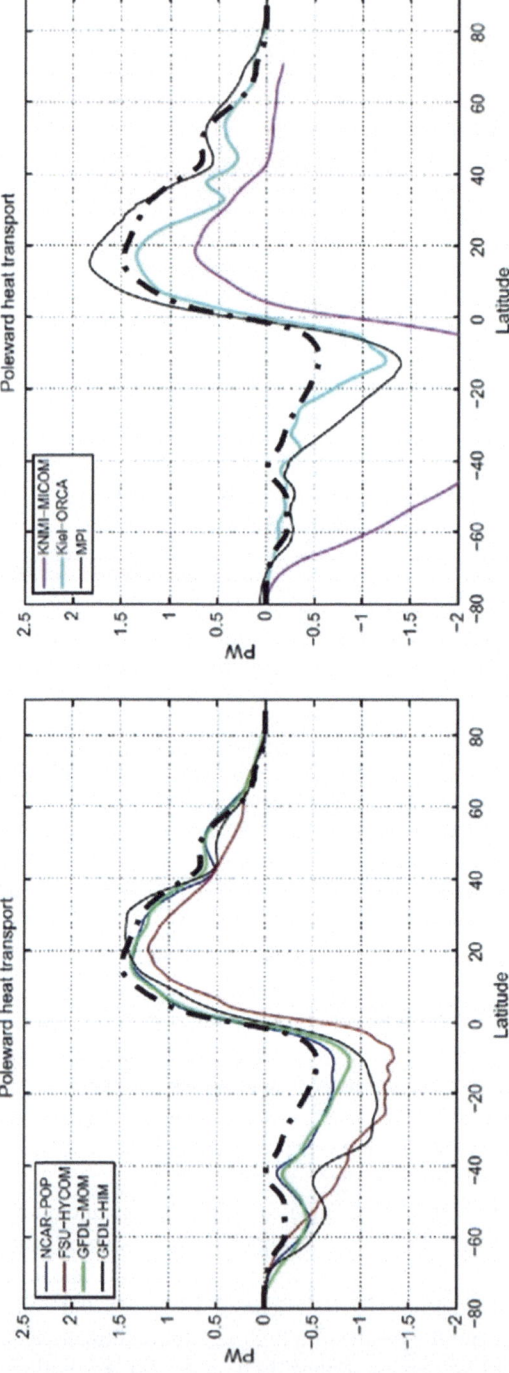

Fig. 1.22 Meridional heat transport in the global ocean averaged over the 491st–500th years of the simulation (dash-dotted thick black curve in each figure) compared with the results of seven OGCMs presented in Fig.22 of Griffies et al. (2009). (Reprinted with permission from Canuto et al. (2019). © NASA all rights reserved)

1.10 OGCM Results

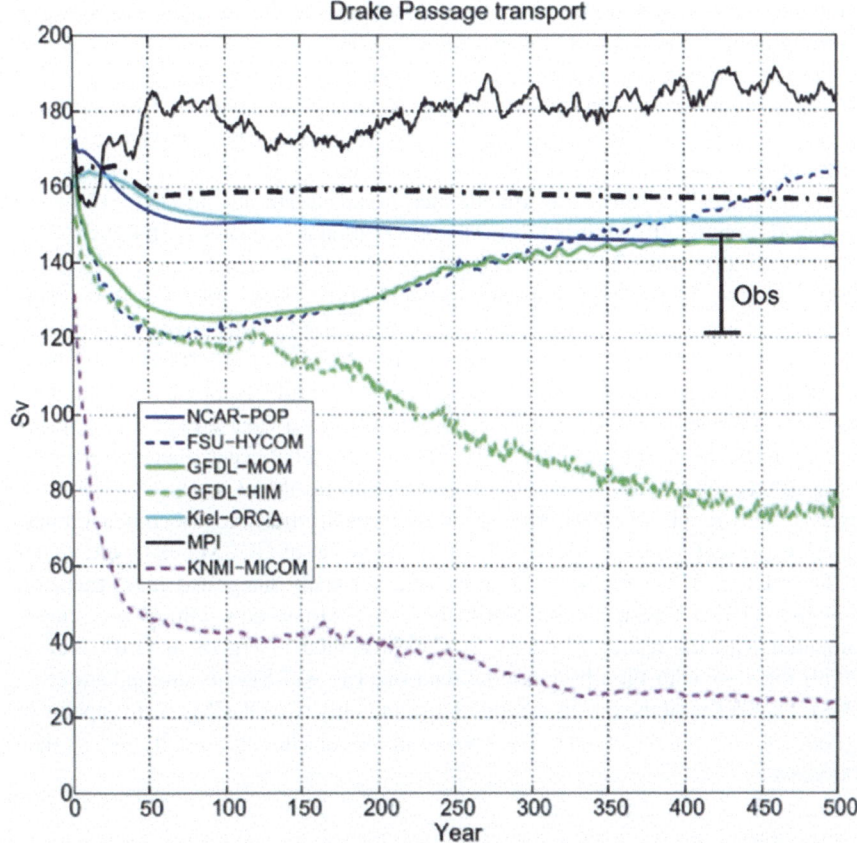

Fig. 1.23 Vertically integrated mass transport through the Drake Passage (dash dotted thick black curve) compared with the results of seven OGCM presented in Fig.18 of Griffies et al. (2009) with the observational data of (137 ± 7.8)Sv, (Cunningham et al. 2003). (Reprinted with permission from Canuto et al. (2019). © NASA all rights reserved)

the equator, 13.8 km at 60^0 North and 9 km at 70^0 N/S). These results are the only case in which we can compare model results with both T/P data and numerical simulations. On average, the simulation results are smaller than those of the present model that are closer to the data. There is an instructive message: at $\pm 40^0$ and $\pm 60^0$ relation (1.86) under predicts the surface eddy kinetic energy with respect to the T/P data. The suggested shear contribution represented by the blue-red dots brings the model results into better agreement with the T/P data. Figure 1.19 shows the ACC surface diffusivity (1.79) which compares well with the numerical simulations shown in Fig.3f of Sallée et al. (2008) and in Fig.3 of Le Sommer et al. (2011). Figure 1.20 shows the time series of the globally and annually averaged temperature and salinity. Of the seven OGCMs results shown in Fig.3 of Griffies et al. (2009), two exhibit a clear cooling tendency, one reaches stationarity only after 500 years while the other exhibits no tendency toward stationarity; the other five cases exhibit warming in time and reach stationarity after approximately 250 years. The result of the present model (black dash-dotted curve) exhibits a warming with a magnitude in the middle of the range of the other warming models. Of the seven OGCMs shown in Fig.4 of Griffies et al. (2009), two have large fresh drifts, one has a moderate salinity drift and the rest have small drifts. The present model salinity drift is among the smallest. Figure 1.21 shows the Atlantic meridional stream function ($Sv=10^6 m^3 s^{-1}$) computed with the diffusivity (1.79). The observational estimates is $16\pm2Sv$ (48^0N, Ganachaud 2003; Lumpkin et al. 2008), $15\pm2Sv$ (42^0N, Ganachaud and Wunsch 2000) and $13\pm2Sv$ (42^0N, Lumpkin and Speer 2003). Since this is a key oceanic feature, it is important to assess how well is reproduced by different parameterizations. Of the seven OGCMs shown in Fig.23 of Griffies et al. (2009), only three, NCAR-POP, GFDL-MOM and MPI OGCMs, yield results comparable to the data. The present model yields about 20Sv within the observed values. Figure 1.22 shows the meridional heat transport ($PW = 10^{15}W$) in the global ocean averaged over the 298th–300th years of the simulation together with results of seven OGCMs from Fig.22 of Griffies et al. (2009). Among the latter results there are discernable outliers while the present model result is well within the group of OGCMs that yield values of the order of 1PW. Figure 1.23 shows the vertically integrated mass transport through the Drake Passage with the diffusivity (1.79) (dash dotted thick black curve) compared with the results of seven OGCM presented in Fig.18 of Griffies et al. (2009) together with the observational data of (137 ± 7.8)Sv (Cunningham et al. 2003). The spread of the results is rather large and only two OGCMs seem capable of reproducing the observed data. The present model reaches stationarity in less than 100 years.

1.11 Subduction by Mesoscales

Subduction is an irreversible process that transfers water masses from the bottom of the mixed layer H to the interior thermocline. Given the observed subduction rate of ±200 m yr.$^{-1}$ (Mazloff et al. 2010) from the Southern Ocean State Estimate (SOSE), it is important to assess if the present mesoscale model is able to reproduce such value. The rate of subduction is given by (Cushman-Roisin 1987; Marshall 1997):

$$S_b = \frac{\partial H}{\partial t} + \underbrace{\overline{\mathbf{u}}\cdot\nabla H + \overline{w}}_{mean} + \underbrace{\mathbf{u}^+\cdot\nabla H + w^+}_{eddy} \qquad (1.88)$$

Sloyan and Rintoul (2001), Sallée et al. (2010) and Sallée and Rintoul (2011, SR11) concluded that the eddy component "*plays an order one role in the overall subduction in the Southern Ocean.*" SR11 employed a model in which the A-regime was treated with the stream function (1.56) with $T(z) = 1$ corresponding to the GM model while in the D-regime, they employed a tapering function $T(x, y, z)$ that depended

1.11 Subduction by Mesoscales

only z only with the boundary conditions $T(0) = 0$ and T(A-D interface) = 1. Finally, the A-D interface was assumed to be at H *and* the mesoscale diffusivity was taken as:

$$\kappa_M = 10^3 \text{m}^2\text{s}^{-1} \tag{1.89}$$

and the 2D form of Visbeck et al. (1997). *Fig.4 of SR11 shows that the resulting subduction rates were an order of magnitude smaller than SOSE data.* On the other hand, use of the diffusivities derived by Sallée et al. (2008) from surface drifter observations, yielded subduction rates that compared significantly better to the data. On those grounds, SR11 suggested a ten-fold increase of the mesoscale diffusivity (1.89). Since the mesoscale diffusivity is but one component of a larger mesoscale parameterization, Canuto and Cheng (2019, CC19) suggested to first assess the latter's overall performance followed by an assessment of the predicted Antarctic Circumpolar Current (ACC) subduction rate. CC19 employed the diffusivity (1.79) which, as shown in Fig. 1.19, compares well with the drifter data of Sallée et al. (2008). The resulting ACC subduction rate reproduced satisfactorily SOSE data without introducing ad hoc changes.

New D-regime mesoscale model. To parameterize the D-regime, SR11 employed the general relation $\nabla_H \cdot \mathbf{u}^+ + \partial_z w^+ = 0$, whereby the eddy-induced velocities read as follows:

$$\mathbf{u}^+ = -\partial_z[\kappa_M T(x,y,z)\mathbf{s}], \qquad w^+ = \nabla_H \cdot [\kappa_M T(x,y,z)\mathbf{s}] \tag{1.90}$$

Although T(x, y, z) depends on *x*, *y*, *z,* it was taken to be a function of *z* only, an assumption that has the following implication. Consider the second relation in Eq. (1.90):

$$w^+ = \kappa_M \mathbf{s} \cdot \nabla_H T(x,y,z) + T(x,y,z)\nabla_H \cdot \mathbf{s}\kappa_M \tag{1.91}$$

Under the assumption that T(x, y, z) depends only on *z,* the first term on the right-hand side of Eq. (1.91) vanishes which affects the subduction rates. As for *T(z),* SR11 adopted a straight line with $T(0) = 0$ to ensure that $w^+(0) = 0$. If the A–D interface is denoted by h_*, matching the fluxes requires $T(h_*) = 1$, and the choice of h_* is not trivial. For example, Gnanadesikan et al. (2007) concluded that a tapering approach yielded OGCMs results that were "*disconcerting*" because of the strong dependence on the isopycnal slope at h_*. Lastly, since tapering functions are a numerical device, the physical content of the D regime can hardly be represented that way since, as discussed in sec. 1b of C18, the A-D regimes satisfy different conservation laws that make use of tapering function a proposition of doubtful validity.

Subduction rates. Since the present subduction rates shown in Fig. 1.24 reproduce satisfactorily SOSE data, it remains to be seen if the mesoscale diffusivities predicted by the present model also reproduces the surface drifter data. Before we do

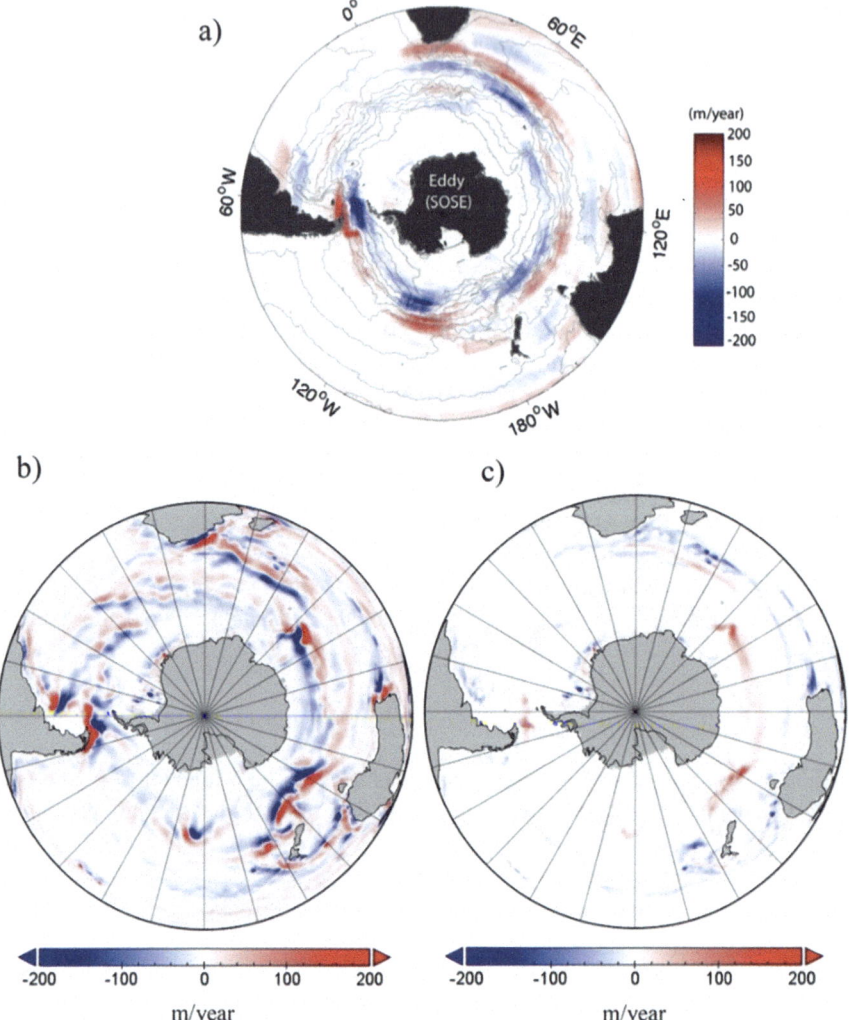

Fig. 1.24 (a) Subduction rates from SOSE (reproduced from SR11), (b) subduction rates from the present mesoscales model, and (c) subduction rates from the present model with $w^+ = 0$. The results correspond to an average of the last 20 winters (July–September) of a 500-yr OGCM run. (From Reprinted with permission from Canuto and Cheng (2019). © NASA all rights reserved)

so, we need to remark that the reason to study the case in Fig. 1.24 with $w^+ = 0$ was to highlight the contribution of w^+ since Hirake et al. (2016), using an eddy-resolving simulation, reported that the w^+ contribution is large; indeed, Fig. 1.24 shows that if $w^+ = 0$, the subduction rates do not reproduce SOSE data.

Next, consider Fig. 1.25. The 3D diffusivities of the present model shown in the left panel compare well to the results in Fig.3 of Sallée et al. (2008); for completeness, the right panel shows the 2D diffusivities used in SR11.

1.12 Upwelling by Mesoscales

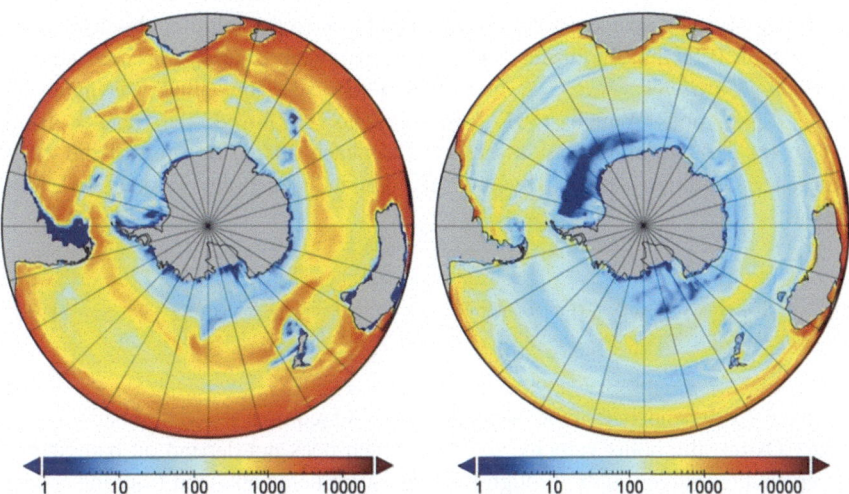

Fig. 1.25 Surface κ_M (m^2 s^{-1}) from the (left) 3D and (right) 2D model. The results correspond to an average of the last 20 winters (July–September) of a 500-yr OGCM run. (Reprinted with permission from Canuto and Cheng (2019). © NASA all rights reserved)

Conclusions. The two models for κ_M employed by SR11 yielded subduction rates smaller than SOSE data by an order of magnitude. On the other hand, the κ_M derived by Sallée et al. (2008) from surface drifter data were larger and reproduced more closely the data. Thus, SR11 suggested boosting the diffusivity Eq. (1.89) ten times. On the other hand, the new mesoscale model reproduces satisfactorily topology (subduction equatorward and obduction poleward) and magnitudes of SOSE data without introducing ad hoc changes.

1.12 Upwelling by Mesoscales

Upwelling in the ACC is an important process for multiple reasons, (a) it closes the global overturning circulation, (b) it upwells cold water from the depth which is the primary process responsible for approximately 70% of the ocean heat uptake (Marshall and Speer 2012) and (c) is responsible for 40% of the anthropogenic carbon uptake. Recent mesoscale numerical simulations with a 0.1^0 resolution by Barthel et al. (2022) and Yung et al. (2022), showed that in the Southern Ocean upwelling is not zonally uniform, as commonly assumed, but exhibits *"upwelling hotspots"*, e.g., Macquarie Ridge, Pacific-Antarctic Ridge, Drake Passage, Southwest Indian Ridge, Kerguelen Plateau, e.g., Fig.7A, 8D of Yung et al. (2022). Though the upwelling hotspots cover only about 30% of the area, they account for more than 70% of the upwelling transport highlighting their relevance. A new paradigm is emerging whereby eddy induced upwelling is better viewed as a topography-induced feature rather than of high eddy kinetic energy regions. Since the commonly used coarse resolution ocean codes have a resolution about 10 times coarser than 0.1^0 thus requiring a mesoscale parameterization, *can such models reproduce the above results?* A first attempt to assess upwelling was made by Liang, Spall and Wunch (2017, LSW) who employed the results from ECCOv4r1 at 1^0 resolution (Wunsch and Heimbach 2013). Because of the zero-divergence condition satisfied by the 3D eddy induced bolus velocity $(\mathbf{u}^+, w^+)$, one has:

$$\nabla_h \cdot \mathbf{u}^+ + \partial_z w^+ = 0, \qquad w^+(z) = -\int_{-H(x,y)}^{z} dz' \nabla_h \cdot \mathbf{u}^+(z') \qquad (1.92)$$

where the x,y dependence of the ocean depth H(x,y) requires a model of bathymetry. In ECCOv4r1, the bolus velocity was GM which does not depend on bathymetry that enters the problem only through the large-scale fields obtained from the coarse resolution OGCMs.

By contrast, in the present case bathymetry affects both large-scale fields and the mesoscale parameterization. Since the A-D regimes have different $\mathbf{u}^+$, the vertical eddy velocity is different in the A-D regimes, specifically:

$$A\text{-}regime: h_* \leq |z| \leq H, \qquad w_A^+(z) = -\int_{-H}^{z} dz' \nabla_h \cdot \mathbf{u}_A^+(z') \qquad (1.93)$$

where H is the ocean depth. The second term is the sum of three contributions:

$$W_1(z) = \int_{-H}^{z} dz' \nabla_h \cdot \mathbf{u}_{GM}^+(z')$$

$$W_2(z) = \left(\text{fr}_d^2\right)^{-1} \int_{-H}^{z} dz' \nabla_h \cdot \Lambda_2(z'), \qquad \Lambda_2 \equiv \kappa_M \mathbf{e}_z \times \mathbf{\bar{u}}$$

$$W_3(z) = \frac{1}{\text{fr}_d^2} (\mathbf{e}_z \times \mathbf{\bar{u}}_d) \cdot \int_{-H}^{z} dz' \nabla_h \kappa_M \qquad (1.94)$$

The LSW model corresponds to the case $W_{2,3} = 0$.

D-regime. The vertical buoyancy vertical flux $F_v(\overline{b})$ is shown in Fig.2 of Appendix C in four locations and the largest values occur in the ACC where its value is $\sim 5 \cdot 10^{-8} \text{m}^2\text{s}^{-3}$. The 2D bolus velocity is then given by:

$$\mathbf{u}_D^+ = -\frac{\partial}{\partial z}[\kappa_M T(z)\mathbf{s}] \qquad (1.95)$$

Since at the bottom of the D-regime $z = -h_*$, the diapycnal flux $F_d = F_v - \mathbf{s} \cdot \mathbf{F}_H$ vanishes and as shown in Canuto and Dubovikov (2011), the horizontal buoyancy flux is $\mathbf{F}_H(\overline{b}) = \kappa_M N^2 \mathbf{s}$, one has:

$$z = -h_* : \quad F_v(\overline{b}) = \kappa_M N^2 s^2 \quad T_* = 1 \qquad (1.96)$$

The D-regime vertical velocity is then given by:

$$D\text{-regime}: \quad 0 \leq |z| \leq h_*, \quad w_D^+ = -\int_{-h_*}^{0} dz \nabla_h \cdot \mathbf{u}_D^+(z) = -\nabla_h \cdot \kappa_{M^*} \mathbf{s}_* \qquad (1.97)$$

1.13 Sub-Mesoscales, SM

In the past few years, it was found that SM are relevant in different quarters especially in biogeochemical studies (McWilliams 2016; Mahadevan 2016; McGillicuddy 2016; Levy et al. 2001, 2013, 2018) that treat the time evolution of tens of passive tracers, Carbon Cycle studies that treat the export or subduction from the surface layers to a depth of (100–1000), e.g., Karleskind et al. (2011), Sanders et al. (2014), and Bopp et al. (2015) and very high-resolution numerical simulations at mid-latitude North Atlantic suggesting that SM may affect the large-scale structure of the upper ocean (Sinha et al. 2022) and the suggestions that SM may be a key component of the global heat budget (Su et al. 2018). A key ingredient is the *vertical buoyancy flux* that we treat next.

Using linear stability arguments and numerical simulations, Fox-Kemper et al. (2008, 2011, FK) proposed the following parameterization of the SM vertical buoyancy flux:

$$F_v(\overline{b}) = -4C_e \sigma(z) h^2 |f|^{-1} |\nabla_H \overline{b}|^2, \quad C_e = 0.06 \qquad (1.98)$$

Here, $z < 0$, $\sigma(z) = z/h(1 + z/h) < 0$, h is the vertical extent of the SM regime, f is the Coriolis parameter and $\nabla_H \overline{b}$ is the horizontal buoyancy gradient representing baroclinic instabilities. Figure 1.26 shows the comparison of (1.98) vs. the numerical simulation data by Capet et al. (2008) for the summer California Current. The FK8 model (dashed line) does not reproduce the data. The new model given in Eq. (1.104) and (1.105) provides a better representation of the data.

The flux (1.98) lacks two ingredients, a dependence on the *SM, kinetic energy* EKE and *wind stresses*. Since the latter are the largest in the ocean upper layers where SM reside, it seems difficult to justify their absence.

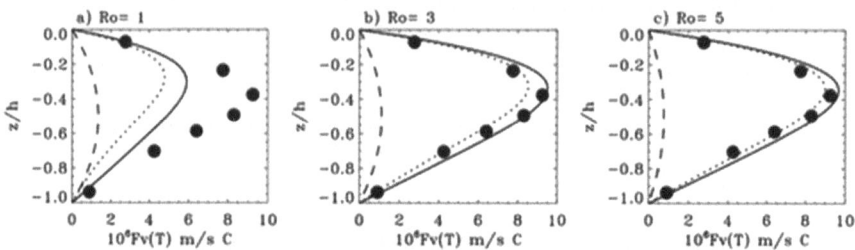

Fig. 1.26 Comparison of the FK8 model of the vertical temperature flux $F_v(T)$ given by Eq. (1.98) dashed line vs. the numerical simulation data of Capet et al. (2008) black dots. The dotted line representing the new flux, Eq. (1.104) and (1.105) provides a closer representation of the data; the solid line is the sum of the dotted and dashed lines. (Reprinted with permission from Canuto and Dubovikov (2010). © NASA all rights reserved)

SM kinetic energy. As shown in CD10, a feature unique to SM[1] is their dependence on the Rossby number Ro:

$$\frac{1}{2}\text{Ro} = \frac{K_{SM}}{r_s^2 f^2} \tag{1.99}$$

where K_{SM} is the SM eddy kinetic energy and $r_s = (0.1\text{–}10)$km are the horizontal sizes of SM. Since when EKE vanishes, the flux must also vanish, one has the relations:

$$C_e \to C_e(\text{Ro}), \qquad C_e(0) = 0 \tag{1.100}$$

In addition, at some Ro, $C_e(\text{Ro})$ must reproduce the value 0.06 of the FK model. The CD10 model derived the following relation:

$$C_e(\text{Ro}) = \frac{1}{8} \frac{\text{Ro}}{1 + \text{Ro}^2} \tag{1.101}$$

that satisfies (1.100) and is such that:

$$\text{Ro} = 1, \quad C_e = 0.06 \tag{1.102}$$

The FK models assume Ro $=1$ everywhere in the ocean which is not consistent with available data (Rosso et al. 2014; Thompson et al. 2016) showing the Ro is location

---

[1] In Appendix C2 of CD10, it was shown that if one separates the eddy velocity into rotational (divergent free) and divergent (curl free), the former is given by $u_R = -ikp'(1 + \frac{1}{2}\text{Ro}^2)^{-1}$. In treating mesoscales for which Ro < 1, one recovers the geostrophic relation $u_R \to -ikp' = u_g$. However, in the SM case when Ro > 1, one must use the complete form of the rotational velocity given by Eq.C12.

1.13 Sub-Mesoscales, SM

dependent. In addition, the solution of the EKE equation shows values of Ro > 1, Figs. 1.30, 1.31 and 1.32.

Wind stresses and EBF. To include wind stresses in (1.98), one can add a term of the form $\tau_w \cdot \nabla_H \bar{b}$; on the other hand, since in the common case of down-front winds, wind stresses are along the geostrophic velocity, $\tau_w \parallel \mathbf{u}_g$, $\mathbf{u}_g \perp \nabla_H \bar{b}$, one has $\tau_w \cdot \nabla_H \bar{b} = 0$. One may then consider the vector product $\nabla_H \bar{b} \times \tau_w$ whose vertical component is $(\nabla_H \bar{b} \times \tau_w) \cdot \mathbf{e}_z$, where $\mathbf{e}_z = (0,0,1)$, Since $\nabla_H \bar{b} \times \tau_w$ is not a vector but a pseudo-vector, it becomes a vector when multiplied by a pseudo-scalar like the Coriolis parameter that yields the *Ekman buoyancy flux*[2]:

$$\rho EBF = f^{-1}(\nabla_H \bar{b} \times \tau_w) \cdot \mathbf{e}_z \tag{1.103}$$

Eq. (1.98) then acquires the new form:

$$F_v(\bar{b}) = -4C_e(Ro)\sigma(z)h^2|f|^{-1}|\nabla_H \bar{b}|^2 + \tilde{C}_e(z, Ro)EBF \tag{1.104}$$

where $\tilde{C}_e(z, Ro)$ satisfies the same condition as $C_e(z, Ro)$, it must vanish at $z = (0, -h)$ and Ro = 0. CD10 derived the following results:

$$C_e(Ro) = \frac{A(Ro)}{8Ro}, \quad \tilde{C}_e(z, Ro) = A(Ro)\lambda(z)$$

$$A(Ro) = \frac{Ro^2}{1 + Ro^2}, \quad \lambda(z) = 1 + z/h \text{-}\alpha(z)\text{-}Ro^{-1}\beta(z) > 0 \tag{1.105}$$

Contrary to the first positive term in (1.104), the EBF term is negative in the ACC (f < 0, τ_x > 0; Hellerman and Rosenstein 1983):

$$EBF \cong -\rho^{-1}|f|^{-1}N^2\tau_x s_y \tag{1.106}$$

In summary one can draw the following conclusions: *first,* without EBF, it is not possible to reproduce Capet et al. data (2008); *second,* Ro is not unity, as in the FK model, but in agreement with the values derived in sec. 1.17; *third,* it is important to recall that in the ACC, EBF is negative thus acting as a sink of EKE. As shown below, this is the reason why in the ACC, SM *de-stratify* the flow (thus favoring C-sequestration and enhancing subduction) which is a novel feature since it is usually thought that SM *re-stratify* the flow.

---

[2] We used the Ekman form of the stresses $\tau(z)$(Chereskin 1995):

$$\tau(z) = \alpha(\zeta)\tau_w + f|f|^{-1}\beta(\zeta)\mathbf{e}_z \times \tau_w, \quad \alpha(z), \quad \beta(z) \equiv e^\zeta(\cos\zeta, \sin\zeta), \quad \zeta = z/\delta_E$$

$\delta_E = (2\nu_t|f|^{-1})^{1/2}$ is the Ekman depth, ν_t is the momentum diffusivity and $E = (\delta_E/h)^2$ is the Ekman number.

1.14 SM Contribution to Subduction, Re-Stratification and Eddy Compensation

The contributions of small-scale turbulence (ss) and mesoscales (M) were analyzed by Karleskind et al. (2011), Rodgers et al. (2014) and Gnanadesikan et al. (2015). Bopp et al. (2015, Fig. 4) showed that in the Southern Ocean, mesoscales contribute - 0.08PgC/yr. while (ss) contribute +0.69PgC/yr. playing a leading role. Rodgers et al. (2014) focused on wind stirring and suggested new forms of the mixed layer depth MLD. However, wind stirring also contributes to the vertical flux generated by baroclinic instabilities that create sub-mesoscales SM that were not included in that study and the goal here is to quantify their contribution. Vertical integration of the transport Eq. (1) of Levy et al. (2013) results in Eq. (2) of that paper but in which we consider only two eddy terms, vertical 1D mixing and SM (all the variables are computed at the bottom of the mixed layer z = -H):

$$h\partial_t C + (\partial_t h + \bar{\mathbf{u}} \cdot \nabla_H h + \bar{w})C - k_v \partial_z C + \underbrace{\int_0^h F_v(z)dz}_{SM} = \underbrace{\int_0^h S_C(z)dz}_{EXT}$$

(1.107)

where the last term represents external sources while the SM vertical flux is given below.

SM vertical tracer flux. In contrast to the vertical mixing scheme like KPP in which a vertical shear instability produces vertical mixing, baroclinic instabilities represented by horizontal gradients, are known to produce vertical fluxes for which CD10 derived the following form:

$$F_v(\bar{c}) = -\boldsymbol{\kappa}_H \cdot \nabla_H \bar{c} \ , \qquad \boldsymbol{\kappa}_H \equiv A(Ro)\left[\mathbf{u}_* \cdot Ro^{-1}\frac{f}{|f|}\mathbf{e}_z \times \mathbf{u}_*\right]$$

$$\mathbf{u}_* = \int_0^z dz\tilde{\mathbf{u}} \ , \quad \tilde{\mathbf{u}} \equiv \bar{\mathbf{u}} - h^{-1}\int_{-h}^0 dz\bar{\mathbf{u}}$$

(1.108)

If one writes the 2D mean velocity $\bar{\mathbf{u}}$ as the sum of geostrophic and a-geostrophic (wind stresses) components, $f\partial_z \bar{\mathbf{u}}_g = \mathbf{e}_z \times \nabla_H \bar{b}$, $f\mathbf{u}_{ag} = -\mathbf{e}_z \times \partial_z \boldsymbol{\tau}/\rho$, one obtains after some algebraic steps:

$$geostrophy: \quad F_v(\bar{c}) = -\mathbf{K}_g \cdot \nabla_H \bar{c} \ ,$$

1.14 SM Contribution to Subduction, Re-Stratification and Eddy Compensation

$$\mathbf{K}_g \equiv \frac{A(Ro)}{2Ro}\sigma(z)h^2|f|^{-1}\left[\nabla_H \overline{b} + Ro\frac{f}{|f|}\mathbf{e}_z \times \nabla_H \overline{b}\right] \quad (1.109)$$

a-geostrophy : $F_v(\overline{c}) = -\mathbf{K}_{ag} \cdot \nabla_H \overline{c}$,

$$\mathbf{K}_{ag} \equiv A(Ro)\left[\lambda(z)f^{-1}\mathbf{e}_z \times \boldsymbol{\tau}_w + \mu(z)|f|^{-1}\boldsymbol{\tau}_w\right]$$

$$\mu(z) = \beta(z) + Ro^{-1}\left[1 + \frac{z}{h} - \alpha(z)\right]$$

Some comments may be useful. The first is that at the bottom of the diabatic regime, $z = -h$, the vertical fluxes vanish. On the other hand, it would be incorrect to identify such depth with the mixed layer depth MLD. This is because below the latter, the flow is still diabatic, as shown by the numerical simulations results of Mensa et al. (2013), Veneziani et al. (2014), Ramachandran et al. (2014), Luneva et al. (2015) and Buckingham et al. (2017). At this point, the sign of the vertical flux (1.109) cannot yet be decided, it can be positive or negative. Positive values would *lower* the contribution of 1D vertical mixing, which, as shown in fig.4 of Bopp et al. (2015), is the largest contribution in the SOC; by contrast, negative values would *enhance* the contribution of vertical mixing. In either case, it seems that the SM contribution ought to be included in biogeochemical studies. In sec. 5 of C19, the suggestion was made that h_* should be less than the depth of the thermocline since at that depth the stratification is too strong for the D regime to exist. The following relation:

$$h_* = \frac{1}{2}\left[HBL + maxN^2(z)\right] \quad (1.110)$$

is a heuristic expression halfway between HBL and the thermocline. We constructed maps of h/HBL and h/MLD, where MLD is computed using the potential density criterion $0.03 kgm^{-3}$. We obtained the following results: maximum values: $h_*/HB = 3.7$, $h_*/MLD = 2.9$, mean values: $h_*/HBL = 1.5$, $h_*/MLD = 1.2$, which confirm the results of the numerical simulations of Mensa et al. (2013), Veneziani et al. (2014), Ramachandran et al. (2014), and Luneva et al. (2015). Buckingham et al. (2017) also suggested a similar relation for h_* but the first term was the MLD instead of HBL. To compare the fluxes (1.109) with the vertical flux models usually expressed as:

$$F_v(\overline{c}) = -K_v\frac{\partial \overline{c}}{\partial z} \quad (1.111)$$

we introduce the slopes:

and rewrite the first of (1.109) in the compact form:

$$\mathbf{s} = -\frac{\nabla_H \bar{b}}{\partial \bar{b}/\partial z}, \quad \mathbf{s}_c = -\frac{\nabla_H \bar{c}}{\partial \bar{c}/\partial z} \tag{1.112}$$

$$F_g(\bar{c}) = K_g \frac{\partial \bar{c}}{\partial z}, \quad K_g \equiv -\frac{A(Ro)}{2Ro}\sigma(z)N^2 h^2 |f|^{-1}\left[\mathbf{s}\cdot\mathbf{s}_c + Ro\frac{f}{|f|}(\mathbf{e}_z \times \mathbf{s})\cdot\mathbf{s}_c\right] \tag{1.113}$$

where $\sigma(z)<0$. Analogously, for the second relation in (1.109), we have:

$$F_{ag}(\bar{c}) = K_{ag}\frac{\partial \bar{c}}{\partial z}, \quad K_{ag} \equiv A(Ro)\left[\lambda(z)f^{-1}(\mathbf{e}_z \times \boldsymbol{\tau}_w)\cdot\mathbf{s}_c + \mu(z)|f|^{-1}\boldsymbol{\tau}_w\cdot\mathbf{s}_c\right] \tag{1.114}$$

which exhibit the combinations:

$$\mathbf{s}\cdot\mathbf{s}_c, \quad (\mathbf{e}_z \times \mathbf{s})\cdot\mathbf{s}_c, \quad \boldsymbol{\tau}_w\cdot\mathbf{s}_c, \quad (\mathbf{e}_z \times \boldsymbol{\tau}_w)\cdot\mathbf{s}_c. \tag{1.115}$$

For example, Omand et al. (2015) assumed $\nabla_H \bar{b} \parallel \nabla_H \bar{c}$, in which case the second term in (1.113) and the ag-fluxes make no contribution. Finally, FK suggested the model:

$$F_v(\bar{c}) \propto \nabla_H \bar{b}\cdot\nabla_H \bar{c} \propto \mathbf{s}\cdot\mathbf{s}_c \tag{1.116}$$

that represents only the first term in the g-component in (1.113) while the other three terms are missing. *Both O15 and FK are incomplete representations of the tracer fluxes.*

1.15 SM Contribution to Re-Stratification and Eddy Compensation

The total rate of stratification by mean advection and SM is given by the following expression:

$$p_g \frac{\partial N^2}{\partial t}\bigg|_{total} = (1+p)Ro-p$$

$$p_g \equiv (1+Ro^2)|f||\nabla_H \bar{b}|^{-2}, \quad p \equiv \frac{|f|EBF}{E^{1/2}h^2|\nabla_H \bar{b}|^2} \tag{1.117}$$

where the subscript g stands for geostrophy, p_g has units of (time)^3 and p is dimensionless. While the above relations are valid for arbitrary Ro and EBF, in

1.15 SM Contribution to Re-Stratification and Eddy Compensation

the FK assumption of Ro = 1, p cancels out and *only re-stratification is possible*. On the other hand, if one considers values of Ro other than unity, the results are quite different:

$$\text{EBF} > 0 : \text{Ro} > \frac{p}{1+p} \text{ re-stratification}, \quad \text{Ro} < \frac{p}{1+p}, \text{de-stratification} \quad (1.118)$$

Since Ro > 1, de-stratification is not allowed. In the ACC where:

$$f < 0, \text{EBF} = -|f|^{-1} N^2 \tau_x s_y, \quad p = -E^{-1/2} \frac{\tau_x/\rho}{h^2 N^2} \frac{s_y}{s^2} \quad (1.119)$$

the results are:

$$\text{Ro} < \frac{|p|}{|p|-1} \text{ re-stratification}, \quad \text{Ro} > \frac{|p|}{|p|-1}, \text{de-stratification} \quad (1.120)$$

Re-stratification is not allowed, the model predicts de-stratification that favors C-sequestration. The result is unexpected since SM are usually viewed as an agent of re-stratification (Oschlies, 2002). Moreover:

$$h = 10^2 m, \quad \text{EBF} = 10^{-8} m^2 s^{-3}, \quad \nabla_H \bar{b} = 10^{-8} s^{-2}, \quad |p| \simeq E^{-1/2} > 1 \quad (1.121)$$

Eddy compensation. Hallberg and Gnanadesikan (2006, HG) were the first to show that in the ACC, *adiabatic mesoscales exhibit a negative stream function* Ψ_e *that partially compensates the positive Ekman stream function* Ψ_{Ek}:

$$\rho \Psi_{Ek} = -f^{-1} \tau, \quad \Psi_{Ek} = \frac{\tau_x/\rho}{|f|} = O(1) m^2 s^{-1} \quad (1.122)$$

where the second relation corresponds to the Southern Ocean where f < 0 and for a wind stress of $0.1 Nm^{-2}$. To estimate the stream function due to SM, we employ relation (1.56):

$$\Psi_e = -\kappa_e s \times e_z, \quad \kappa_e \equiv \frac{F_v(\bar{b})}{N^2 |s^2|} \quad (1.123)$$

The stream function due to the *a-geostrophic* EBF term in (1.104), is given by:

$$\Psi_{ag} = -\kappa_{ag} s_y, \quad \kappa_{ag} \equiv A(\text{Ro}) \lambda(z) \frac{\text{EBF}}{N^2 s^2} \quad (1.124)$$

In the ACC where $\text{EBF} \cong -\rho^{-1} |f|^{-1} N^2 \tau_x s_y$, we have:

$$\Psi_{ag} = A(Ro)\lambda(z)\Psi_{Ek} \qquad (1.125)$$

Contrary to the case of mesoscales, the SM stream function is now positive. If we further depth average $< . >$ the function $\lambda(z)$ and use the relations $2 < \alpha > \, = E^{1/2}$, $2 < \beta > \, = -E^{1/2}$ of Appendix B, (1.125) becomes:

$$\Psi_{ag} \simeq \frac{1}{2}A(Ro)\left[1 - E^{1/2}(1 - Ro^{-1})\right]\Psi_{Ek} \qquad (1.126)$$

For $Ro > 1$, $A(Ro) \simeq 1$ and since $E < 1$, we finally obtain:

$$\Psi_{ag} \simeq \frac{1}{2}\Psi_{Ek} \qquad (1.127)$$

The wind-driven SM stream function is *positive, opposes compensation and enhances the mean Ekman stream function up to 50%*. By comparison, the first, geostrophic term in (1.104) yields the stream function:

$$\Psi_g = -\, 4C_e|\sigma(z)||f|^{-1}h^2N^2s_y \simeq -\, 4s_y \qquad (1.128)$$

It is negative, favors eddy compensation but is negligible compared to the Ekman term (1.122).

1.16 PV Destruction by SM

Numerical simulations by Thomas (2005) showed that a *positive surface value of the vertical frictional PV flux induces PV destruction by down-front winds*. The numerical resolution of 12 km did not capture SM and their effect on PV destruction could not be assessed. Canuto and Cheng (2021, CC21) studied the problem with the following results. While in the absence of SM, there were no restrictions on where PV destruction may occur, in the presence of SM, the condition of a positive surface vertical frictional flux is limited to SM regimes of either $Ro < 1$ or $Ro > 10$. Since the SM kinetic energy is location dependent, so is PV destruction. The overall conclusion is that PV destruction is a much rarer occurrence than previously thought. To prove it, we use the conservation law for potential vorticity Q (Haynes and McIntyre 1987):

1.16 PV Destruction by SM

$$\frac{\partial Q}{\partial t} + \nabla \cdot \mathbf{J} = 0 \qquad (1.129)$$

where $\mathbf{J} = \mathbf{J}_a + \mathbf{J}_d + \mathbf{J}_f$ are the *advective* (a), *diabatic* (d) and *frictional* (f) fluxes. The vertical component of the *frictional flux* $J_f = \mathbf{J}_f \cdot \mathbf{e}_z$ is given by:

$$J_f = \underbrace{\left(\frac{\partial}{\partial z}\overline{\mathbf{u}'w'} \times \nabla_H \overline{b}\right) \cdot \mathbf{e}_z}_{SM} + \underbrace{\left(\nabla \overline{b} \times \frac{\partial \boldsymbol{\tau}}{\partial z}\right) \cdot \mathbf{e}_z}_{Ekman} \qquad (1.130)$$

where $\overline{w'\mathbf{u}'}$ are the SM induced momentum fluxes given by:

$$\frac{\overline{w'\mathbf{u}'}}{2|f|A(Ro)} = \frac{A(Ro)}{Ro}\boldsymbol{\kappa}(z) + \frac{f}{|f|}\boldsymbol{\kappa}(z) \times \mathbf{e}_z$$

$$\boldsymbol{\kappa}(z) = \int_0^z dz'[\overline{\mathbf{u}}(z') - <\overline{\mathbf{u}}(z)>], \quad <\overline{\mathbf{u}}> = h^{-1}\int_{-h}^0 \overline{\mathbf{u}}(z)dz \qquad (1.131)$$

Using the Ekman form of $\boldsymbol{\tau}$ in footnote #2 in sec. 1.13, the second term in (1.130) becomes:

$$z = 0: \quad J_f(Ek) = \frac{f}{hE^{1/2}}EBF \qquad \alpha(0) = 1, \quad \beta(0) = 0. \qquad (1.132)$$

Using (1.131), the SM frictional J_f then acquires the simple form:

$$\frac{J_f(0)}{J_f(Ek)} = 1 - Q_{BI}\frac{Ro^3}{(1+Ro^2)^2} + Q_{WS}\frac{Ro^2}{1+Ro^2} \qquad (1.133)$$

$$Q_{BI} = \frac{h^2|\nabla_H b|^2}{|f|EBF}E^{1/2}, \quad \frac{1}{2}Q_{WS} = 1 - E^{1/2} - Ro^{-1}A(Ro) \qquad (1.134)$$

which are shown below (Fig. 1.27).

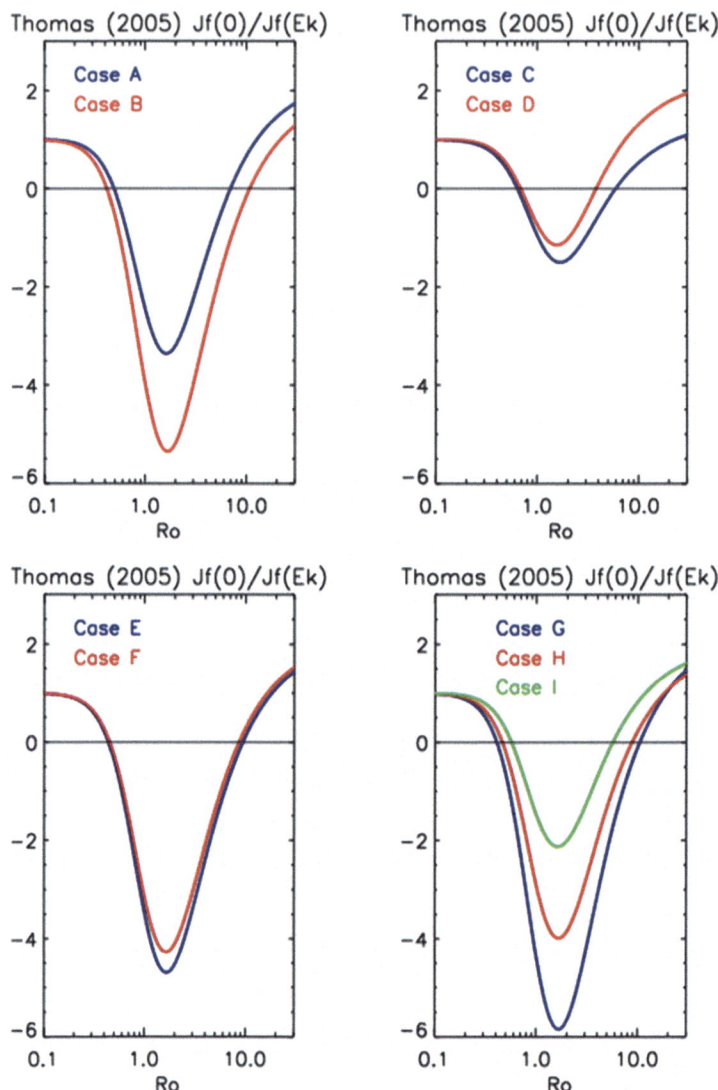

Fig. 1.27 The ratio (1.133) vs. Ro (inputs from Thomas 2005). (Reprinted with permission from Canuto and Cheng (2021). © NASA all rights reserved)

In Table 1 of Thomas (2005), the horizontal buoyancy gradients in Q_{BI} are given in the second column, EBF is given by second, third and fourth columns and rewriting the fifth column as $f^2 \delta_E^{-1} EBF$, one obtains $\delta_E = hE^{1/2}$ which, together with the values $h = 80$ m, yield the Ekman number E for each of T5 nine cases A-I. The condition for down-front winds to destroy PV, positive *surface vertical frictional PV flux,* is satisfied only in the regimes:

$$\text{Ro} < 1, \quad \text{Ro} > 10 \qquad (1.135)$$

corresponding to SM with low and high kinetic energies. Not only has PV destruction become location dependent but it also appears to be less frequent than previously thought Thomas and Taylor (2010).

1.17 SM Kinetic Energy

The stationary, local limit of the SM kinetic energy equation becomes the production = dissipation relation:

$$<F_v(\overline{b})> + \; <P_s(z)> \; = \; <\varepsilon>, \qquad P_s(z) = - \overline{w'\mathbf{u}'} \cdot \frac{\partial \overline{\mathbf{u}}}{\partial z} \qquad (1.136)$$

where $<..>$ stands for depth average, primes indicate SM fields, an overbar represents time or ensemble average that, under ergodic conditions, yield the same result, $\overline{w'\mathbf{u}'}$ are the SM momentum fluxes, $\overline{\mathbf{u}}$ is the 2D mean velocity, and ε is the rate of EKE dissipations. Since the FK model discussed in sec. 1.13, did not provide a parameterization of $\overline{\mathbf{u}'w'}$, Eq. (1.136) could not be solved which we do next. To compute the dissipation term, we employ Kolmogorov's kinetic energy spectrum $E(k) = K_o \varepsilon^{2/3} k^{-5/3}$, where K_o is the Kolmogorov constant that in 2D turbulence is $4 < K_o < 8$ (Danilov and Gurarie 2000). Integrating $E(k)$ over all wavenumbers from $k_0 = \pi \; r_s^{-1}$ to infinity, one obtains K_{SM}. Using the definition of Ro in Eq. (1.99), one obtains:

$$\frac{<\varepsilon>}{h^2 N^2 |f|} = (q_0 \pi^2)^{-1} \text{Ro}^3, \qquad q_0 \equiv \left(\frac{3}{2} K_o\right)^{3/2} \qquad (1.137)$$

Next, using the momentum fluxes Eq. (1.131), one derives the relation:

$$P_s(z) = \underbrace{p_0(z) h^2 |f|^{-1} |\nabla_H \overline{b}|^2}_{\text{GSP}} + \underbrace{p_1(z) \text{EBF}}_{\text{ASP1}} + \underbrace{p_2(z) \left(h^2 |f|^{-1}\right)^2 (\tau_w/\rho)^2}_{\text{ASP2}} \qquad (1.138)$$

where the dimensionless, z-dependent functions $p_{0,1,2}$ are given in Appendix B. GSP and ASP stand for geostrophic/a-geostrophic shear production. Figures 1.28 and 1.29 show the vertical profiles of the buoyancy flux, shear production and their sum.

Fig. 1.28 Vertical profiles of the terms in Eq. (1.28) for Ro = 1. We used the following input from Thomas et al. (2013): h = 170 m, EBF = $6.5*10^{-7} m^2 s^{-3}$, $|\nabla_H \bar{b}| = 1.3*10^{-7} s^{-2}$, $\tau_w^{x,y} = -(0.2 - 0.48) Nm^{-2}$

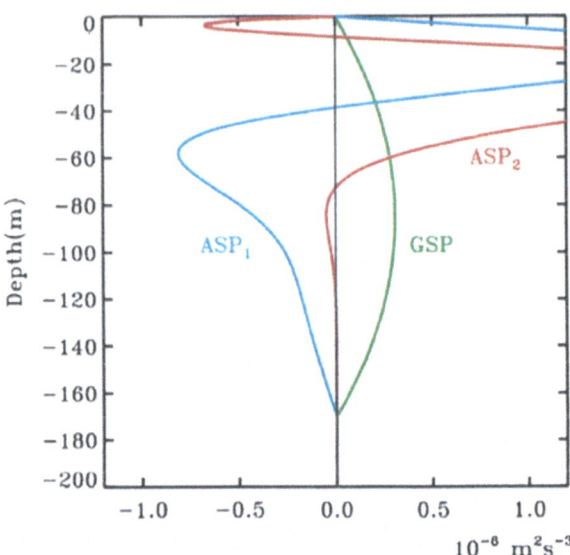

Fig. 1.29 Vertical profiles of the vertical buoyancy flux, shear production and their sum

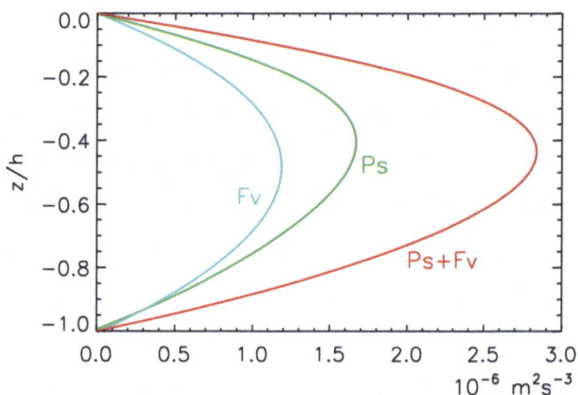

Using the depth averages relations given in Appendix B, one obtains:

$$\frac{<F_b(\bar{b})>}{|f|h^2 N^2} = aRi_g^{-1} + be_*, \quad a = \frac{A(Ro)}{12Ro}, \quad b = \frac{1}{2}\left[1 - E^{1/2}(1-Ro^{-1})\right] \quad (1.139)$$

$$\frac{<P_s(z)>}{|f|h^2 N^2} = cRi_g^{-1} + de_* + e\tau_*, \quad c = \frac{A(Ro)^2}{6Ro}, \quad d = 12cE^{1/2}, \quad e = 12cE^{-1/2}\left(1 - E^{1/2}\right) \quad (1.140)$$

where we have introduced the following dimensionless variables, *geostrophic Richardson number* Ri_g, *interaction of buoyancy gradients with wind stress* e_* *and wind stress* τ_*:

1.17 SM Kinetic Energy

$$Ri_g = \frac{f^2 N^2}{|\nabla_H \bar{b}|^2}, \qquad e_* = \frac{EBF}{h^2 N^2 |f|}, \qquad \tau_* = \frac{(\tau_w/\rho)^2}{h^4 N^2 f^2} \qquad (1.141)$$

Using the above relations in the first of (1.136), algebraic steps yield the 6-th order equation:

$$Ro^6 + 2Ro^4 + d_1 Ro^3 + d_2 Ro^2 + d_1 Ro = d_0 \qquad (1.142)$$

where we have defined the dimensionless variables:

$$12 d_0 = q_o \pi^2 \left(Ri_g^{-1} + 6 E^{1/2} e_* \right), \qquad 2 d_1 = -q_o \pi^2 \left(1 - E^{1/2} \right) e_*$$

$$d_2 = 1 - \frac{1}{4} q_o \pi^2 \left[Ri_g^{-1} + 10 E^{1/2} e_* + 8 E^{-1/2} \left(1 - E^{1/2} \right) \tau_* \right] \qquad (1.143)$$

No wind stress, $e_* = \tau_* = 0$. In this case, Eq. (1.142) becomes a cubic equation in Ro^2:

$$Ro^6 + 2 Ro^4 + d_2 Ro^2 + \frac{1}{3}(d_2 - 1) = 0, \qquad d_2 = 1 - \frac{1}{4} q_o \pi^2 Ri_g^{-1} \qquad (1.144)$$

Figure 1.30 shows that Ro decreases with increasing Ri_g since the larger Ri_g, the weaker the baroclinic instability that sustains K_{SM} (Fig. 1.31).

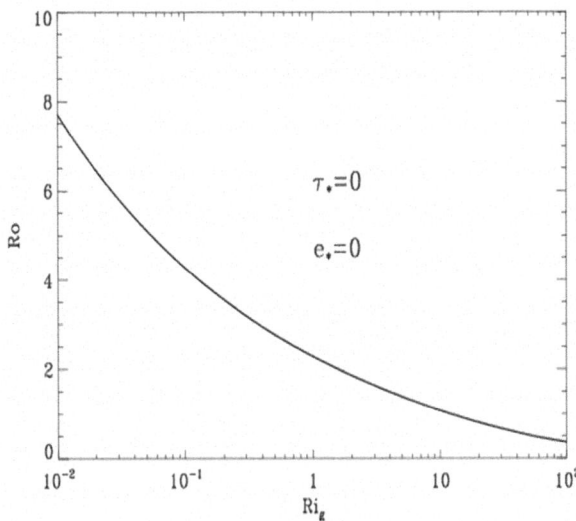

Fig. 1.30 Solution of Eq. (1.142) in the no-wind stress case

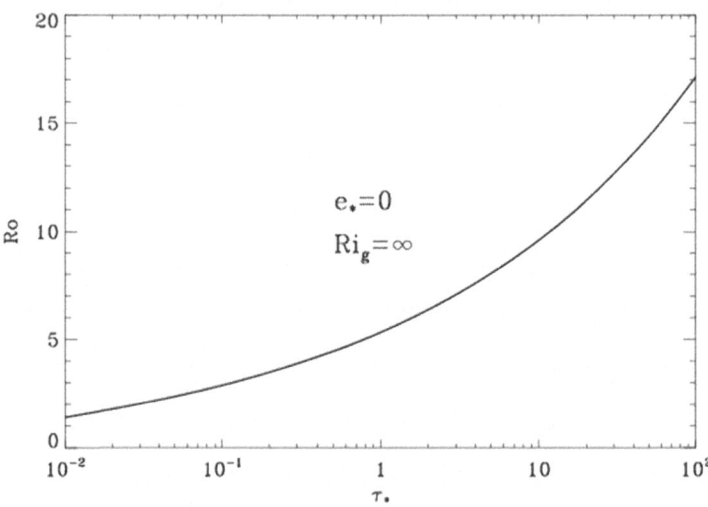

Fig. 1.31 Solution of (1.142) in the wind stress only case

Wind stress only, $\text{Ri}_g = \infty$, $e_* = 0$. In this case we have:

$$\text{Ro}^2 = (\alpha\tau_*-1)^{1/2} - 1, \qquad \alpha = 2q_0\pi^2\left(E^{-1/2} - 1\right) \tag{1.145}$$

For large values of τ_*, the SM kinetic energy grows linearly with wind stress:

$$\text{Ro}^2 \propto \tau_w, \qquad K_{SM} \propto \tau_w \tag{1.146}$$

In the Ro > 1 regime, the coefficient $C_e = \frac{A(\text{Ro})}{8\text{Ro}}$ given by Eq. (1.101), scales like:

$$C_e \propto \tau_w^{-1/2} \tag{1.147}$$

which shows that in the limit of large EKE, the geostrophic contribution to the flux becomes less important than the wind stress term EBF.

C8 case. Using $\tau_w = 0.05 \text{Nm}^{-2}, \tau_* = 0.1, h = 40\text{m}, N^2 = 10^{-6}\text{s}^{-2}, \text{EBF} = 3*10^{-8}\text{m}^2\text{s}^{-3}, e_* = 0.2, \text{Ri}_g = 2.5$ from C8, in Fig. 1.32 we show the solution of Eq. (1.142). The results show that without the contribution of τ_* the value of Ro from C8 simulation data Fig. 1.26, could not be reproduced. In the absence of winds, Ro decreases monotonically with increasing Ri_g as in Fig. 1.30. In the presence of wind, the decrease of Ro is reduced since there is now a source of kinetic energy. The *black dot* is the Ro derived from the C8 data which is close to the value predicted by the model for $\tau_* = 0.1$. In the absence of τ_*, the value of Ro is half the simulation result.

1.18 Vertical Velocities

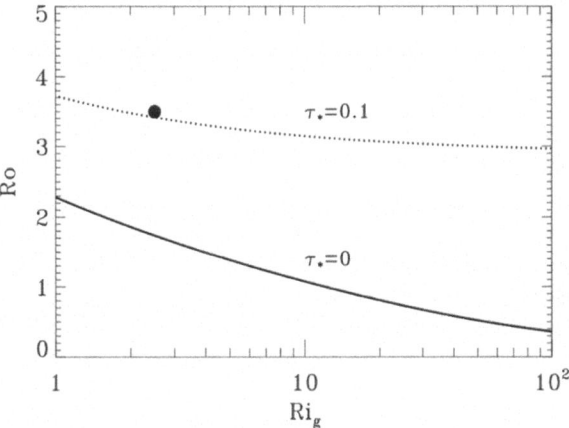

Fig. 1.32 Solution of (1.142) for the Capet et al. (2008) data

1.18 Vertical Velocities

High resolution simulation of SM (e.g., Levy et al. 2001) have highlighted the large vertical velocities of several tens of m/day generated by SM that are instrumental to the transport of nutrients to oligotrophic regimes (e.g., Mahadevan 2016). The vertical velocity has traditionally been computed by solving the equation (Hoskins 1974; Polland and Regier 1992):

$$f^2 \frac{\partial^2 w}{\partial z^2} + \nabla_H \cdot (N^2 \nabla_H w) = \nabla \cdot \mathbf{Q} \quad (1.148)$$

Koszalka et al. (2009) pointed out however that this equation underestimates the vertical velocity by an order of magnitude with respect to the results of their Fig.6 from a regional ocean code at 1 km resolution. Since the latter represents the SM regime, their result casts doubts about the ability of (1.148) to describe SM. Pinot et al. (1996) discussed different forms of the source term **Q**, pointing out the limitations of the original QG form and suggesting the need for a more general form of it. The generalization of (1.148) suggested by Giordani et al. (2006) included a *vertical flux and wind stresses* ($\rho = 1$):

$$f^2 \frac{\partial^2 w}{\partial z^2} + \nabla_H \cdot (N^2 \nabla_H w) = \nabla \cdot \mathbf{Q} + \underbrace{|f| \frac{\partial}{\partial x} \left(\frac{\partial^2 \tau_{yz}}{\partial z^2} \right) + \frac{\partial^2}{\partial x^2} \left(\frac{\partial F_v}{\partial z} \right)}_{\text{new terms}} \quad (1.149)$$

Canuto and Cheng (2017, CC17) identified F_v with the SM vertical buoyancy flux given by Eq. (1.104). Since the solution of the full Eq. (1.149) can only be obtained numerically, CC suggested it would be of interest to quantify the role of the new terms as the only source of w. A major advantage is that the resulting model:

$$f^2 \frac{\partial^2 w}{\partial z^2} = |f| \frac{\partial}{\partial x}\left(\frac{\partial^2 \tau_{yz}}{\partial z^2}\right) + \frac{\partial^2}{\partial x^2}\left(\frac{\partial F_v}{\partial z}\right) \quad (1.150)$$

can be solved analytically. The solution was written as the sum of wind and SM terms:

$$w(x,z) = w_{wind}(z)\cos\theta_1 + w_{SM}(z)\sin\theta_2 \quad (1.151)$$

where $\theta_1 = \kappa_1 x$, $\theta_2 = \kappa_2(x - x_{min})$; $\kappa_1 = \pi/x_{mx}$, $\kappa_2 = \pi/(x_{max} - x_{min})$ represent the horizontal extents of the region under consideration. It was shown that:

$$w_1(z) = \kappa_1 |f|^{-1} T(z) \quad (1.152)$$

where, using the Ekman form of the wind stresses, algebraic steps lead to:

$$T(z) = T_0(z)\tau_y + T_1(z)\frac{f}{|f|}(\mathbf{e}_z \times \boldsymbol{\tau})_y \quad T_0(z) = \alpha(\zeta)\text{-}1 + \frac{z}{h}$$
$$\times (\alpha_* - 1), \quad T_1(z) = \beta(\zeta) + \frac{z}{h}\beta_* \quad (1.153)$$

where $\zeta = z/\delta_e$, $(\alpha, \beta)_* = (\alpha, \beta)_{-E^{-1/2}}$. The wind contribution to the vertical velocity is thus complete.

The SM contribution yields the two g-ag terms:

$$w_{SM}(z) = w_g I_g + w_{ag} I_{ag}, \quad 4Row_g = A(Ro)h^2|f|^{-3}\kappa_2^2|\nabla_H \overline{b}|^2, \quad w_{ag}$$
$$= -A(Ro)\kappa_2^2 f^{-2} h EBF \quad (1.154)$$

where, using the notation $\sigma(z) = z/h$, we have:

$$I_g(\sigma) \equiv \sigma\left(\frac{1}{3} + \sigma + \frac{2}{3}\sigma^2\right), \quad I_{ag}(\sigma) = I(\sigma) + \sigma I(-1),$$
$$I(\sigma) = \sigma\left(1 + \frac{1}{2}\sigma\right) - \frac{1}{2}E^{1/2}(\alpha + \beta\text{-}1) - \frac{1}{2Ro}E^{1/2}(\beta\text{-}\alpha + 1) \quad (1.155)$$

To exhibit the contributions from wind and SM, CC17 used:
$x_{min} = 200\text{km}$, $x_{max} = 300\text{km}$, $h_0 = 50\text{m}$, $\Delta h = 100\text{m}, x_1 = 0$ and $x_2 = 500\text{km}$ (D'Asaro et al. 2011; Giordani et al. 2006), together with, EBF = $10^{-4}\text{m}^2\text{s}^{-3}$, $d\overline{b}/dx = 0$, $d\overline{b}/dy = -10^{-5}\text{s}^{-2}$, $\tau_{w,x}/\rho = 10^{-3}\text{m}^2\text{s}^{-2}$, $\tau_{w,y}/\rho = 0$ Ro = 3.5.

In summary, though the total vertical velocity can only be obtained by solving the new ϖ-Eq. (1.149), CC17 considered it instructive to separate and assess the magnitude of the new terms that had not been considered before, *vertical SM buoyancy flux and wind stresses*. The results in Figs. 1.33, 1.34 and 1.35 show

1.18 Vertical Velocities

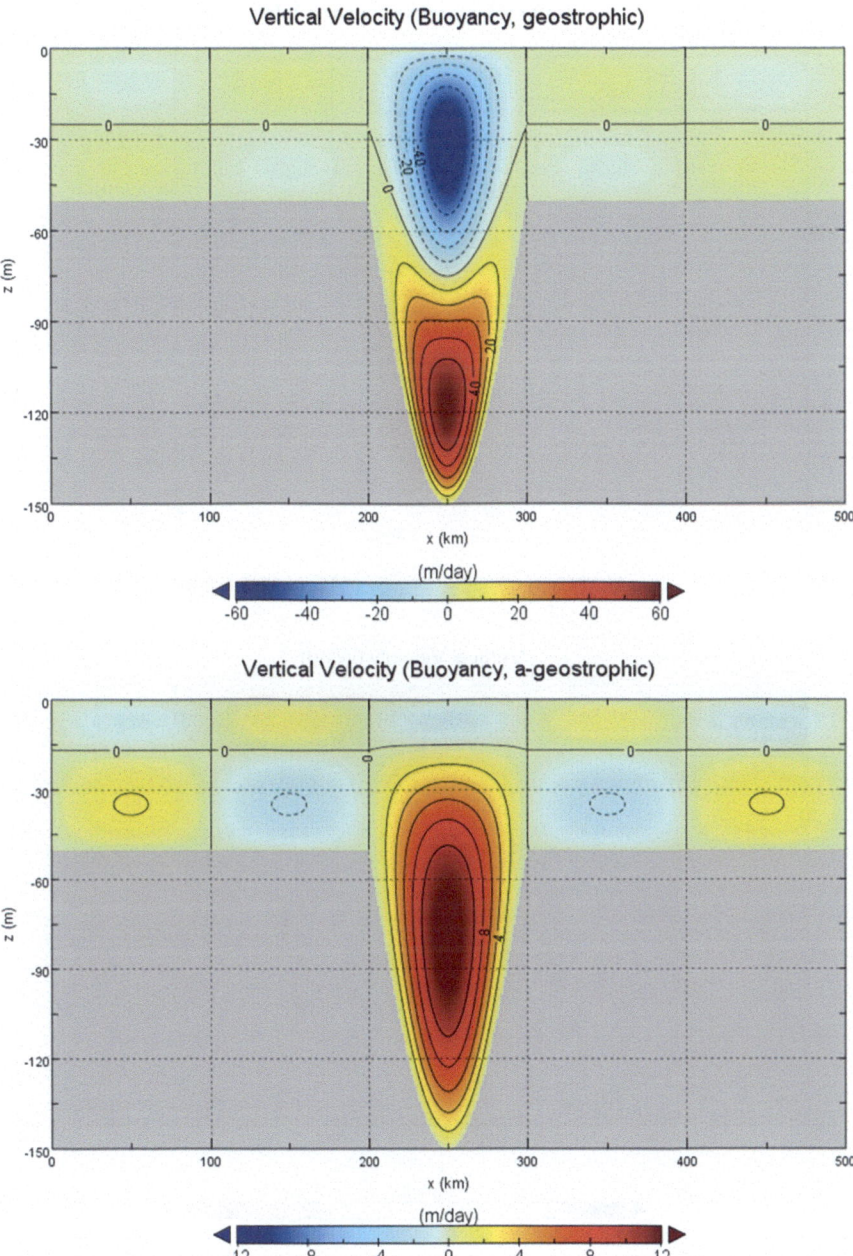

Fig. 1.33 Upper panel. Vertical velocity (m/day) contour map of the geostrophic contribution. Lower panel. Contour map of the a-geostrophic $w_2(ag)$. (Reprinted with permission from Canuto and Cheng (2017). © NASA all rights reserved)

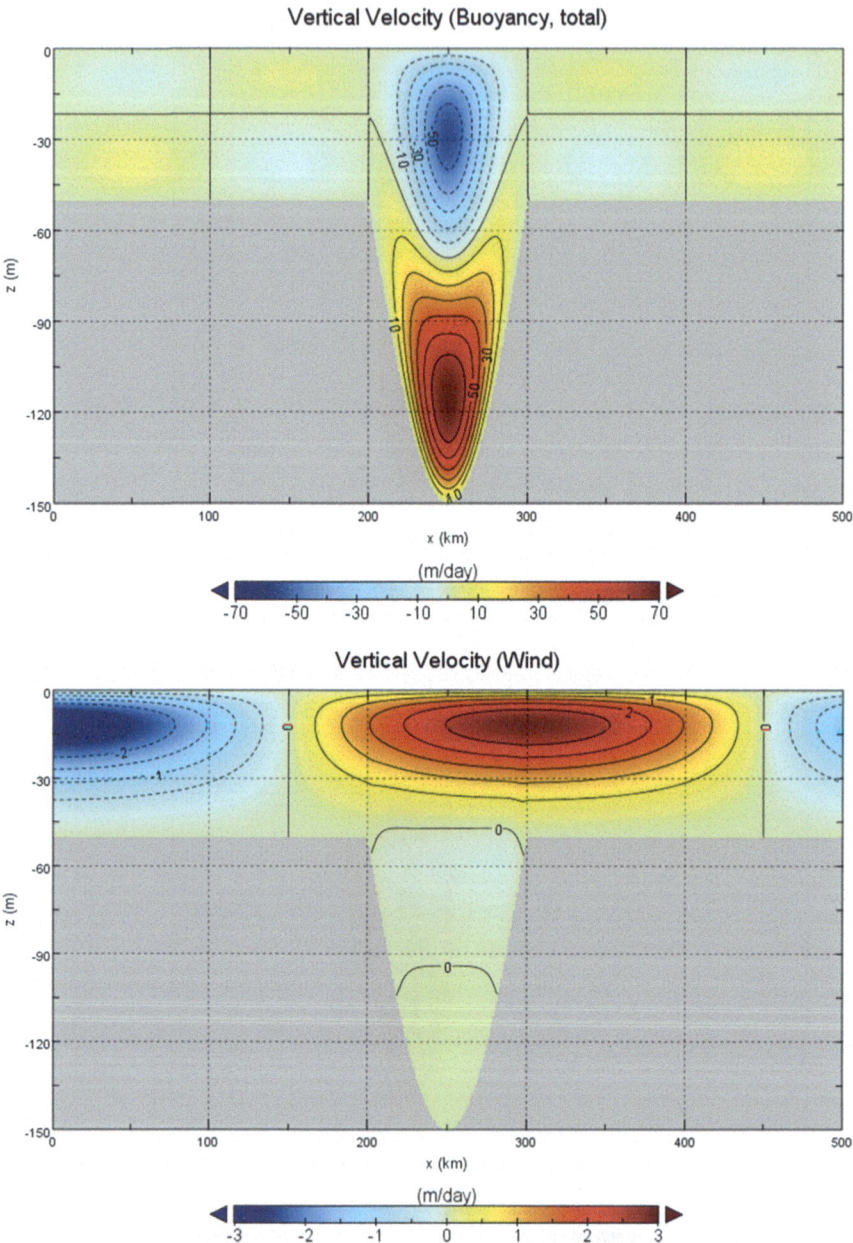

Fig. 1.34 (upper panel) shows the total SM induced vertical velocity) $w_2(g) + w_2(ag)$, the lower panel shows the vertical velocity due to wind. (Reprinted with permission from Canuto and Cheng (2017). © NASA all rights reserved)

1.19 Sub-Mesoscales and Mesoscale Interaction

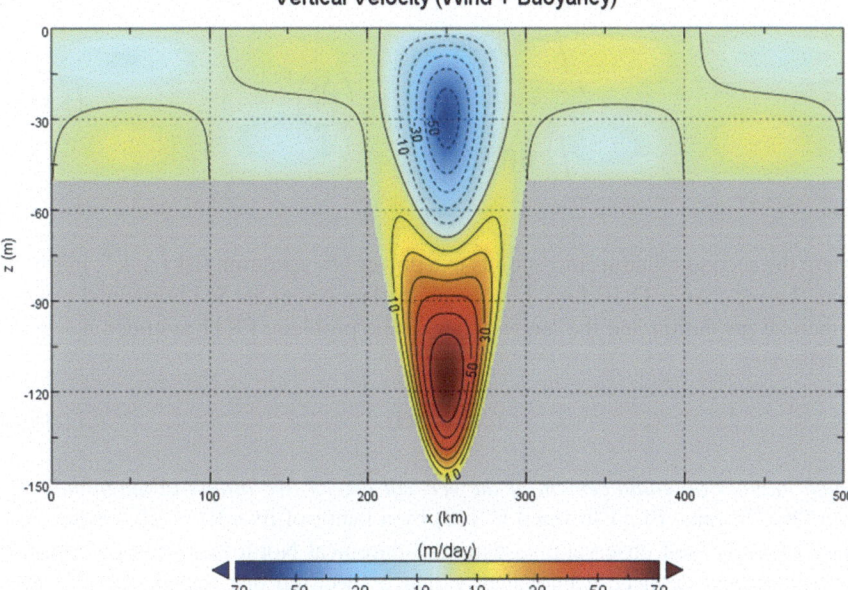

Fig. 1.35 Total velocity $w_2(g) + w_2(ag) + w_1(wind)$. (Reprinted with permission from Canuto and Cheng (2017). © NASA all rights reserved)

that the SM contributions to the vertical velocity are sufficiently large to suggest that future solutions of the ϖ-equation ought to include the SM contribution.

1.19 Sub-Mesoscales and Mesoscale Interaction

Since in the diabatic regime SM coexist with mesoscales, FK11 showed that the latter change the SM flux by introducing a dimensionless factor R_b so that the new flux reads as follows:

$$F_v(\overline{b}) \propto (1 + R_b) h^2 |f|^{-1} \sigma(z) |\nabla_H \overline{b}|^2 \qquad (1.156)$$

where R_b is defined in terms of the *buoyancy variance spectrum B(k):*

$$R_b = \frac{\int_{1/\Delta s}^{1/L_s} B(k)dk}{\int_{1/L}^{1/\Delta s} B(k)dk}, \quad R_b \propto \left(\frac{\Delta s}{L_s}\right)^{n+1} - 1 \quad (1.157)$$

where the second relation corresponds to a power law spectrum $B(k) \propto k^{-n}$. Furthermore, $L_s = r_s$, $\Delta s = O(10^2 \text{km})$, $L = O(10^3 \text{ km})$ represent the SM horizontal extent, numerical resolution and the largest scale of the problem. FK11 assumed $n = 0$ in which case:

$$R_b = 100 \quad (1.158)$$

which induces an enhancement of the vertical flux of two orders of magnitude. In their OGCM runs, FK11 lowered (1.158) by a factor of five, $R_b = 20$. Callies and Ferrari (2013) used observational data in Subtropical North Pacific in the interval (1–200)km and deduced a buoyancy variance spectrum that scaled like $k^{-1/2}$ that yields $R_b = 10$; Zhong and Bracco (2013) studied the Northwestern gulf of Mexico in the interval $k = (10^{-6} - 4*10^{-4})\text{m}^{-1}$. Using the spectrum in their Fig.8e at a depth of 5 m, one obtains $R_b = 2$. Since high resolution numerical simulations (Siegelman 2020; Siegelman et al. 2020; Su et al. 2018) concluded that the buoyancy flux is about seven times larger than the FK8 flux, yielding $R_b = 5$ - 6. An even more stringent value of $R_b = O(1)$ has recently been derived by Sinha et al. (2022) in a high-resolution numerical simulation of mid-latitude North Atlantic. These results lead to the conclusion that the FK11 OGCMs results overestimated the effect of mesoscales on the SM vertical buoyancy flux. Finally, the a-geostrophic component of the SM buoyancy flux is not affected by mesoscales since wind stresses operate on scales much larger than those on which the mesoscale averaging operates.

1.20 Conclusions

In today global circulation models, ocean mixing processes are parameterized using the KPP model for vertical shear, the GM model for mesoscales and the FK8–11 model for sub-mesoscales. At first sight, such parameterizations seem to have little in common which is the likely reason why no attempt was made thus far to unify the three processes. However, a closer look reveals that these mixing processes have in common a basic feature, they are all macroscopic manifestations of *non-linearity under different external forcings*. On the basis of such commonality, we have suggested a unified approach to mixing processes. The classical Brownian motion offered the conceptual template of a motion under the influence of an external force and a fluctuating one. The turbulence version is that of an eddy whose velocity is acted upon by external forces such as shear, buoyancy etc., and turbulence forces by eddies larger and smaller to the one under consideration. The latter forcing is represented by a turbulent viscosity whose form was obtained heuristically in this paper but dynamically in C1 using the RNG (renormalization group). The forcing due to large eddies cannot be obtained using RNG and that is why it was not derived in the 1986 Yakhot-Orszag YO first application of RNG to turbulence. The absence of such a term made the YO model incomplete precluding the application to realistic flows. The RNG formulation thus remained an interesting but incomplete scheme to study turbulence. To overcome this limitation, it was imperative to parameterize such a force which is the main contribution of this work. It was done in sec. 1.3 leading to the kinetic energy Eq. (1.26) that includes the contributions of both large and small eddies. The next step was to assess the overall performance of the complete LSE on 80 + turbulence statistics representing different turbulent flows, e.g., buoyancy, shear, 2D turbulence, effect of rotation, subduction by M and SM, re-stratification by SM in the ACC, SM opposition to eddy compensation, PV destruction by SM, vertical velocity induced by SM and derivation of the algebraic equation that yields the SM kinetic energy in terms of large scale variables. Taken together, the results presented here show that the LSE represent a viable, operationally deterministic approach to treat turbulent flows.

Appendices

Appendix A

We begin with the NSE equations in which we use the notation $\kappa = (\mathbf{k}, \omega)$:

$$(-i\omega + \nu k^2)u_i(\kappa) = f_i^{ext}(\kappa) + P_{ijl}(\mathbf{k})M_{ij}(\kappa) \qquad (1.159)$$

where the non-linear term is given by:

$$M_{ij}(\kappa) = \int d\kappa' u_i(\kappa') u_j(\kappa - \kappa'), \quad P_{ij}(k) = \delta_{ij} - k^{-2}k_ik_j, \quad 2iP_{ijl}(\mathbf{k}) = k_j P_{il} + k_l P_{ij} \qquad (1.160)$$

The correlation function of the external forces are given by:

$$< f_i^{ext}(\kappa) f_j^{ext}(\kappa') > \equiv \phi_{ij}(\kappa, \kappa') = P_{ij}(k)\phi(\kappa)\delta(\kappa + \kappa') \qquad (1.161)$$

To the lowest order of the expansion of Q_{ij} in powers of the non-linear term M:

$$Q(\kappa) \to Q_0(\kappa) = G_0(\kappa)\phi(\kappa)G_0(\kappa)^* = \frac{\phi(\kappa)}{\omega^2 + \nu^2 k^4} \quad (1.162)$$

where the zeroth-order propagator is defined as follows:

$$G_0(\kappa) = \left(-i\omega + \nu k^2\right)^{-1} \quad (1.163)$$

Appendix B: The p(z) Functions in (1.138)

The dimensionless functions p's in Eq. (17.3) are given by:

$$\frac{Ro}{A(Ro)^2} p_0(z) = -\sigma(z), \quad \frac{Ep_2(z)}{4A(Ro)} = \alpha^2 + \beta^2 - \left(\alpha + \frac{A(Ro)}{Ro}\beta\right)(1 + z/h),$$

$$\frac{p_1(z)}{2A(Ro)} = -\left(1 + z/h - \alpha + \frac{A(Ro)}{Ro}\beta\right) - E^{-1}\sigma(z)\left(\frac{A(Ro)}{Ro}\alpha - \beta\right)$$

$$(1.164)$$

Depth averages:

$$2<\alpha,\beta> = \left(E^{1/2}, -E^{1/2}\right), \quad 8<\alpha^2,$$
$$<\beta^2> = (3,1)E^{1/2}, \quad <\alpha z/h> = 0, \quad 2<\beta z/h> = E,$$
$$2<(\alpha,\beta)z^2/h^2> = -E^{3/2}$$
$$6<p_0> = a(Ro), \quad <p_1> = 2E_k^{1/2}a(Ro), \quad <p_2> = 2E^{-1/2}\left(1 - E^{1/2}\right)a(Ro),$$
$$a(Ro) = Ro^{-1}A(Ro)^2$$

$$(1.165)$$

Appendices

Appendix C: Mesosale Vertical Buoyanc Flux. Extent of the D-Regime (Figs. 1.36 and 1.37)

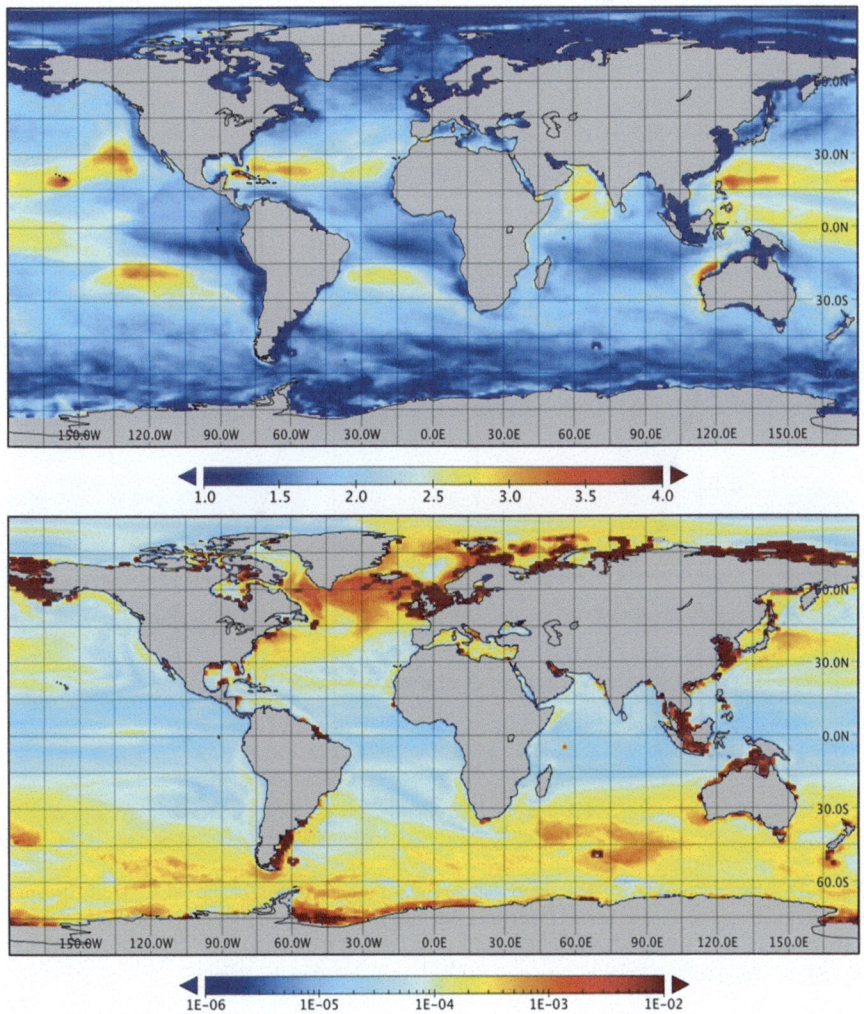

Fig. 1.36 *Upper panel.* Map of h_* (1.110) in units of the MLD. In all regions $h_* \geq$ MLD indicating that below the mixed layer the flow is still diabatic. *Lower panel:* s_* at h_*. The value of s_* is no longer arbitrary as in tapering schemes, it is now given by the model. It is not a universal constant but a location dependent variable

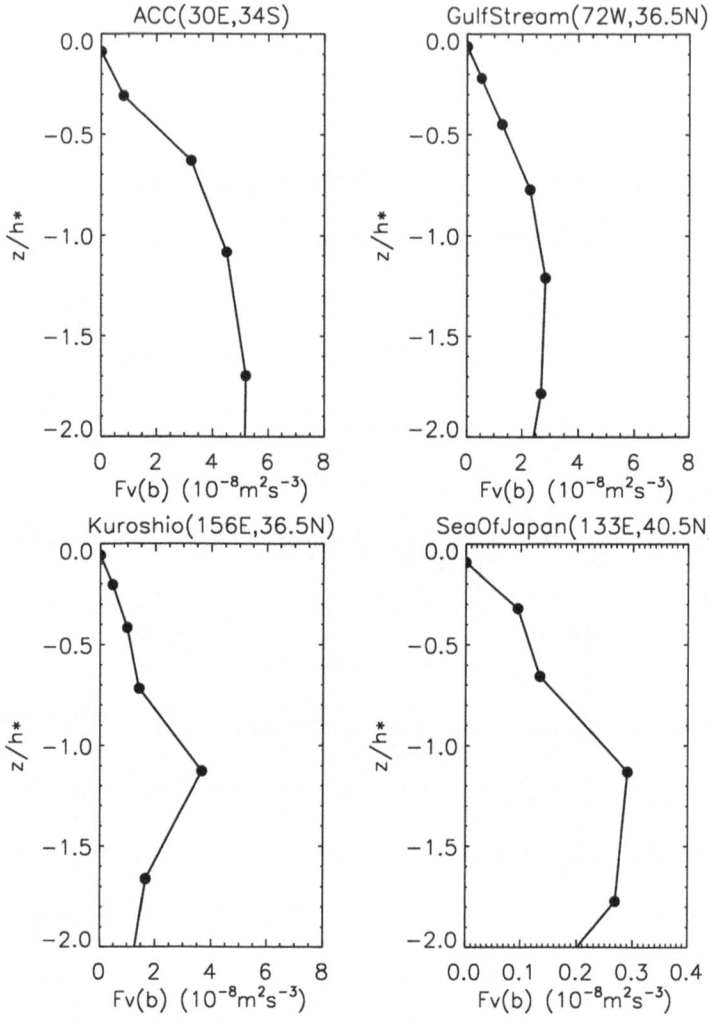

Fig. 1.37 Vertical profiles of the mesoscale vertical buoyancy flux in the ACC, Gulf Stream, Kuroshio and Sea of Japan. In the D-regime, the flux is given by (1.104) while in the A-regime $F_v(b) = \kappa_M N^2 \text{ s·}\xi$. The depth z is in units of h_*. To convert the flux to Wm^{-2}, multiply $F_v(b)$ by $\rho c_p/g\alpha_T = 42\alpha_0^{-1}\left(10^8 s^3 m^{-2}\right) Wm^{-2}$, $\alpha_T = \alpha_0 10^{-4} K^{-1}$, α_0 is location dependent

Appendix D: Volume and Ensemble Averages

In a fully developed turbulent flow, the largest scales cannot exhibit a universal spectrum since they depend on the type of stirring while intermediate scales that are no longer affected by geometry and/or type of stirring, exhibit a universal character that does not depend on the specific type of energy input but only on the amount of energy input. This part of the energy spectrum has been successfully described by the Kolmogorov law, Eq. (1.32). The small-scale portion of the spectrum contain little energy but large vorticity which is the opposite of the features that characterize the largest scales which contain most of the energy to the point of having given rise to the one-eddy model in which it is assumed that bulk properties such as heat fluxes can be represented by a single large eddy, e.g., the Prandtl's mixing length model. The ratio of the largest to the smallest scales like the 3/4 power of the Reynolds number Re which in a 3D case, means that the number of grid points or degrees of freedom, scales like the 9/4 power of Re (see textbooks on turbulence, e.g., Lesieur 1991; Pope 2000; McComb 1992). In geophysical and astrophysical regimes (e.g., Re = 10^8 for the planetary boundary layer, Re = 10^{14} for the sun's interior), the number of grid points easily exceeds the computational capability of modern computers. Moreover, since the time scale of evolution of a turbulent flow is of the order of the turnover time of the energy containing eddies, one needs to perform Re $^{3/4}$ time steps to represent a significant time evolution. The requirement of computer storage would then be O(Re$^{9/4}$). Historically, there have been two approaches to describe turbulence.

Volume average approach. In this approach, one carries out a large eddy simulation LES by numerically resolving the largest scales while the remaining smaller, unresolved scales are treated with a SG (sub-grid) model. Results from the LES + SG approach, as used in different fields, have shown that its reliability depends on the reliability of the SG model for both numerical stability and physical completeness. Since there is a documented history of different SG models suggested over the years, we limit the discussion to the well-known Smagorinsky (1963) and Lilly (1966) models for shear driven flows. It yielded the following turbulent viscosity $\nu_t = (C_s\Delta)^2 S$, where S is the mean shear, Δ is the smallest resolved scale and $C_s = 0.20 - 0.22$ It turned out that the latter value was twice as large as what was needed to reproduce channel flow data (Deardorff 1970). The discrepancy was discussed in Canuto and Cheng (1997) who employed a form of the Reynolds stress vs. strain rate tensor (see their eq. 3 g) more general than the traditional Fickian model in which the Reynolds stresses depended linearly on the strain rate tensor. A SG model was developed in Canuto (1994, C94) in which a general form of SG included shear, vorticity, rotation and buoyancy, the complete SG model being given by Eq. (22)–(26). Such equations became considerably simpler in the stationarity and local limit, see Eq. (34a)-(35b). The Smagorinsky-Lilly model was recovered as a particular limit in the absence of buoyancy. Though this SG model represents the most complete form one can construct, there two important limitations. The first is that the derivation of the model equations in Appendix A of C94, shows that closures of the pressure gradients terms were necessary which entailed a set of constants that had to be specified which naturally introduces some uncertainty.

Ensemble averages. It is represented by the Reynolds Stress Model RSM whose first application dates to 1945 when P.Y. Chou (1945) published his pioneering work on the application to shear driven flows. Given the flexibility of the model to incorporate external forcing such as buoyancy, the RSM became quite popular. Application of the RSM to geophysical flows began with the pioneering work of

Mellor and Yamada (1982), followed by several others, e.g., Canuto et al. (2001, 2002, 2010). It is fair to conclude that after several years of efforts by many groups, the local SOC (second order closure) models have achieved a considerable level of sophistication. The models by Cheng et al. (2020) represent the most up to-date SOC models presently available. A limitation that affects the RSM is that turbulence not only generates finite second-order fluxes (e.g., of heat, salt and momentum) but also transports them to regions where locally there is no source of mixing, an effect usually referred to as *non-locality*. This is accomplished by the TOMs (third-order moments) representing the fluxes of second-order fluxes. In Canuto et al. (2005), a model for a non-local convective PBL (planetary boundary layer) was proposed that contained third and fourth order moments and in Canuto et al. (2007) an attempt was made to recast the mixing model in a "plume form" reminiscent of the well-known plume model by Morton et al. (1956). The attempt was moderately successful. Though parameterizations of non-locality by third and fourth-order models are available, their complexity is quite apparent as is the need to justify terminating the chain of moments at the fourth order. The most common models, the SOC models, have no third-order moments and are local in the sense that turbulent mixing generated at a given location is dissipated there without being transported elsewhere.

References

Aiki, H., et al.: Parameterizing Ocean eddy transport from surface to bottom. Geophys. Res. Lett. **31**, L19302 (2004)

Barthel, A., et al.: Baroclinic control of Southern Ocean eddy upwelling near topography. J. Geophys. Res. **49** (2022). https://doi.org/10.1029/2021GL097491

Batchelor, G.K.: The Theory of Homogeneous Turbulence. Cambridge University Press, Cambridge (1971)

Bates, M., et al.: Rationalizing the spatial distribution of mesoscale eddy diffusivity in terms of mixing length theory. J. Phys Oceanogr. **44**, 1523–1540 (2014). https://doi.org/10.1175/JPO-D-13-0130.1

Bopp, L., et al.: Pathways on anthropogenic carbon subduction in the global ocean. Geophys. Res. Lett. **42**, 6416–6423 (2015). https://doi.org/10.1002/2015GL065073

Buckingham, C.E., et al.: Testing Munk's hypothesis for sub-mesoscale eddy generation using observations in the North Atlantic. J. Geophys. Res. Oceans. **122**, 6725–6745 (2017). https://doi.org/10.1002/2017JC012910

Busecke, J.J.M., Abernathey, R.P.: Ocean mesoscale mixing linked to climate variability. Sci. Adv. **5**, eaav5014 (2019)

Callies, J., Ferrari, R.: Interpreting energy and tracer spectra of upper ocean turbulence in the sub-mesoscale range. J. Phy. Oceanogr. **43**, 2456–2474 (2013)

Canuto, V.M.: Large eddy simulation of turbulence: a sub-grid model with shear and buoyancy. Astrophys. J. **428**, 729–752 (1994)

Canuto, V.M., Cheng, Y.: Determination of the Smagorinsky-Lilly constant. Phys. Fluids. **9**, 1368–1378 (1997)

Canuto, V.M., Cheng, Y.: Contribution of sub-mesoscales to the vertical vorticity: the ϖ-equation. Ocean Model. **115**, 70–76 (2017). https://doi.org/10.1016/j.ocemod.2017.05.004

Canuto, V.M., Cheng, Y.: Subduction by mesoscales. J. Phys. Oceanogr. **49**, 3263–3272 (2019)

References

Canuto, V.M., Cheng, Y.: Do sub-mesoscales inhibit destruction of PV? J. Geophys. Res. Oceans. **126**(11), e2020JC016991 (2021). https://doi.org/10.1029/2020JC016991

Canuto, V.M., Dubovikov, M.S.: 1. A dynamical model for turbulence. General formalism. Phys. Fluids. **8**, 571–586 (1996a)

Canuto, V.M., Dubovikov, M.S.: 2. A dynamical model for turbulence. Shear driven flows. Phys. Fluids. **8**, 587–598 (1996b)

Canuto, V.M., Dubovikov, M.S., Cheng, Y., Dienstfrey, A.: 3. Numerical results. Phys. Fluids. **8**, 599–613 (1996c)

Canuto, V.M., Dubovikov, M.S., Dienstfrey, A.: 4. A dynamical model for turbulence. Buoyancy driven flows. Phys. Fluids. **9**, 2118–2131 (1997a)

Canuto, V.M., Dubovikov, M.S.: 5. A dynamical model for turbulence. The effect of rotation. Phys. Fluids. **9**, 2131–2140 (1997b)

Canuto, V.M., Dubovikov, M.S., Wielaard, D.J.: 6. A dynamical model for turbulence. Two-dimensional turbulence. Phys. Fluids. **9**, 2141–2147 (1997c)

Canuto, V.M., Dubovikov, M.S.: 7. A dynamical model for turbulence. Complete system of 5 orthogonal tensors; shear driven flows. Phys. Fluids. **11**, 659–664 (1999a)

Canuto, V.M., Dubovikov, M.S., Yu, G.: 9. A dynamical model for turbulence. Reynolds stresses for shear driven flows. Phys. Fluids. **11**, 678–691 (1999b)

Canuto, V.M., Dubovikov, M.S., Yu, G.: 8. A dynamical model for turbulence. IR and UV Reynolds stress spectra for shear driven flows. Phys. Fluids. **11**, 665–677 (1999c)

Canuto, V.M., Dubovikov, M.S.: Modeling mesoscale eddies. Ocean Model. **8**, 1–30 (2005)., CD5

Canuto, V.M., Dubovikov, M.S.: Dynamical model of mesoscale eddies in z- coordinates. Ocean Model. **11**(28–46), CD6 (2006)

Canuto, V.M., Dubovikov, M.S.: Mixed layer sub-mesoscales parameterization. Part I. Derivation and Assessment. Ocean Sci. **6**, 679–593 (2010)

Canuto, V.M., Dubovikov, M.S.: Comparison of four mixed layer mesoscale parameterizations and the equations for an arbitrary tracer. Ocean Model. **39**, 200–207 (2011)

Canuto, V.M., Cheng, Y., Howard, A.: New third-order moments for the convective boundary layer. J. Atmos. Sci. **58**, 1169–1172 (2001)

Canuto, V.M., et al.: Ocean Turbulence. Part 1: one-point closure model. Momentum and heat vertical diffusivities. J. Phys Oceangr. **31**, 1413–1426 (2002)

Canuto, V.M., Cheng, Y., Howard, H.: Non-local convective PBL model based on new third and fourth order moments. J. Atmos. Sci. **62**, 2189–2204 (2005)

Canuto, V.M., Cheng, Y., Howard, H.: Non-local ocean mixing model and a new plume model for deep convection. Ocean Model. **17**, 172–181 (2007)

Canuto, V.M., et al.: Ocean turbulence III, New GISS vertical mixing scheme. Ocean Model. **34**, 70–91 (2010)

Canuto, V.M., et al.: Parameterization of mixed layer and deep-ocean mesoscales including nonlinearity. J. Phys. Oceanogr. **48**, 555–572 (2018a)

Canuto, V.M., et al.: Subduction by sub-mesoscales. J. Geophys. Res. Oceans. **123**, 8688–8700 (2018b)

Canuto, V.M., et al.: 3D space dependent mesoscale diffusivity. Derivation and implications. J. Phys. Oceanogr. **49**, 1055–1074 (2019)

Capet, X., et al.: Mesoscale to sub-mesoscale transition in the California current system. Part I: flow structure, eddy flux, and observational tests. J. Phys. Ocean. **38**, 29–43 (2008)

Castaing, B., et al.: Scaling of hard thermal turbulence in Rayleigh-Benard convection. J. Fluid Meh. **204**, 1-X (1989)

Cessi, P.: An energy constrained parameterization of eddy buoyancy flux. J. Phys. Ocean. **38**, 1897–1819 (2008)

Chandrasekhar, S.: A theory of turbulence. Proc. Roy. Soc. A. (1955). https://doi.org/10.1098/rspa.1955.0070

Chelton, D.B., Schlax, M.G., Samelson, R.M.: Global observations of non-linear mesoscale eddies. Prog. Oceanogr. **91**, 167–216 (2011)

Cheng, Y., et al.: A second-order closure turbulence model. New heat flux equation and no critical Richardson number. J. Atmos. Sci. **77**(8), 2743–2759 (2020)

Chereskin, T.K.: Direct evidence for an Ekman balance in the California Current. J. Geophys. Res. **100**(C9), 18261–18269 (1995)

Chou, P.Y.: On the velocity correlations and the solutions of the equation of turbulent fluctuations. Quart. J. Applied Math. **3**, 38–54 (1945)

Coffey, W.T., Kalmykov, Y.P.: Chapter 1, Historical Background and introductory concepts. In: The Langevin Equation, 4th edn, pp. 1–180. Word Scientific (2017)

Cunningham, S.A., et al.: Transport and variability of the Antarctic Circumpolar Current in the Drake Passage. J. Geophys. Res. **108**, 8084 (2003)

Cushman-Roisin, B.: Dynamics of the ocean surface mixed layer. http://www.soest.hawaii.edu/pubservices/1987pdfs/cushman_roisin.pdf (1987)

D'Asaro, E., et al.: Enhanced turbulence and energy dissipation at ocean fronts. Science. **332**, 318–322 (2011). https://doi.org/10.1126/science.120155

Danasaboglu, G., Marshall, J.: Effects of vertical variations of thickness diffusivity in an ocean general circulation model. Ocean Model. **18**, 122–141 (2007)

Danilov, S.D., Gurarie, D.: Quasi-2D turbulence. Soviet Phys. Uspekhi. **43**, 863–890 (2000)

Deardorff, J.W.: A numerical study of three-dimensional turbulent channel flow at large Reynolds numbers. J. Fluid Mech. **41**(2), 453–480 (1970)

Farneti, R.T.L.D., Rosati, A.J., Griffies, S.M., Zeng, F.: The role of mesoscale eddies in the re-stratification of the Souter Ocean response to climate change. J. Phys. Oceanogr. **40**, 1539–1557 (2010)

Ferrari, R., Nikurashin, M.: Suppression of eddy diffusivity across Jets in the SouthernOcean. J. Phys. Oceanogr. **40**, 1501–1519 (2010)

Ferrari, R., Polzin, K.I.: Finescale structure of the T-S relation in the eastern North Atlantic. J. Phys. Oceanogr. **35**, 1437–1454 (2005). https://doi.org/10.1175/JPO2763.1

Ferrari, R., et al.: A boundary-value problem for the parameterized mesoscale eddy transport. Ocean Model. **32**, 143–156 (2010)

Fox-Kemper, B., Ferrari, R., Hallberg, R.: Parameterization of mixed layer eddies. Part I: theory and diagnosis. J. Phys. Oceanogr. **38**, 1145–1165 (2008)

Fox-Kemper, B., et al.: Parameterization of ML eddies: implementation and impact in global ocean climate simulations. Ocean Model. **39**, 61–78 (2011)

Frenger, I., et al.: Southern Ocean eddy phenomenology. J. Geophys. Res. **120**, 7413–7449 (2015). https://doi.org/10.1002/2015JC011047

Fu, L.L.: Pattern and velocity of propagation of the global ocean eddy variability. J. Geophys. Res. **114**, C11017 (2009). https://doi.org/10.1029/2009JC005349

Ganachaud, A.: Large-scale mass transport, water mass formation and diffusivities estimated from World Ocean Circulation Experiment (WOCE) hydrographic data. J. Geophys. Res. **108** (2003). https://doi.org/10.1029/2002JC001565

Ganachaud, A., Wunsch, C.: Improved estimates of global circulation, heat transport and mixing from hydrographic data. Nature. **408**, 453–456 (2000)

Gent, P.R.: The Gent-McWilliams parameterization: 20/20 hindsight. Ocean Model. **39**, 2–9 (2011). https://doi.org/10.1016/j.ocemod.2010.08.002

Gent, P.R., McWilliams, J.C.: Isopycnal mixing in ocean circulation models. J. Phys. Oceanogr. **20**, 150–155 (1990)

Gille, S.T., Davis, R.: The influence of mesoscales on coarse resolution overturning circulation of subgrid-scale eddy parameterization. J. Phys. Oceanogr. **20**, 1109–1123 (1999)

Giordani, H., Prieur, L., Caniaux, G.: Advances insights into sources of vertical velocity in the ocean. Ocean Dyn. **56**, 513–524 (2006)

Gnanadesikan, A., et al.: How does ocean ventilation change under global warming? Ocean Sci. **3**(1), 43–53 (2007)

References

Gnanadesikan, A., Pradal, M.A., Abernathey, R.: Isopycnal mixing by mesoscale eddies significantly impacts oceanic anthropogenic carbon uptake. Geophys. Res. Lett. **42**, 4249 (2015). https://doi.org/10.1002/2015GL064100

Griffies, S.M., et al.: Coordinated Ocean-ice Reference Experiments (COREs). Ocean Model. **26**, 1–46 (2009)

Hallberg, R., Gnanadesikan, A.: The role of eddies in determining the structure and response of the wind-driven Southern Ocean overturning. J. Phys. Oceanogr. **36**, 2232–2252 (2006)

Hamilton, K.: At the dawn of global climate modeling: the strange case of the Leith atmospheric model. Hist. Geo. Space Sci. **11**, 93–103 (2020). https://doi.org/10.5194/hgss-11-93-2020

Haynes, P., McIntyre, M.: On the evolution of vorticity and potential vorticity in the presence of diabatic heating and frictional or other forces. J. Atmos. Sci. **44**, 828–841 (1987)

Hellerman, S., Rosenstein, M.: Normal monthly wind stress over the world ocean with error estimates. J. Phys. Oceanogr. **13**, 1093–1104 (1983)

Hirake, Y., et al.: Subduction of Pacific Antarctic water in an eddy-resolving model. J. Geophys. Res, Oceans. **121**, 133–147 (2016)

Hoskins, B.J.: The role of PV in symmetric stabilities and instabilities. Q.J.R. Meteorol. Soc. **100**, 480–482 (1974)

Karleskind, P., Levy, M., Memory, L.: Subduction of carbon, nitrogen and oxygen in the northeast Atlantic. J. Geophys. Res. (2011). https://doi.org/10.1029/2010JC006446

Kerr, R.M.: Rayleigh number scaling in numerical convection. J. Fluid Mech. **310**, 139 (1996)

Killworth, P.D.: On parameterization of eddy transport. J. Mar. Res. **55**, 1171–1197 (1997)

Klocker, A., Abernathey, R.: Global patterns of mesoscale eddy properties and diffusivities. J. Phys. Oceanogr. **44**, 1030–1046 (2014)

Klocker, A., Marshall, D.P.: Advection of baroclinic eddies by depth mean flow. Geophys. Res. Lett. **41**, 3517–3521 (2014). https://doi.org/10.1002/2014GL060001

Koszalka, L., et al.: Dynamics of wind forced coherent anticyclones in the upper ocean. J. Geophys Res. **114**, C08001 (2009)

Kraichnan, R.H.: Lagrangian-history closure approximation for turbulence. Phys. Fluids. **8**, 575 (1965)

Kraichnan, R.H.: An almost Markovian Galilean -invariant turbulence model. J. Fluid Mech. **47**, 513 (1971)

Langevin, P.: Sur la theory du movement Brownian. Comptes Rendus Acad. Sci. Paris. **146**, 530–533 (1908)

Large, W.G., McWilliams, J.C., Doney, S.C.: Oceanic vertical mixing: a review and a model with a nonlocal boundary layer parameterization. Rev. Geophys. **32**, 363–403 (1994). https://doi.org/10.1029/94RG01872

Le Sommer, J., et al.: Parameterization of subgrid stirring in eddy resolving ocean models. Part 1: theory and diagnostics. Ocean Model. **39**(3–4), 154–169 (2011)

Leith, C.E.: Atmospheric predictability and two-dimensional turbulence. J. Atmos. Sci. **8**, 145 (1971)

Lemons, D.S.: Paul Langevin 1908 paper "on the theory of Brownian motion". Am. J. Phys. **65**(11) (1997)

Lesieur, M.: Turbulence in Fluids. Kluwer, Boston, MA (1991)

Levy, M., Klein, P., Treguier, A.M.: Impact of SM physics on production and subduction of phytoplankton in an oligotrophic regime. J. Mar. Res. **59**, 535–565 (2001)

Levy, M., et al.: Physical pathways for carbon transfer between the surface layer and the ocean interior. Global Biogeoch. Cycle. **27**, 1001–1012 (2013). https://doi.org/10.1002/gbc.20092

Levy, M., et al.: The role of sub-mesoscales currents in structuring marine ecosystems. Nat. Commun. **9**, 4758 (2018). https://doi.org/10.1038/s41467-018-07059-3

Liang, X., Spall, M., Wunsch, C.: Global ocean vertical velocity from a dynamically consistent ocean state estimate. J. Geophys. Res. **122**, 8208 (2017). https://doi.org/10.1002/2017JC012985

Lilly, D.K.: On the application of the eddy viscosity concept in the inertial subrange of turbulence. NCAR Ms.123, Boulder, Co. (1966)

Liu, J., et al.: Sensitivity of sea ice to physical parameterizations in the GISS global climate model. J. Geophys. Res. **108**(C2), 3053 (2003). https://doi.org/10.1029/2001JC001167

Lumley, J.L., Yaglom, A.: A Century of Turbulence. In: Flow, Turbulence and Combustion, pp. 241–286. Kluwer Academic Publishers (2001)

Lumpkin, R., Speer, K.: Large scale vertical and horizontal circulation in the North Atlantic Ocean. J. Phys. Oceanogr. **33**, 1902–1920 (2003)

Lumpkin, R., Speer, K., Kolterman, K.: Transport across 28N in the Atlantic Ocean. J. Phys. Oceanogr. **38**, 733–752 (2008)

Luneva, M.V., et al.: Effects of mesoscale eddies in the active mixed layer: test of the parameterization in eddy resolving simulations. Geophys. Astrophys. Fluid Dyn. (2015). https://doi.org/10.1080/03091929.2015.1041023

Mahadevan, A.: The impact of sub-mesoscale physics on primary production of plankton. Ann. Rev. Mar. Sci. **8**, 161–184 (2016)

Malkus, W.V.R.: Discrete transition in turbulent convection. Proc. R. Soc. London Ser. A. **225**, 196 (1954)

Marshall, D.: Subduction of water masses in an eddying ocean. J. Mar. Res. **55**, 201–222 (1997)

Marshall, P., Speer, N.: Closure of the meridional overturning circulation through Southern Ocean upwelling. Nat. Geosci. **5**(3), 171–180 (2012)

Mazloff, M.R., Heimbach, P., Wunsch, C.: An eddy-permitting Southern Ocean state estimate (SOSE). J. Phys. Oceanogr. **40**, 880–888 (2010)

McComb, W.D.: The Physics of Fluid Turbulence. Oxford Science Publications (1992)., 72pp

McGillicuddy, D.J.: Mechanisms of physical, biological, bio-geochemical interaction at the ocean mesoscale. Annu. Rev. Mar. Sci. **8**, 125–159 (2016)

McWilliams, J.C.: Sub-mesoscale currents in the ocean. Proc. R. Soc. **A472**, 290160117 (2016). https://doi.org/10.1098/rspa.2016.0117

Mellor, G.L., Yamada, T.: Development of a turbulence closure model for geophysical fluid problems. Rev. Geophys. Space Phys. **20**, 851–875 (1982)

Mensa, J.A., et al.: Seasonality of sub-mesoscale dynamics in the Gulf Stream region. Ocean Dyn. **63**, 923–941 (2013)

Meredith, N.P., et al.: Global model of lower band and upper band chorus from multiple satellite observations. J. Geophys. Res. Space Phys. **117**(A10) (2012)

Monin, A.S., Yaglom, A.M.: Statistical Fluid Mechanics. MIT Press, Cambridge, MA (1971)

Morton, B.R., Tayor, G.I., Turner, J.S.: Turbulent gravitational convection from maintained and isolated sources. Proc. Roy. Soc. London. **A234**, 1–23 (1956)

Ogura, Y.: A consequence of the zero-fourth-order-cumulant approximation in the decay of isotropic turbulence. J. Fluid Mech. **16**, 33–40 (1963)

Omand, M.M., et al.: Eddy-driven subduction exports particulate organic carbon from the spring bloom. Science. **348**, 222–225 (2015)

Orszag, S.A.: Analytic theories of turbulence. J. Fluid Mech. **41**, 363–386 (1970)

Oschlies, A.: Improved representation of upper-ocean dynamics and mixed layer depth in a model of the North Atlantic on switching from eddy permitting to eddy resolving grid resolution. J. Phys. Oceanogr. **32**, 2277–2298 (2002)

Pinot, J.-M., et al.: A study of the equation for diagnosing vertical motion at ocean fronts. J. Mar. Res. **54**, 239–259 (1996)

Polland, R.T., Regier, R.A.: Vorticity and vertical circulation at the ocean front. J. Phys. Oceanogr. **22**, 609–625 (1992)

Pope, S.B.: Turbulent Flows. Cambridge University Press, Cambridge (2000)., 771pp

Prandtl, L.: Bericht uber Untersuchungen zur ausgebildeten Turbulenz. Z. Angew. Math. Mech. **5**, 136–139 (1925)

Radko, T., Marshall, J.: Eddy induced diapycnal fluxes and their role in the maintenance of the thermocline. J. Phys. Oceanogr. **34**, 372–383 (2004)

Ramachandran, S., Tandon, A., Mahadevan, A.: Enhancement in vertical fluxes at a front by mesoscale-submesoscale coupling. J. Geophys. Res. Oceans. **119**, 8495–8511 (2014)

Redi, M.H.: Oceanic isopycnal mixing by coordinate rotation. J. Phys. Oceanogr. **12**, 1154–1158 (1982)

Roach, C.J., Balwada, D., Speer, K.: Global observations of horizontal mixing from Argo floats and surface drifter trajectories. J. Geophys. Res. Oceans. **123**, 4560–4575 (2018)

Robert, M.J., Marshall, D.P.: On the validity of downgradient eddy closure in ocean models. J. Geophys. Res. **105**, 28613–28627 (2000)

Rodgers, K.B., et al.: Strong sensitivity of Southern Ocean carbon uptake and nutrient cycling to wind stress. Biogeosciences. **11**, 4077–4098 (2014)

Rosso, I., et al.: Vertical transport in the ocean due to mesoscale structure. Impacts in the Kerguelen region. Ocean Modell. **80**, 10–23 (2014)

Russell, G.L., Miller, J.R., Rind, D.H.: A coupled atmosphere-ocean model for transient climate change. Atmosphere-Ocean. **33**, 683–730 (1995)

Russell, G.L., et al.: Comparison of model and observed regional temperature changes during the past 40+ years. J. Geophys. Res. **105**, 14891–14898 (2000)

Sallée, J.B., Rintoul, S.R.: Parameterization of eddy-induced subduction in the Southern Ocean surface layer. Ocean Model. **39**, 146–153 (2011)

Sallée, J.B., et al.: An estimate of Lagrangian eddy statistics and diffusion in the mixed layer of the Southern Ocean. J. Mar. Res. **66**, 441–463 (2008)

Sallée, J.B., et al.: Southern Ocean thermocline ventilation. J. Phys. Oceanogr. **40**, 509–529 (2010)

Sanders, R., et al.: The biological carbon pump in the North Atlantic. Prog. Oceanogr. **129**, 200–218 (2014)

Schaffenberg, M.G., Stammer, D.: Seasonal variation of the geostrophic flow field and of eddy kinetic energy inferred from TOPEX/Poseidon and Jason-I tandem mission dada. J. Geophys. Res. **115**, C02008 (2010)

Schmidt, G.A., et al.: Configuration and assessment of the GISS ModelE2 contributions to the CMIP5 archive. J. Adv. Model. Earth Syst. **6**(1), 141–184 (2014). https://doi.org/10.1002/2013MS000265

Shraiman, B.P., Siggia, E.D.: Heat transfer in high Rayleigh number convection. Phys. Rev. A. **42**, 3650 (1990)

Siegelman, L.: Energetic sub-mesoscale dynamics in the ocean interior. J. Phys. Oceanogr. **50**, 727–749 (2020)

Siegelman, L., et al.: Enhanced upward heat transport at deep sub-mesoscale ocean fronts. Nat. Geosci. **13**, 50–55 (2020)

Sinha, A., Callies, J., Menemenlis, D.: Do sub-mesoscales affect the large-scale structure of the upper ocean? J. Phys. Oceanogr. (2022). https://doi.org/10.1175/JPO-D-22-0129

Sloyan, B.M., Rintoul, S.R.: Circulation, renewal, and modification of Antarctic Mode Water and Intermediate Water. J. Phys. Oceanogr. **31**, 1005–1030 (2001)

Smagorinsky, J.: General circulation experiments with the primitive equations: I. The basic equations. Mon. Weather Rev. **91**, 99–164 (1963)

Su, Z., et al.: Ocean sub-mesoscales as a key component o the global heat budget. Nat. Commun. **9**(1), 775 (2018)

Thomas, L.N.: Destruction of PV by winds. J. Phys. Oceanogr. **35**, 2457–2466 (2005)

Thomas, L.N., Taylor, J.R.: Reduction of the usable wind-work on the general circulation by forced symmetric instability. Geophys. Res. Lett. **32**, L18606 (2010). https://doi.org/10.1029/2010GL044680

Thomas, L.N., et al.: Symmetric instability in the Gulf Stream. Deep-Sea Res. II. **91**, 96–110 (2013)

Thompson, A.F., et al.: Open-ocean sub-mesoscale motions: a full seasonal cycle of mixed layer instabilities from gliders. J. Phys. Oceanogr. **46**, 1285–1306 (2016)

Treguier, A.M.: Evaluating eddy mixing coefficient from eddy-resolving ocean model: a case study. J. Mar. Res. **57**, 89–108 (1999)

Tulloch, R., et al.: Direct estimates of lateral eddy diffusivity upstream of Drake Passage. J. Phys. Oceanogr. **44**, 2593–2616 (2014)

Veneziani, M., et al.: Barrier layers in the tropical South Atlantic: mean dynamics and submesoscale effects. J. Phys. Oceanogr. **44**, 265–288 (2014)

Visbeck, M.J., et al.: On the specification of eddy transfer coefficients in coarse resolution ocean circulation models. J. Phys. Oceanogr. **27**, 381–402 (1997)

Wang, S., et al.: Ocean eddy energetics in the spectral space as resolved by high-resolution general circulation models. J. Phys. Oceanogr. **49**, 2815–2827 (2019)

WOCE Data Products Committee. WOCE Global Data, Version 3.0, WOCE International Project Office, WOCE Report No. 180/02, Southampton, UK (2002)

Wu, X.Z., Libchaber, A.: Scaling relations in thermal convection. The aspect ratio dependence. Phys. Rev A. **45**, 842 (1990)

Wunsch, C.: The vertical partition of oceanic horizontal kinetic energy. J. Phys. Oceanogr. **27**, 1770–1794 (1997)

Wunsch, C., Heimbach, P.: Bidecadal thermal changes in the abyssal ocean. J. Phys. Oceanogr. **43**(8), 1750–1766 (2013)

Wyld, H.W.: Formulation of the theory of turbulence in an incompressible fluid. Ann. Phys. **14**, 143–165 (1961)

Yakhot, V., Orszag, S.A.: Renormalization group analysis of turbulence. I. Basic theory. J. Sci. Comput. **1**, 3–51 (1986)

Yung, C.K., et al.: Topographic hotspots of Southern Ocean eddy upwelling. Front. Mar. Sci. **9**, 855785 (2022). https://doi.org/10.3389/fmars2022.855785

Zhong, Y., Bracco, A.: Submesoscale impacts on horizontal and vertical transport in the Gulf of Mexico. J. Geophys. Res. Oceans. **118**(10), 5651–5668 (2013)

Chapter 2
Vertical Mixing in Oceans, PBL and Stars

Abstract We apply the Langevin Stochastic Equations LSE to study vertical mixing in the ocean, PBL and stars. In the first case, we derive the algebraic form of two key variables, *heat and momentum turbulent diffusivities*. We also derive the 3D equations governing the case of mixing due to shear, vorticity, heat and salt. In the 1D case, we present the explicit solutions for the important case of *Salt Fingers* for which there are data to assess the model predications. The case of *Diffusive Convection* is also discussed. Carbon-cycle studies that are an important ingredient of future climate studies, require the vertical diffusivity of a passive tracer that, not being available thus far, was often identified with that of salt which we show is not correct. A passive tracer diffusivity is then presented. As for stars, we consider thermal convection, overshooting and the angular momentum problem that has risen from helioseismology. Overflows by gravity currents and deep convection are discussed for which the present analysis offers improvements. In the case of the PBL, we present the results of the most updated turbulence model.

2.1 LSE Equations for Buoyancy

Using the results of Chap. 1, we have the following LSE equations:
Velocity:

$$\frac{\partial}{\partial t} u_i(k) = f_i(\text{ext}) + f_i^<(\text{turb}) + f_i^>(\text{turb})$$

$$E = \int dk E(k), \quad \partial_t E(k) = A_t(k) - 2\nu_d k^2 E(k) + A_{\text{ext}}$$

$$A_t(k) = -\ r(k)\partial_k E(k), \quad \frac{1}{2}r(k) = \int_0^k q^2 \nu_t(q) dq \quad (2.1)$$

Temperature:

$$\frac{\partial \theta(k)}{\partial t} = f_\theta(\text{ext}) + f_\theta^<(\text{turb}) - \chi_d(k) k^2 \theta(k)$$

$$\frac{1}{2}\overline{\theta^2} = \int E_\theta(k) dk, \quad \partial_t E_\theta(k) = A_\theta^t - 2k^2 \chi_d(k) E_\theta(k) + A_{\text{ext}}$$

$$A_\theta^t = r_\theta(k) \partial_k E_\theta(k), \quad \frac{1}{2}r_\theta(k) = \int_0^k q^2 \chi_t(q) dq, \quad \chi_d(k) = \chi + \chi_t(k) \quad (2.2)$$

Dynamical viscosity, $\nu_d(k) = \nu + \nu_t(k)$ ***and conductivity***:

$$\nu_d(k) = \left(\nu^2 + \frac{2}{5}\int_k^\infty q^{-2} E(q) dq\right)^{1/2}, \quad \frac{d\chi_d}{d\nu_d} = \frac{10}{3} \frac{\nu_d}{\nu_d + \chi_d} \quad (2.3)$$

Next, we introduce the spectral densities:

$$e(k) = \frac{1}{2} <u_i(\mathbf{k}) u_i(\mathbf{k})>, \quad e_{\theta,s}(k) = \frac{1}{2}(<\theta(\mathbf{k})\theta(\mathbf{k})>, \quad <s(\mathbf{k})s(\mathbf{k})>)$$
$$(2.4)$$
$$j(k) = \frac{1}{2} <w(\mathbf{k})\theta(\mathbf{k})>, \quad e_z(k) = \frac{1}{2} <w(\mathbf{k})w(\mathbf{k})>$$

and the spectral functions:

$$E(k), \ E_\theta(k) = k^2 \int d\Omega_k [e(k), e_\theta(\mathbf{k})], \quad [J(k), \ E_z(k)]$$
$$= k^2 \int d\Omega_k [j(\mathbf{k}), e_z(\mathbf{k})] \quad (2.5)$$

Using the above dynamic equations, one then derives the equations for heat flux, buoyancy variance e_θ, z-component of the kinetic energy e_z and kinetic energy e:

2.1 LSE Equations for Buoyancy

$$\partial_t j + D_f(j) = 2g\alpha P_{zz}e_\theta + 2\beta e_z - 2k^2(\nu_d + \chi_d)j$$

$$\partial_t e_\theta + D_f(e_\theta) = a_\theta^t + \beta j - 2k^2\chi_d e_\theta$$

$$\partial_t e_z + D_f(e_z) = \frac{1}{2}P_{zz}\, a^t + g\alpha P_{zz}j - 2k^2\nu_d e_z \quad (2.6)$$

$$\partial_t e + D_f(e) = a^t + g\alpha j - 2k^2\nu_d e$$

where the functions a_θ^t, a_t represent the work done by the turbulent forces:

$$4\pi k^2 a^t = A^t(k), \quad 4\pi k^2 a_\theta^t = A_\theta^t(k) \quad (2.7)$$

together with the third relation in (2.1) and (2.2). Let us note that Eq. (2.6) do not introduce adjustable parameters and that heat flux and z-kinetic energy are given by:

$$\overline{w\theta} = J = \int dk J(k), \quad \frac{1}{2}\overline{w^2} = \int dk E_z(k) \quad (2.8)$$

Equation (2.6) must be compared with RSM results derived in Canuto (1992, Eqs.57–60):

Convective flux $J = \overline{w\theta}$: $\quad \dfrac{\partial}{\partial t} J + D_f = \beta \overline{w^2} + (1-c_7)\alpha_T g\overline{\theta^2} - 2c_6\tau^{-1}J \quad (2.9)$

Temperature variance $\overline{\theta^2}$: $\quad \partial_t \dfrac{1}{2}\overline{\theta^2} + D_f = \beta J - e_\theta, \quad e_\theta = 2\chi \int k^2 E_\theta dk \quad (2.10)$

Vertical KE $\frac{1}{2}\overline{w^2}$:

$$\frac{\partial}{\partial t}\frac{1}{2}\overline{w^2} + D_f = -c_4\tau^{-1}\left(\overline{w^2} - \frac{2}{3}K\right) + \left(1 - \frac{2}{3}c_5\right)\alpha_T gJ - \frac{1}{3}\varepsilon \quad (2.11)$$

Kinetic energy K : $\quad \partial_t K + D_f = g\alpha_T J - \varepsilon, \quad \varepsilon = 2\nu \int k^2 K(k) dk \quad (2.12)$

Here, D_f represents diffusion, $\beta = -\,dT/dz. - g/c_p$, α_T is the thermal expansion coefficient and ν, χ are the kinematic viscosity and conductivity. Since the RSM cannot determine $c_{4,\,5,\,6,\,7}$, the following heuristic values were suggested (C92, Eq. 44d):

$$2c_4 = \frac{7}{2}, \quad c_5 = \frac{3}{10}, \quad 2c_6 = 7.5, c_7 = 1/3, 1/5 \tag{2.13}$$

which in the next section we show to be derivable from the LSE.

2.2 Derivation of the RSM Equations from the LSE

We integrate the fourth of (2.6) over $d\Omega_k$, use (2.5) and the first of (2.6). Since $\nu_t > > \nu$, we have:

$$\partial_t E + D_f E = A^t(k) + g\alpha J - 2k^2 \nu_t E \tag{2.14}$$

After integrating (2.4) over k, the result coincides with the RSM Eq. (2.12). Using a similar procedure, the second Eq. (2.6) becomes the RSM Eq. (2.10). Next, from the first of (2.6), we recover the RSM Eq. (2.9) with:

$$1 - c_7 \equiv \left[\int dk e_\theta(k) \right]^{-1} \int dk P_{zz}(k) e_\theta(k), \quad 2c_6 \tau^{-1} \equiv \left[\int dk J(k) \right]^{-1} \int dk k^2 (\nu_d + \chi_d) j(k) \tag{2.15}$$

Next, we integrate over k the third equation in (2.6). Using the first of (2.7), we obtain:

$$\int dk a^t(k) P_{zz} = \frac{2}{3} \int dk A_t(k) \tag{2.16}$$

Since $\nu_t > > \nu$, the conservation law for the transfer T(k), first of Eq. (1.35) of Chap. 1, gives:

$$\int dk T(k) = \int dk \left[A^t(k) - 2k^2 \nu_t(k) E(k) \right] = 0 \tag{2.17}$$

and thus:

$$\frac{\partial}{\partial t} \frac{1}{2} \overline{w^2} + D_f = \alpha g \int dk j(k) P_{zz} - 2\nu \int dk k^2 E_z(k) - 2 \int dk k^2 \nu_t(k) \left(E_z - \frac{1}{3} E \right) \tag{2.18}$$

The last term in (2.18) differs from zero only for small k since for large k, eddies become isotropic and $E_z = \frac{1}{3} E$. We can then take $k^2 \nu_t(k)$ out of the integral and compute it at $k = k_0$. By the same token, since the second term peaks at large k, we take the same limit and Eq. (2.18) becomes:

2.2 Derivation of the RSM Equations from the LSE

$$\frac{\partial}{\partial t}\frac{1}{2}\overline{w^2} + D_f = \alpha g J <P_{zz}> - \nu_t(k_0)k_0^2\left(\overline{w^2}-\frac{2}{3}K\right) - \frac{1}{3}\varepsilon \quad (2.19)$$

where we have defined the dimensionless variable:

$$<P_{zz}> = \frac{\int dk j(k) P_{zz}(k)}{\int dk j(k)} \quad (2.20)$$

Comparing (2.19) with the RSM Eq. (2.11), we obtain the relations:

$$c_4 \tau^{-1} = \nu_t(k_0)k_0^2, \quad 1-\frac{2}{3}c_5 = <P_{zz}> \simeq \frac{2}{3} \quad (2.21)$$

Assuming a Kolmogorov spectrum (Chap.1, Eq. 1.32), the first of (2.21) gives:

$$\nu_t \equiv \nu_t(k_0) = \left(\frac{3 Ko}{20}\right)^{1/2} \varepsilon^{1/3} k_0^{-4/3}, \quad K = \frac{3}{2} Ko \varepsilon^{2/3} k_0^{-2/3} \quad (2.22)$$

Using the definition $\tau = 2K/\varepsilon$ and the first of (2.21), we obtain:

$$c_4 = \left(\frac{27}{20}K_o^3\right)^{1/2} = \frac{5}{2}, \quad Ko = \frac{5}{3} \quad (2.23)$$

to be compared with (2.13). Next, we consider $2c_6$ whose value depends on the *efficiency of convection* determined by the Peclet number:

$$Pe = \frac{\tau_\chi}{\tau} = \frac{c_\varepsilon^2}{2c_\mu}\frac{\nu_t}{\chi}, c_\mu = c_e(10\pi^2)^{-1/2}, \quad c_e = \pi\left(\frac{2}{3Ko}\right)^{3/2} \quad (2.24)$$

Using these results, Canuto and Dubovikov (2008) showed that for both $\chi_t > \chi$ and $\chi > \chi_t$ one has:

$$\frac{\tau_{pv}}{\tau} = \frac{2}{5}$$

$$\frac{\tau_{p\theta}}{\tau} = \frac{1}{4\pi^2}Pe\left(1+\frac{1}{2}\frac{\nu_t}{\chi}+\frac{1}{2}\frac{\chi_t}{\chi}\right)^{-1} \quad (2.25)$$

$$\frac{\tau_\theta}{\tau} = \frac{4}{7\pi^2}Pe\left(1+\frac{8}{35}\frac{\chi_t}{\chi}\right)^{-1}$$

In conclusion, the RSM equations can be derived from the LSE and the parameters that enter the RSM are determined in the LSE framework with results comparable to the heuristic values (2.13).

2.3 Heat and Momentum Turbulent Diffusivities

As discussed in Part I, the LSE for the velocity and buoyancy turbulent fields contain an external forcing and two types of turbulent forcings. The former have the following forms:

$$\text{Velocity field}: f_i^{ext} : -g^{-1}g_j P_{ij}b' - ik_j \bar{u}_j u_i' - S_{ij}^{\perp} u_j' + k_j S_{jm} P_{i\ell} \frac{\partial u_\ell'}{\partial k_m} \quad (2.26)$$

where the turbulent buoyancy field b' is given in terms of the fluctuating temperature θ and salinity s:

$$b' = g(\alpha_\theta \theta - \alpha_s s) \quad (2.27)$$

with $\alpha_{\theta, s}$ are thermal expansion and haline contraction coefficients. The other terms in (2.26) are:

$$S_{ij} = \bar{u}_{i,j}, \quad S_{ij}^{\perp} = P_{mi} S_{m\ell} P_{\ell j}, \quad P_{ij} = \delta_{ij} - k_i k_j / k^2 \quad (2.28)$$

where for simplicity, we omitted some of the k-dependences; primes denote turbulent fields and an overbar denotes a mean variable.

Buoyancy field:

$$f^{ext}(b) = -\frac{\partial \bar{b}}{\partial x_i} u_i' \quad (2.29)$$

Momentum structure functions. To obtain LSE for the Reynolds stresses in **k**-space $R_{ij}(\mathbf{k}) = \overline{u_i' u_j'}$, we substitute (2.26), (2.27) and (2.28) into the LSE for the turbulent velocity:

$$\frac{\partial u_i'(\mathbf{k})}{\partial t} = f_i(\text{ext}) + f_i^{<} + f_i^{>} \quad (2.30)$$

multiply the result by $u_j'^*$ and ensemble average the function $\text{Re } \overline{a(\mathbf{k})b^*(\mathbf{k}')} = \overline{ab}(\mathbf{k})\delta(\mathbf{k} - \mathbf{k}')$. The result that includes external and non-linear terms:

2.3 Heat and Momentum Turbulent Diffusivities

$$\frac{1}{2}\partial_t \mathbf{R}(\mathbf{k}) = Non-linear + Ext.$$

$$Non-linear = -(8\pi k^2)^{-1} r(k) \frac{\partial E(k)}{\partial k} \mathbf{P}(\mathbf{k}) - k^2 \nu_d(k) \mathbf{R}(\mathbf{k})$$

(2.31)

The first term of the non-linear forcing is due to the IR (infrared) term $f_i^<$ in (2.30) while the second term is due to the UV forcing $f_i^>$. The external forcing is as follows:

$$Ext. = -(\mathbf{S}^\perp \cdot \mathbf{R})_s + [(\mathbf{k} \cdot \mathbf{S} \cdot \partial_\mathbf{k})\mathbf{R}]^\perp - g^{-1}(\mathbf{g}^\perp \mathbf{J})_s \qquad (2.32)$$

where $\mathbf{J}$ is the buoyancy flux we used the notation $(\mathbf{T})_s \equiv \frac{1}{2}(T_{ij} + T_{ji})$ to symmetrize a second rank tensor. Since OGCMs require terms in physical space, we need to average Eq. (2.31) over directions and magnitude of $\mathbf{k}$. Though the general result is rather complex, we can derive simpler results in the IR and UV regimes corresponding to small and large k. In the UV case, one has the smallness parameters:

$$\left|\frac{k^2 \nu_d}{S_{ij}}\right| << 1, \qquad \left|\frac{\mathbf{J} \cdot \nabla_H b}{k^2 \nu_d e_b}\right| << 1 \qquad (2.33)$$

and we expand all the spectra using (2.33) the latter and the UV energy spectrum results in the Kolmogorov spectrum, see below Eq. (2.40). The same procedure gives the following evolution equation for the spectrum of the traceless RS defined as follows $\widehat{\mathbf{R}}(\mathbf{k}) = \mathbf{R}(\mathbf{k}) - \frac{2}{3}E(k)\boldsymbol{\delta}$:

$$UV\ regime: \frac{D\widehat{\mathbf{R}}}{Dt} = c_s(k)\mathbf{S} + c_V(k)[\mathbf{V}, \mathbf{S}] - 2k^2 \nu_d \widehat{\mathbf{R}} + \frac{4}{3}\left[\left(\mathbf{e}_z \mathbf{j}(k) - \frac{1}{3}\mathbf{j}(k)\boldsymbol{\delta}\right)\right]_s$$

(2.34)

where, when not strictly necessary, we omitted the k-dependence. We further have the relations:

$$c_s(k) = -\frac{20}{27}\varepsilon^{2/3} k^{-5/3}, \qquad c_V(k) \approx \varepsilon^{1/3} k^{-7/3}, \qquad 2\nu_d(k) = \varepsilon^{1/3} k^{-4/3} \qquad (2.35)$$

Here, $\mathbf{S}$, $\mathbf{V}$ are the mean shear and vorticity tensors, $[\mathbf{V}, \mathbf{S}] = \mathbf{V} \cdot \mathbf{S} - \mathbf{S} \cdot \mathbf{V}$ is the commutator of those tensors, $\overline{\mathbf{j}(k)} = \mathrm{Re}\,\overline{b'^* \mathbf{u}'}(k)$ is the spectrum of the horizontal buoyancy flux and, $j(k) = \mathrm{Re}\,\overline{b'^* w'}(k)$ is its vertical counterpart. To obtain (2,34) in physical space, we integrate over the whole k region, i.e., $\widehat{\mathbf{R}} \equiv \int d\mathbf{k}\widehat{\mathbf{R}}(\mathbf{k})$. A reasonable approximation may be obtained from its UV part for $k_* < k < \infty$ and using a cut-off at $k < k_*$ i.e., and using relations:

$$K = \int_{k_*}^{\infty} E(k)dk, \quad k_* = \left(\frac{5}{2K}\right)^{3/2} \varepsilon, \quad \widehat{\mathbf{R}} \approx \int_{k_*}^{\infty} \widehat{\mathbf{R}}(k)dk, \quad \mathbf{J} \approx \int_{k_*}^{¥} \mathbf{J}(k)dk \quad (2.36)$$

in which the integrands are assumed to be in the *UV regime*. Integrating Eq. (2.34) over k and using the asymptotic relation $\widehat{\mathbf{R}}(k) \sim k^{-7/3}$, the evolution equation for the one-point traceless component $\widehat{\mathbf{R}}$ is given by:

$$\frac{D\widehat{\mathbf{R}}}{Dt} = -\frac{4}{9}KS - \frac{27}{20}[\mathbf{V},\widehat{\mathbf{R}}] + \frac{4}{3}\left(\mathbf{e}_z\mathbf{J} - \frac{1}{3}\mathbf{j}\boldsymbol{\delta}\right)_s - \frac{10}{\tau}\widehat{\mathbf{R}} \quad (2.37)$$

which may be compared with Eqs. (A.6)–(A.8) of C10a. In the stationary limit, from (2.34) and (2.35), we obtain the following expression for the UV spectrum of the traceless Reynolds stress:

$$\textit{UV regime}: \widehat{\mathbf{R}}(k) = -\frac{20}{27}\varepsilon^{1/3}k^{-7/3}\mathbf{S} + k^{-3}[\mathbf{V},\mathbf{S}] + \frac{4}{3}\varepsilon^{-1/3}k^{-2/3}\left[\mathbf{e}_z\mathbf{J}(k) - \frac{1}{3}\mathbf{j}(k)\boldsymbol{\delta}\right]_s$$
$$(2.38)$$

Using relations (2.36), the one-point RS is then found to be:

$$\widehat{\mathbf{R}} \approx -\frac{2}{45}K\tau\mathbf{S} + 8^*10^{-3}K\tau^2[\mathbf{V},\mathbf{S}] + 9^*10^{-2}\tau\left[2\mathbf{j}\left(\mathbf{e}_z\mathbf{e}_z - \frac{1}{3}\boldsymbol{\delta}\right) + \mathbf{e}_z\mathbf{j} + \mathbf{j}\mathbf{e}_z\right] \quad (2.39)$$

Heat structure function. Consider the LSE for an arbitrary tracer:

$$\left(\frac{\partial}{\partial t} + \chi_\tau k^2\right)\tau' = f^{ext} + f^< + f^> \quad (2.40)$$

where χ_τ is the tracer kinematic diffusivity and consider the buoyancy case $\tau' = b'$. We substitute (2.29) into (2.40), multiply the resulting equation $u_i'^*$ and Eq. (2.30) by b*. Summing and averaging the resulting equations, we obtain the following LSE equation for the buoyancy flux **J**:

$$\partial_t J_i = -\overline{b_{,j}u_i'u_j'} + P_{i3}\overline{|b'|^2} - P_{im}S_{mj}J_j - k^2(\nu_d + \chi_d)J_i \quad (2.41)$$

where, as before, for simplicity, we omitted the k-dependence. Since in the ML $|j| < |$ j|, $|w_{,i}| < |u_{i,z}|$ and $\bar{b}_z = N^2 > |\overline{b_{,i}}|$ (i = 1, 2), Eq. (2.41) becomes:

$$\partial_t J_i = -N^2 R_{i3} + 2e_b P_{i3} - P_{im}\overline{u_{m,z}}j - k^2(\nu_d + \chi_d)J_i \quad (2.42)$$

where $e_b = \frac{1}{2}\overline{|b'|^2}$ whose LSE is:

2.3 Heat and Momentum Turbulent Diffusivities

$$\partial_t e_b = A_t^b - \mathbf{J}\cdot\nabla\overline{b} - 2k^2\chi_d e_b \qquad (2.43)$$

Equations (2.42) and (2.43) are however not appropriate for OGCMs that require results in physical space. To do so, we divide the k-space into UV(ultraviolet) and IR (infrared) regimes that represent small and large scales. In the UV regime, one has the small parameters:

$$\left|\frac{k^2\nu_d}{S_{ij}}\right| < <1, \quad \left|\frac{\mathbf{J}\cdot\nabla_H b}{k^2\nu_d e_b}\right| < <1 \qquad (2.44)$$

and we can expand all the terms using (2.44). Furthermore, in the ocean ML, the vertical component of the buoyancy flux is considerably larger than the horizontal component. Then, integrating over the directions of **k**, the UV vertical and horizontal fluxes are given by:

$$\frac{Dj}{Dt} = \frac{2}{3}(2e_b - N^2 e) - k^2\nu_d(1+\sigma_t^{-1})j \qquad (2.45)$$

UV regime:

$$\frac{Dj_i}{Dt} = -\left(\frac{2}{3}j\partial\overline{u}/\partial z + N^2\widehat{R}_{i3}\right) - k^2\nu_d(1+\sigma_t^{-1})j_i, \quad i=1,2 \qquad (2.46)$$

where $\widehat{R} = \mathbf{R} - \frac{2}{3}E(k)\delta$ and j is the z-component of $\mathbf{J}(k)$. In the stationary approximation from (2.45), (2.46), we obtain the expression for the UV spectrum of the buoyancy flux:

UV regime : $j = -\frac{20}{9}\sigma\epsilon^{1/3}k^{-7/3}N^2(1-2\sigma_t\lambda N^{-2}), \quad \sigma \equiv \frac{\sigma_t}{1+\sigma_t}, \quad \lambda = \frac{\epsilon_b}{\epsilon} \quad (2.47)$

$$j_i = -2\sigma\epsilon^{-1/3}k^{-2/3}\left(\frac{2}{3}j\partial\overline{u}_i/\partial z + N^2\widehat{R}_{i3}\right) \qquad (2.48)$$

In deriving these equations, we used the Kolmogorov and Batchelor spectra for EKE and buoyancy variance and the corresponding dissipations $\epsilon(\epsilon_b)$ and the $<P_{ij}> = \frac{2}{3}\delta_{ij}$. To obtain the results in physical space, we integrate the above results over k. A reasonable approximation may be obtained from its UV counterpart at $k_* < k < \infty$ and cutting off $\widehat{R}(k)$ at $k < k_*$, i.e.

$$K = \int_{k_*}^{\infty} E(k)dk, \quad k_* = \left(\frac{5}{2}K\right)^{3/2}\epsilon, \quad \widehat{\mathbf{R}} \approx \int_{k_*}^{\infty}\widehat{\mathbf{R}}(k)dk, \quad \mathbf{J} \approx \int_{k_*}^{\yen}\mathbf{J}(k)dk \quad (2.49)$$

As shown in C8–9, k_* is located in the vicinity of the maximum of the EKE spectrum and thus (2.40) may be interpreted as the definition of k_* provided the UV energy spectrum coincides with the Kolmogorov one $E(k) = \frac{5}{3}\varepsilon^{2/3}k^{-5/3}$. In this case, we obtain:

$$j = -\frac{2}{15}\sigma N^2(1-2\sigma_t\lambda N^{-2})K\tau \tag{2.50}$$

$$j_i = -\frac{4}{15}\sigma\tau\left(\frac{2}{3}j\,\partial\bar{u}_i/\partial z + N^2\widehat{R}_{i3}\right), \quad i=1,2 \tag{2.51}$$

The *evolution equations* for the buoyancy flux in physical space is given by:

$$Dj/Dt = \frac{2}{3}\left(\overline{|b'|^2}\text{-}N^2K\right) - 5(1+\sigma_t^{-1})\tau^{-1}j \tag{2.52}$$

$$Dj_i/Dt = -\frac{2}{3}j\frac{\partial\bar{u}_i}{\partial z} - N^2\widehat{R}_{i3} - 5(1+\sigma_t^{-1})\tau^{-1}j_i, \quad i=1,2 \tag{2.53}$$

which are quite close to Eq. (A.1) and (A.9) of C10a.
Heat structure function.
Next, consider the down-gradient relations:

$$j = \overline{w'b'} = -\kappa_b N^2, \quad \kappa_b = \Gamma_b\frac{\varepsilon}{N^2}, \quad \Gamma_b = \frac{1}{2}S_b(\tau N)^2 \tag{2.54}$$

where $S_b(Ri)$ is a dimensionless *structure function* that depends on Ri. From (2.53) and (2.54), we obtain the following relation:

$$S_b = \frac{2}{15}\sigma(1-2\sigma_t\lambda N^{-2}) \tag{2.55}$$

In Appendix A we derive the following expression for λ:

$$\lambda N^{-2} = -\frac{1}{2}a + \left(\frac{1}{4}a^2 + \frac{1}{2\sigma_t}\right)^{1/2}, \quad a = 1 - \frac{1}{2\sigma_t} + \frac{1}{12}Ri^{-1}\sigma_t^{-1}(1+\sigma_t^{-1}) \tag{2.56}$$

which, once substituted into (2.55), yields:

$$S_b(Ri) = \Phi(\sigma_t)\left[1 + X - (X^2 + 2\sigma_t)^{1/2}\right]$$
$$\Phi(\sigma_t) = \frac{2}{15}\sigma_t(1+\sigma_t)^{-1}, \quad X = \sigma_t - \frac{1}{2} + \frac{1}{12}Ri^{-1}(1+\sigma_t^{-1}) \tag{2.57}$$

Using $\sigma_t \simeq 0.72$, we obtain:

$$S_b(Ri) \approx 0.056\left[1 - 0.22\left[(Y^2 + 30)^{1/2} - Y\right]\right] \quad , \quad Y = 1 + 0.9Ri^{-1} \qquad (2.58)$$

Momentum structure function. Let us consider the down-gradient relations:

$$\widehat{R}_{i3} \equiv \overline{w'u'_i} = -\kappa_m \frac{\partial \bar{u}_i}{\partial z}, \quad \kappa_m = \frac{2K^2}{\varepsilon} S_m \qquad (2.59)$$

To find the structure function S_m, we use Eqs.(2.39) and (2.55) to derive the relation:

$$\widehat{R}_{i3} = -\frac{1}{45}(\tau K + 0.72\sigma\tau^2 j)\frac{\partial \bar{u}_i}{\partial z}\left(1 + \frac{0.36}{15}\sigma\tau^2 N^2\right)^{-1} \qquad (2.60)$$

Using (2.59), we obtain:

$$S_m = \frac{1 - 2\alpha S_b}{45 + 3\alpha}, \quad \alpha = 0.36\sigma\tau^2 N^2 \qquad (2.61)$$

where S_b (Ri) is given in (2.57).

2.4 Shear, Vorticity, Heat and Salt Fluxes

This case was treated in Canuto et al. (2011). Consider the EKE equation:

$$D_t K + \partial_s(\overline{u_i K}) = P - \varepsilon, \quad P = P_m + P_b \qquad (2.62)$$

where the momentum, heat and salt productions are defined as follows:

$$P_m = -R_{ij}\overline{u_{i,j}}, \quad R_{ij} = \overline{u'_i u'_j}, \quad P_{h,s} = g\alpha_{T,s} J_i^{h,s}, \quad J_{i,j}^{h,s} = \overline{u_i(\theta, s)} \qquad (2.63)$$

R_{ij} are the Reynolds stresses, primes represent turbulent fields and an overbar stands for time or ensemble averages. The dynamic equation for the heat flux is given by:

$$D_t J_i^h = R_{ij}\beta_j - J_j^h(S_{ij} + V_{ij}) + g\lambda_i\left(\alpha_T \overline{\theta^2} - \alpha_s \overline{s\theta}\right) - \tau_{p\theta}^{-1} J_i^h \qquad (2.64)$$

where $2S_{ij} = \overline{u_{i,j}} + \overline{u_{j,i}}, 2V_{ij} = \overline{u_{i,j}} - \overline{u_{j,i}}$ are mean shear and vorticity. We further have:

$$\lambda_i = -(g\rho)^{-1}\partial p/\partial x_i, \quad \beta_i = -\partial T/\partial x_i \qquad (2.65)$$

Analogously, the dynamic equation for the salinity flux is given by:

$$D_t J_i^s = -R_{ij} S_{,j} - J_j^s (S_{ij} + V_{ij}) + g\lambda_i \left(\alpha_T \overline{s\theta} - \alpha_s \overline{s^2}\right) - \tau_{ps}^{-1} J_i^s \tag{2.66}$$

where S is the mean salinity. Furthermore, we have the equations:

$$\overline{\theta^2} = \tau_\theta \beta_i J_{,i}^\theta, \quad \overline{s^2} = -\tau_s J_i^h S_{,i} \quad \overline{s\theta} = \tau_{p\theta} \left(\beta_i J_i^s - J_i^\theta S_{,i}\right) \tag{2.67}$$

The solutions of the stationary limits of the above equations yield the algebraic relations:

$$\left(\delta_{ij} + \eta_{ij}\right) J_j^h = \gamma_{ij} \beta_j \quad , \quad \left(\delta_{ij} + \xi_{ij}\right) J_j^s = -d_{ij} S_{,j} \tag{2.68}$$

in which the two tensors in the left-hand sides of (2.68) are given by:

$$\eta_{ij} = \pi_4 \tau (S_{ij} + V_{ij}) - \pi_4 \tau^2 g \lambda_i \left(\pi_5 \alpha_T \beta_j + \pi_2 \alpha_s S_{,j}\right)$$
$$\xi_{ij} = \pi_1 \tau (S_{ij} + V_{ij}) - \pi_1 \tau^2 g \lambda_i \left(\pi_2 \alpha_T \beta_j + \pi_3 \alpha_s S_{,j}\right) \tag{2.69}$$

while the tensors on the right-hand sides are given by:

$$d_{ij} = \pi_1 \tau \left(R_{ij} + \pi_2 g \alpha_T \lambda_i J_j^h\right), \gamma_{ij} = \pi_4 \tau \left(R_{ij} - \pi_2 g \alpha_s \lambda_i J_j^s\right) \tag{2.70}$$

It is quite remarkable that the 3D case that includes *shear, vorticity, heat and salt fluxes* can be represented by *the algebraic relations (2.68)* showing how heat and salinity fluxes depend on each other as do the corresponding heat and salt diffusivities that are often parameterized as they were independent of one another. Finally, the dimensionless functions π's are defined as follows:

$$\pi_{1,2,3,4,5} = \tau_{ps}/\tau, \quad \tau_{p\theta}/\tau, \quad \tau_s/\tau, \quad \tau_{p\theta}/\tau, \quad \tau_\theta/\tau \tag{2.71}$$

and represent the dissipation time scales $\tau_{p\theta}$, τ_θ, τ_{ps}, τ_s, $\tau_{s\theta}$ in units of the turn-over time scale $\tau = 2K/\varepsilon$ for which in C10a relations (A.11, 12) were suggested.

1D case. A key feature is the possibility to account for Double-Diffusion DD processes represented by the following regimes:

$$\begin{aligned}
&\textit{Salt-fingers}: & &T_{,z}, S_{,z} > 0, \quad Ri > 0, 1, \quad R_\rho > 0.6 \\
&\textit{Diffusive-convection}: & &T_{,z}, S_{,z} < 0, \quad Ri > 0, \quad R_\rho > 0 \\
&\textit{Doubly-stable}: & &T_{,z} < 0, \quad S_{,z} > 0, \quad Ri > 0, \quad R_\rho < 0 \\
&\textit{Doubly-unstable}: & &T_{,z} < 0, \quad S_{,z} > 0, \quad Ri < 0, \quad R_\rho < 0
\end{aligned} \tag{2.72}$$

in which the following large-scale variables were defined:

2.4 Shear, Vorticity, Heat and Salt Fluxes

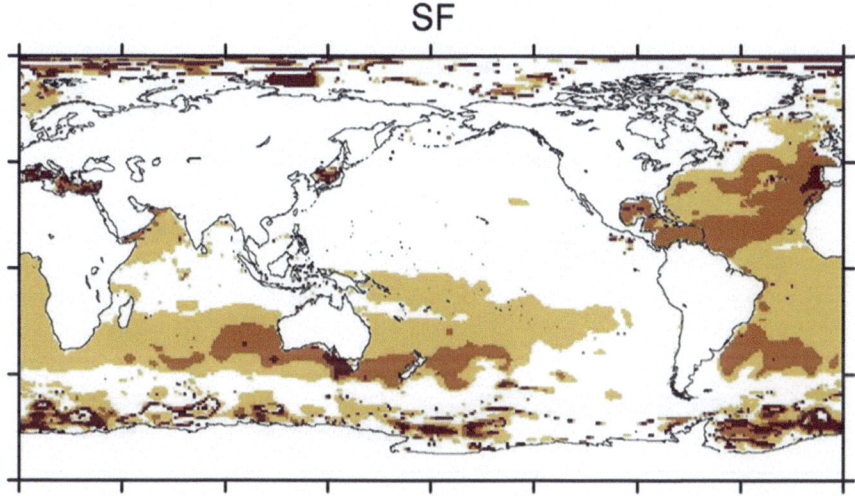

Fig. 2.1 *Salt Fingers*. Regions susceptible to SF; R_ρ^{-1} intervals are: 1–1.5 (red), 1.5–2 (light brown) and 2–3 (yellow). The reddest color has the lowest R_ρ^{-1} favorable to SF. (Reprinted with permission from Canuto et al. (2010a). © NASA all rights reserved)

$$Ri = N^2 \Sigma^{-2}, \quad \Sigma^2 = 2S_{ij}S_{ij}, \quad N_h^2 = g\alpha_T T_{,z}, \quad N^2 = N_h^2(1-R_\rho) \quad (2.73)$$

$$R_\rho = \alpha_S S_{,z}(\alpha_T T_{,z})^{-1} \quad (2.74)$$

As shown below in Eq. (2.75), the 1D case exhibits an interesting structure that simplifies considerably the entire formulation: *the equations for temperature flux, salinity flux, temperature-salinity correlation, temperature variance and salinity variance, do not depend on shear,* which suggests the following strategy (Figs. 2.1 and 2.2). One separates the problem into two parts; first, one solves Eqs. (2.75) analytically and in a second step, one solves the equations for the Reynolds stresses that bring shear into the problem. The equations in question are given by (e.g., Appendix A of Canuto et al. 2010a):

$$
\begin{aligned}
J^h = \overline{w\theta}: &\quad D_t J^h + D_f = -\overline{w^2}T_{,z} + g\left(\alpha_T \overline{\theta^2} - \alpha_s \overline{s\theta}\right) - \tau_{p\theta}^{-1} J^h \\
J^s = \overline{ws}: &\quad D_t J^s + D_f = -\overline{w^2}S_{,z} + g\left(\alpha_T \overline{\theta s} - \alpha_s \overline{s^2}\right) - \tau_{ps}^{-1} J^s \\
\overline{\theta s}: &\quad D_t \overline{\theta s} + D_f = -\left(J^h S_{,z} + J^s T_{,z}\right) - \tau^{-1}\pi_2^{-1}\overline{\theta s} \quad (2.75) \\
\tfrac{1}{2}\overline{\theta^2}: &\quad D_t \tfrac{1}{2}\overline{\theta^2} + D_f = -J^h T_{,z} - \tau_\theta^{-1}\overline{\theta^2}
\end{aligned}
$$

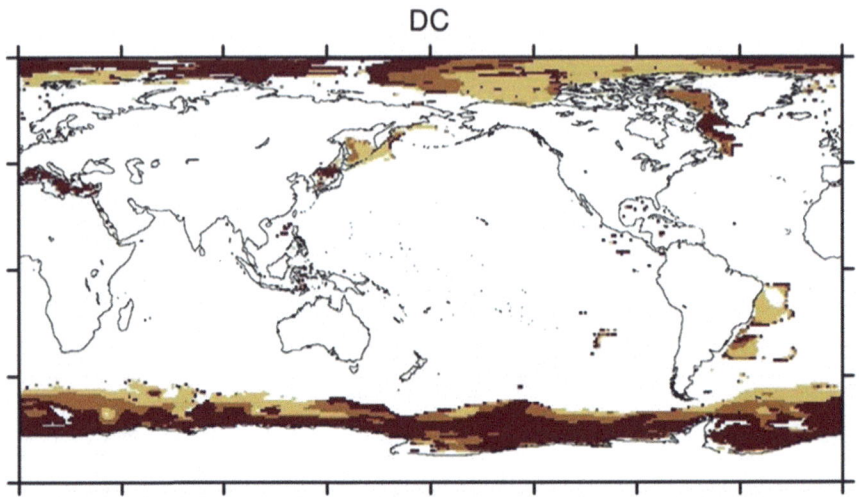

Fig. 2.2 *Diffusive convection*. Regions susceptible to DC; R_ρ intervals 1–3 (red), 3–5 (light brown) and 5–10 (yellow). The reddest color has the lowest R_ρ which is most favorable to DC. (Reprinted with permission from Canuto et al. (2010a). © NASA all rights reserved)

$$\tfrac{1}{2}\overline{s^2}: \qquad D_t \tfrac{1}{2}\overline{s^2} + D_f = -J^s S_{,z} - \tau_s^{-1}\overline{s^2}$$

Reynolds stresses:

$$b_{ij} \equiv \overline{u_i u_j} - \tfrac{2}{3} K \delta_{ij}: \qquad D_t b_{ij} + D_f = -\tfrac{8}{15} K S_{ij} - \tfrac{1}{2} Z_{ij} + \tfrac{1}{2} B_{ij} - 5\tau^{-1} b_{ij} \quad (2.76)$$

where the tensors entering (2.76) are defined as follows:

$$Z_{ij} = b_{ik} V_{jk} + b_{jk} V_{ik}, \quad B_{ij} = g\left(\lambda_i J_j^\rho + \lambda_j J_i^\rho - \tfrac{2}{3}\delta_{ij}\lambda_k J_k^\rho\right), \quad J_i^\rho = \alpha_T J_i^h - \alpha_s J_i^s, \quad g\overline{\rho}\lambda_i = -\overline{p}_{,i}$$
(2.77)

The horizontal T-S fluxes enter the buoyancy tensor B_{ij} are defined as follows ($i = 1,2$):

$$J_i^h = \overline{u_i \theta}: \qquad D_t J_i^h + D_f = -\overline{u_i w}T_{,z} - J^h \partial_z U_i - \tau^{-1}\pi_4^{-1} J_i^h \quad (2.78)$$

$$J_i^h = \overline{u_i \theta}: \qquad D_t J_i^s + D_f = -\overline{u_i w}S_{,z} - J^S \partial_z U_i - \tau^{-1}\pi_1^{-1} J_i^s \quad (2.79)$$

Though we employed the same symbol D_f, the diffusion terms are different.

2.5 Solutions for the Salt-Fingers Case

In the stationary, local limit ($D_f = 0$) limit, Eq. (2.75) become a set of algebraic equations that can be solved analytically with the following results:

$$\overline{wu} = -K_m U_{,z}, \quad \overline{w\theta} = -K_h T_{,z}, \quad \overline{ws} = -K_s S_{,z} \tag{2.80}$$

where the *diffusivities* have the following form:

$$K_\alpha = \frac{2K}{\varepsilon} S_\alpha, \quad S_\alpha(N, Ri, R_\rho) \tag{2.81}$$

and where S_α are dimensionless *"structure functions"* given by:

$$S_{h,s} = A_{h,s} \frac{\overline{w^2}}{K}, \quad A_h = \pi_4 \left(1 + px + \pi_2\pi_4 x(1-r^{-1})\right)^{-1}, \quad A_s = A_h (xR_\rho)^{-1} \tag{2.82}$$

where the variable "r" is commonly used to represent the heat-to-salt flux ratio:

$$r = \frac{\alpha_T J^h}{\alpha_s J^s} = R_\rho^{-1} \frac{K_h}{K_s}, \quad \frac{K_h}{K_s} = \frac{\pi_4}{\pi_1} \frac{1+qx}{1+px}$$

$$x = (\tau N)^2 (1-R_\rho)^{-1}, \quad R_\rho < 1, \quad x > 0 \tag{2.83}$$

$$p = \pi_4\pi_5 - \pi_4\pi_2(1 + R_\rho), \quad q = \pi_1\pi_2(1 + R_\rho) - \pi_1\pi_3 R_\rho$$

There is a form alternative to (2.81) suggested by Osborn and Cox (1972) and Osborn (1980) that reads as follows:

$$K_\alpha = \Gamma_\alpha \frac{\varepsilon}{N^2}, \quad \Gamma_\alpha = \frac{1}{2}(\tau N)^2 S_\alpha \tag{2.84}$$

where the dimensionless functions Γ_α are called *mixing* coefficients, see below.

2.6 Reynolds Stresses

We next consider the algebraic solutions of Eqs.(2.77), (2.78) and (2.79) which were found to be:

$$S_m = A_m \frac{\overline{w^2}}{K}, \quad A_m = \frac{A_{m1}}{A_{m2}}$$

$$A_{m1} = \frac{4}{5} - xA_h\left[\pi_4-\pi_1 + \left(\pi_1-\frac{1}{150}\right)(1-r^{-1})\right]$$

$$A_{m2} = 10 + x(\pi_4-\pi_1 R_\rho) + \frac{1}{50}(\tau\Sigma)^2$$

$$\frac{\overline{w^2}}{K} = \frac{2}{3}\left[1 + \frac{2}{15}Y + \frac{1}{10}A_m\tau\Sigma\right)^2\right]^{-1}, \quad Y = xA_h(1-r^{-1}) \quad (2.85)$$

2.7 The EKE Equation

In the stationary-local limit, the EKE equation becomes the production = dissipation relation $P = \varepsilon$ where the total production by shear and buoyancy is given by:

$$P_s = K_m\Sigma^2, \, P_b = -K_h N^2, \, P = K_m\Sigma^2(1-R_f) \quad (2.86)$$

where R_f is the flux *Richardson number* given by the ratio of buoyancy to shear fluxes:

$$R_f = Ri\sigma_t^{-1}, \, \sigma_t = \frac{K_m}{K_h} \quad (2.87)$$

where σ_t is the *turbulent Prandtl number*. Using $\tau = 2K/\varepsilon$, $P = \varepsilon$ becomes the cubic equation:

$$G_m \equiv (\tau\Sigma)^2, \, c_3 G_m^3 + c_2 G_m^2 + c_1 G_m + 1 = 0 \quad (2.88)$$

where the c_k are defined in relations (B.11–12) of C10a in terms of the variables π_k.

2.8 Results for the SF Case

Using the above framework, below we present the following results (Figs. 2.3, 2.4, 2.5, 2.6, 2.7, 2.8, 2.9, 2.10, 2.11, 2.12 and 2.13).

Effect of shear on heat efficiency vs. NATRE data. Model results dashed and full lines while the data are from St. Laurent and Schmitt (1999). The heat efficiency increases as the shear decreases and Ri increases.

2.9 Diffusive Convective DC

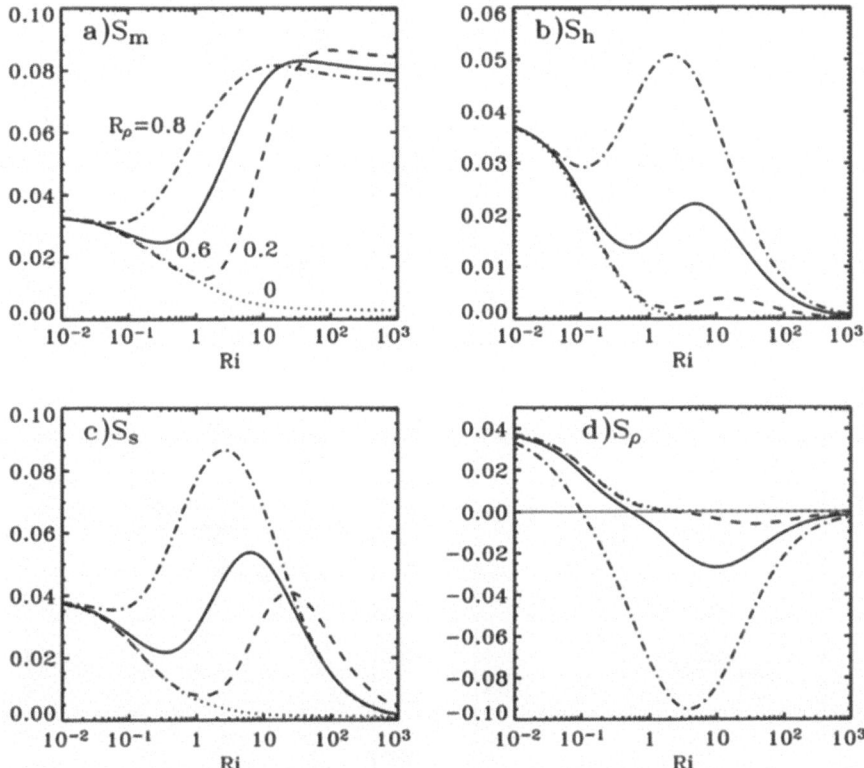

Fig. 2.3 The structure functions S_a for momentum, heat, salt and density vs. Ri for different R_ρ. (Reprinted with permission from Canuto et al. (2010a). © NASA all rights reserved)

2.9 Diffusive Convective DC

Figure 2.2 shows that the DC case (warm-salty waters underneath cold-fresh waters) may play a significant role in the Arctic and Southern oceans, a point discussed in detail by Kelley et al. (2003) who concluded that DC *"could be of major importance"*. Diffusive convection has been the subject of many studies in the past (Stern 1960; Kelley 1984, 1990, 2001; Kelley et al. 2003; Schmitt 1994; Brandt and Fernando 1995) and even more so recently because of a suggestion that DC may play an important role in the ice melting at the bases of ice shelves, a process with clear implications on the sea-level rise problem. For example, Wang et al. (2023) wrote that *"intrusions of circumpolar deep waters into the West Antarctic Peninsula increase the heat content in the cavities, raise the heat delivered to the bases of the ice shelves and drive basal meting"*. Middleton et al. (2021, 2022) wrote that *"the relatively warm and salty circumpolar deep waters is thought to be mixed up to the base of the ice shelves via shear-driven turbulence where it melts the ice"*. These authors studied the EKE dissipation with a new parameterization based on stirring

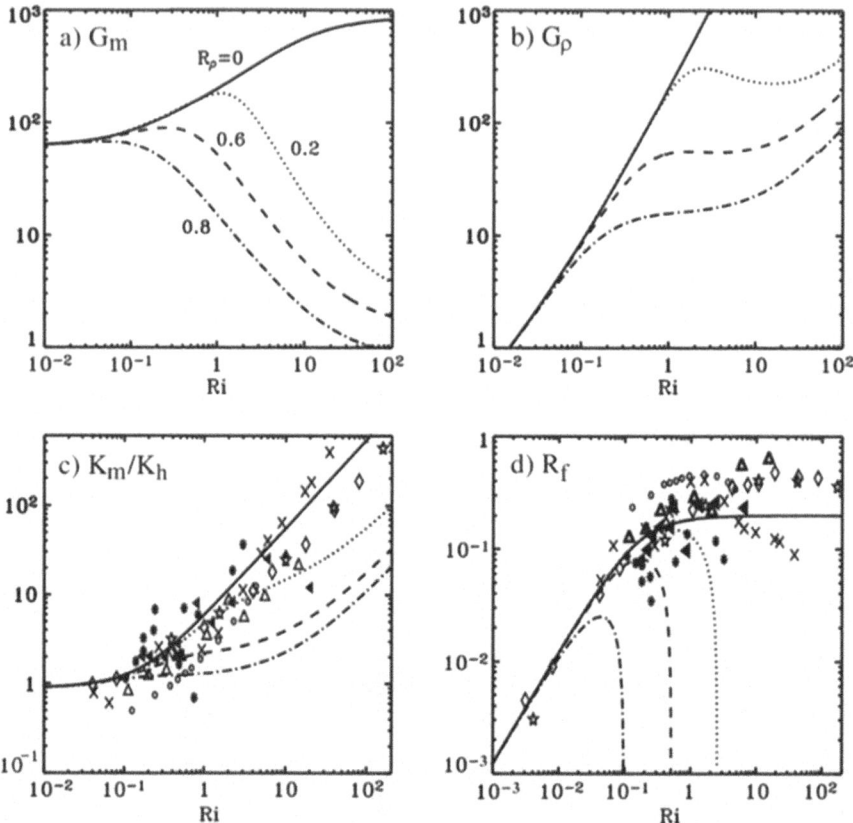

Fig. 2.4 $G_m, G_p = (\tau N)^2$, K_m/K_h, R_f. In (c) and (d) the data correspond to no DD. (Reprinted with permission from Canuto et al. (2010a). © NASA all rights reserved)

along-isopycnals that was shown to reproduce 72% of the data. While the SF case was studied here; in sec. 2.5 with a SOC model, no such treatment is available in the DC case for which we have heuristic relations and the data shown in Figs. 2.14 and 2.15 below but no turbulence-based model has yet predicted them. While the dynamic equations governing the 1D SF case are formally the same as in the DC case, there is an important difference in the definition in one of the variables. Consider the definition of the Brunt-Vaisala frequency N in terms of the mean buoyancy $\bar{b}$:

$$N^2 = \frac{\partial \bar{b}}{\partial z}, \quad \bar{b} = -g\frac{\bar{\rho}}{\rho_0} = g\alpha_T T - g\alpha_s S \tag{2.89}$$

$$N^2 = g\alpha_T T_{,z} - g\alpha_s S_{,z} = g\alpha_T T_{,z}(1 - R_\rho) = N_h^2(1 - R_\rho) \tag{2.90}$$

Since in the DC case one has:

2.9 Diffusive Convective DC

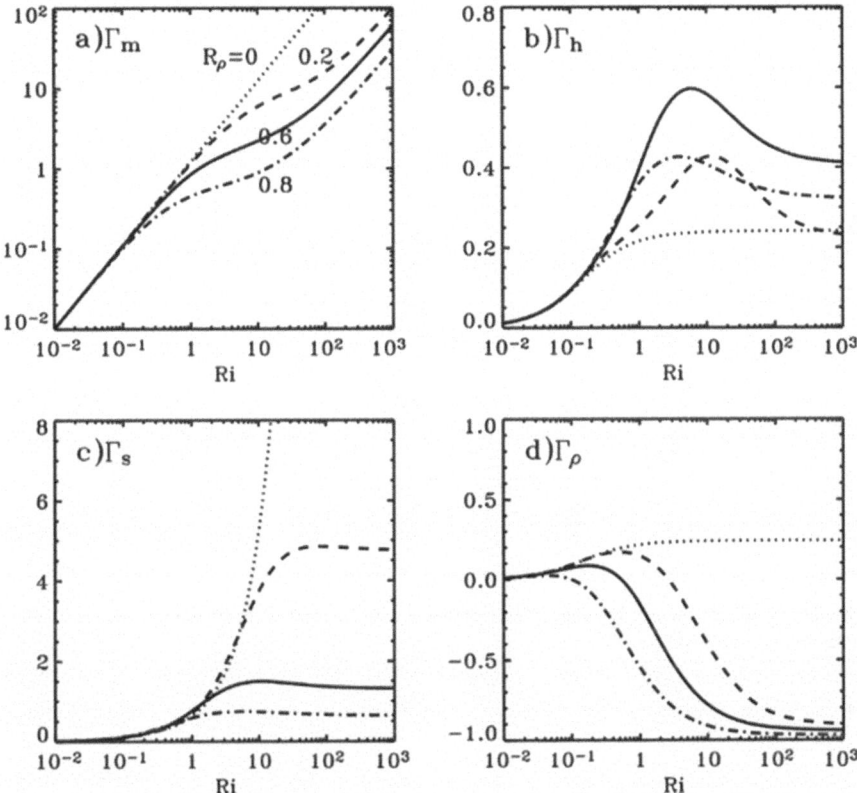

Fig. 2.5 Mixing efficiencies defined in (2.54). (Reprinted with permission from Canuto et al. (2010a). © NASA all rights reserved)

$$T_{,z} < 0, \quad N_h^2 < 0, R_\rho > 1 \tag{2.91}$$

and $N^2 > 0$ while x is negative:

$$x = (\tau N)^2 (1 - R_\rho)^{-1} < 0 \tag{2.92}$$

In the SF case, one has: $T_{,z} > 0$, $N_h^2 > 0$, $R_\rho < 1$, $N^2 > 0$, $x > 0$.

Determination of the functions π defined in (2.71). In their pioneering work, Zeman and Lumley (1982) employed a SOC model to study SF and suggested the following forms:

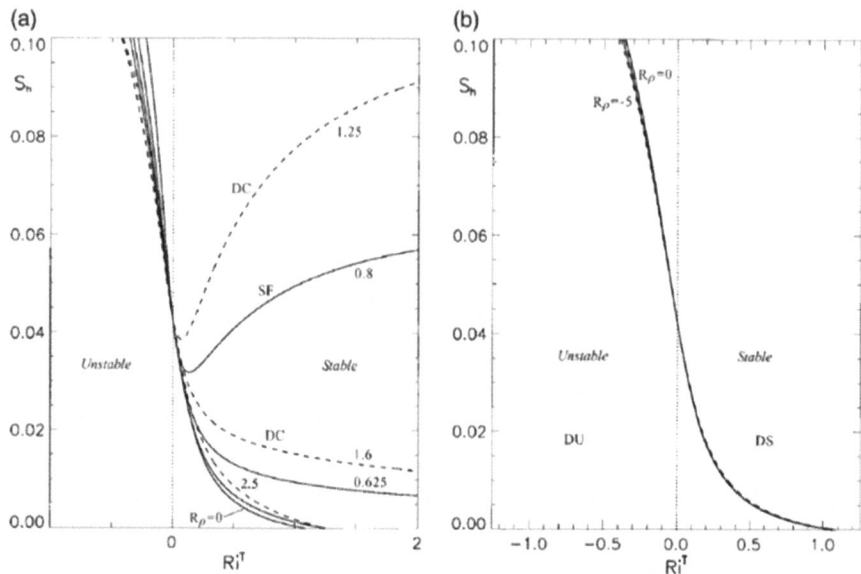

Fig. 2.6 S_h vs. Ri for SF, DC and DU. (Reprinted with permission from Canuto et al. (2010a). © NASA all rights reserved)

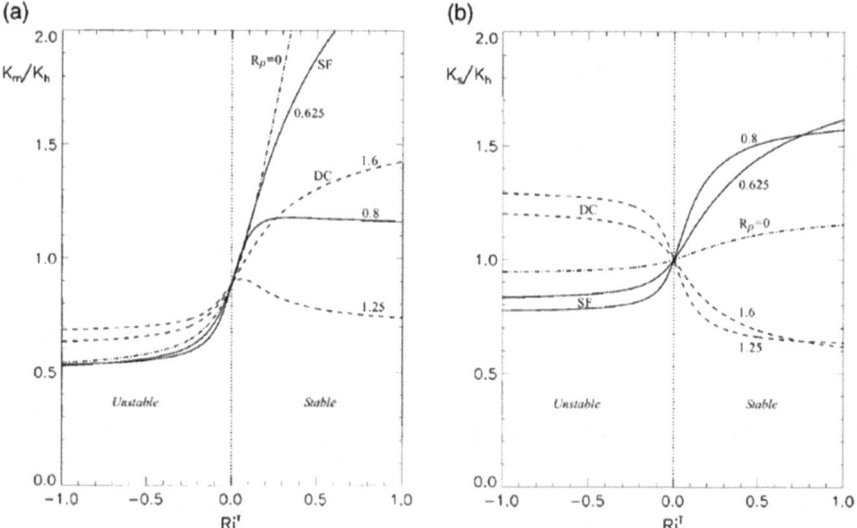

Fig. 2.7 Momentum/heat and salt/heat diffusivities vs Ri for different R_ρ. (Reprinted with permission from Canuto et al. (2010a). © NASA all rights reserved)

2.9 Diffusive Convective DC

Fig. 2.8 Ratio salt to heat diffusivities vs Ri for different R_ρ. (Reprinted with permission from Canuto et al. (2010a). © NASA all rights reserved)

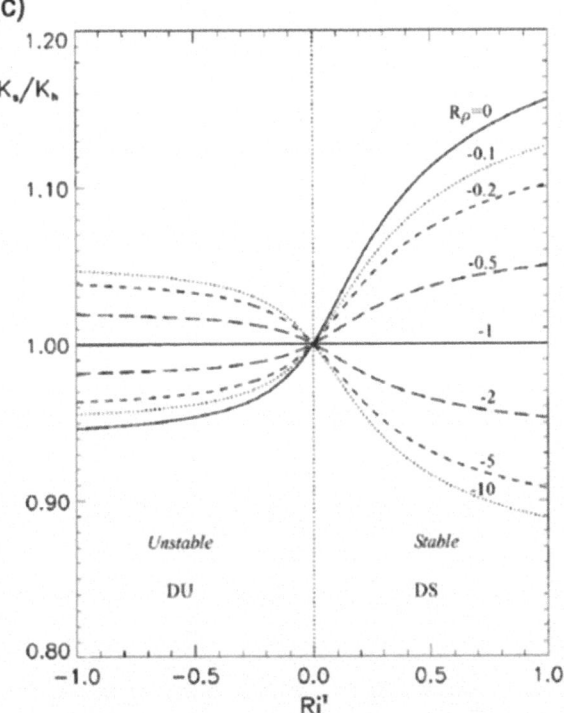

Fig. 2.9 Heat to salt diffusivity ratio vs. R_ρ^{-1}. The results compare well with Walsh and Ruddick (2000). (Reprinted with permission from Canuto et al. (2010a). © NASA all rights reserved)

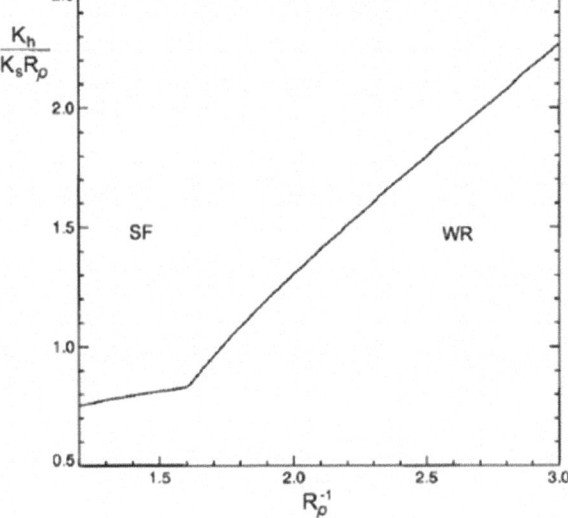

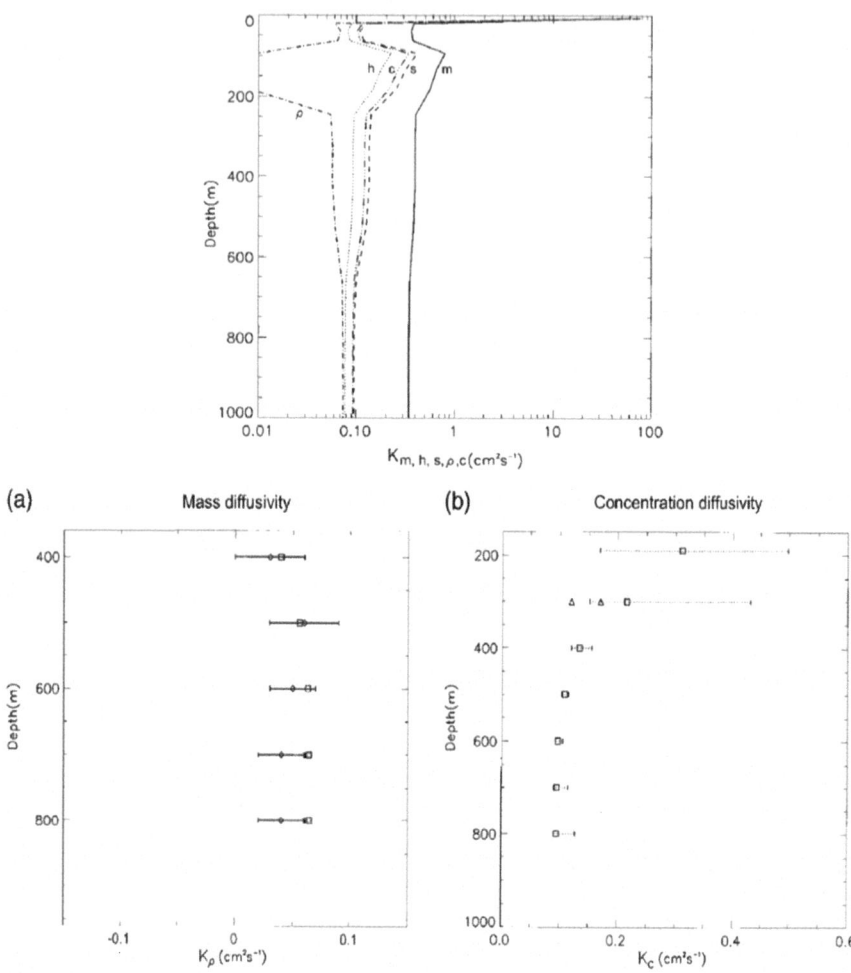

Fig. 2.10 Momentum, heat, salt, mass and concentration diffusivities vs. depth from NATRE. (Reprinted with permission from Canuto et al. (2010a). © NASA all rights reserved)

$$\pi_1^{-1} = \left(1 + \frac{\kappa_s}{\nu}\right), \quad \pi_2^{-1} = \frac{\kappa_\theta + \kappa_s}{\nu}, \quad \pi_3^{-1} = \frac{\kappa_s}{\nu}, \quad \pi_4^{-1} = \left(1 + \frac{\kappa_\theta}{\nu}\right), \quad \pi_5^{-1} = \frac{\kappa_\theta}{\nu}$$
(2.93)

with the values:

$$\nu = 1.35{*}10^{-2} \text{cm}^2\text{s}^{-1}, \quad \nu/\kappa_\theta = 7, \quad \kappa_\theta/\kappa_s = 100 \tag{2.94}$$

However, it not yet known same time scales are able to reproduce the following key data on DC.

2.10 Passive Tracers

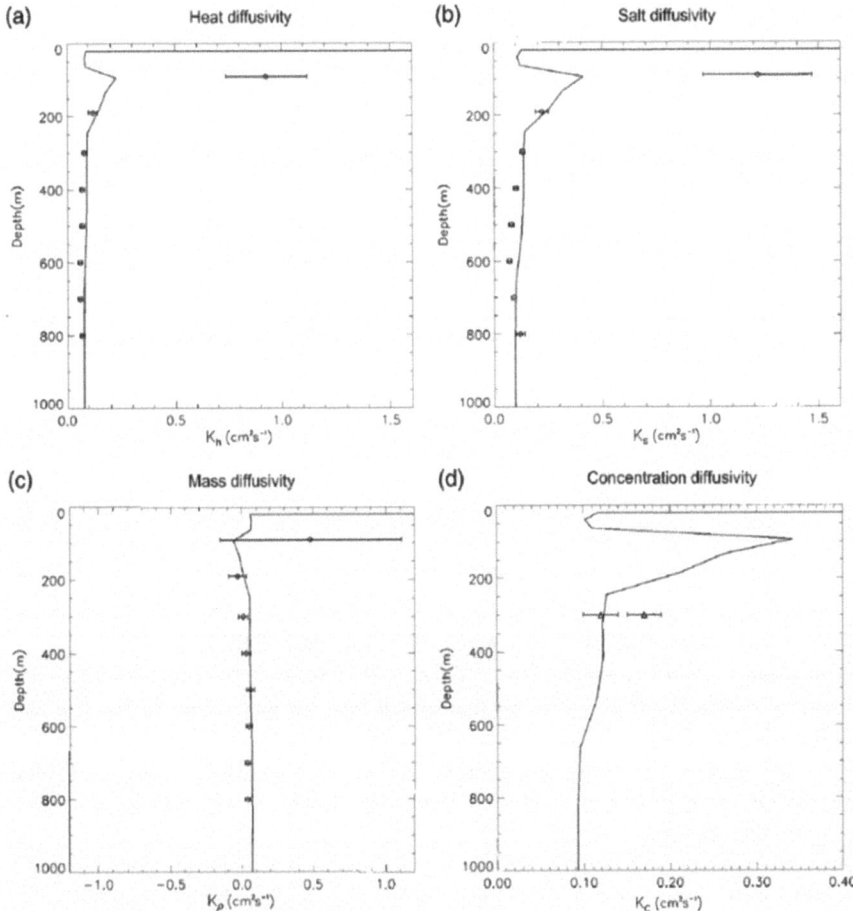

Fig. 2.11 Vertical profiles of different diffusivities. NATRE data. (Reprinted with permission from Canuto et al. (2010a). © NASA all rights reserved)

2.10 Passive Tracers

Climate models that include a Carbon-cycle need the *vertical diffusivity of passive tracers*. Since in the past the latter was not available, it was identified with that of salt. The identification is however incorrect since both T, S are active not passive tracers. Below, we derive the diffusivity of a passive tracer in terms of Ri and R_ρ. The following results emerged:

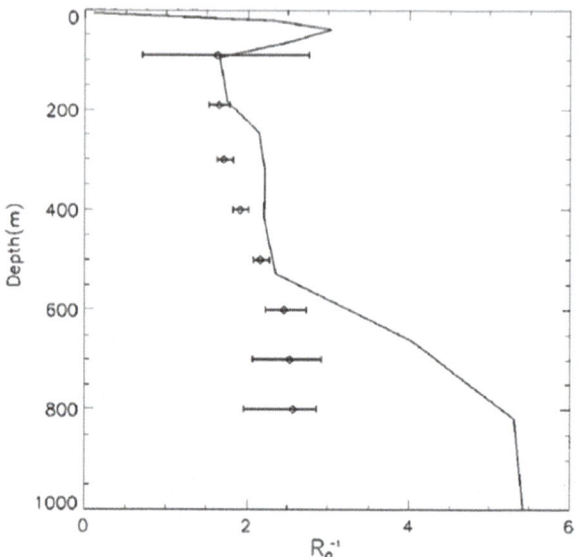

Fig. 2.12 Vertical profile of R_ρ^{-1}. NATRE data (diamonds), model results (full line). (Reprinted with permission from Canuto et al. (2010a). © NASA all rights reserved)

(a) the passive tracer diffusivity is an algebraic function of Ri, R_ρ,
(b) in *doubly stable regimes* (DS: $\partial T/\partial z > 0,) \partial S/\partial z < 0$), the passive scalar diffusivity is nearly the same as that of salt-heat for any values of $R_\rho < 0$ and Ri > 0.
(c) in *DC regimes* (diffusive convection, $\partial T/\partial z < 0, \partial S/\partial z < 0$, $R_\rho > 1$), the passive scalar diffusivity is *larger* than that of salt. At Ri = O(1), it can be more than twice as large.
(d) in *SF regimes* (salt fingers, $\partial T/\partial z > 0, \partial S/\partial z > 0$, $R_\rho<1$), the passive scalar diffusivity is *smaller* than that of salt. At Ri = O(1), it can be less than half of it.
(e) In summary:

$$DC : K_h > K_c > K_s, \quad SF : K_h < K_c < K_s \tag{2.95}$$

(f) a most interesting conclusion is that the identification of the tracer diffusivity with that of salt is valid only in DS regimes, which is not realized in the ocean. In the Southern Ocean, where there is a large CO2 absorption, the dominant regime is DC case where the tracer diffusivity is larger than that of salt.

Formulation of the problem. Ocean general circulation models (OGCMs) solve the time dependent equations for the mean velocity, mean temperature, mean salinity and mean concentration of passive tracers such as CO2, since the Carbon Cycle is an important diagnostic in climate predictions and some studies also include CFC While both temperature and salinity are active tracers because they affect the velocity field via the buoyancy term and thus alter the status of the turbulent flow, tracers such as CO2, CFC, Oxygen, if sufficiently diluted as is ordinarily the case in

2.10 Passive Tracers

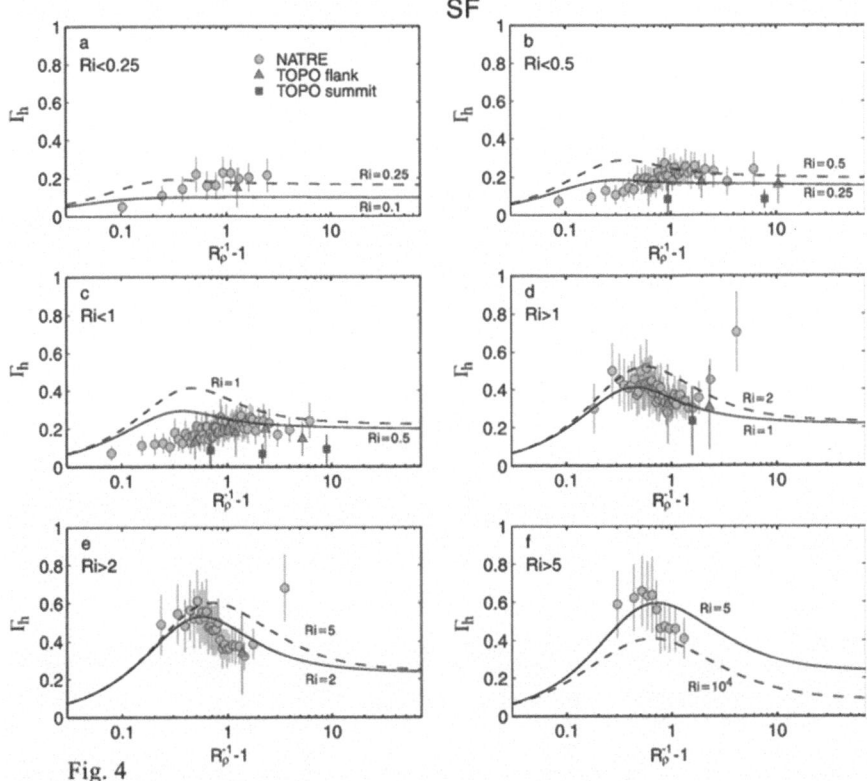

Fig. 2.13 Heat mixing efficiency at different Ri. The canonical value of 0.2 valid only for small Ri; as the latter increases, so does the heat efficiency. (Reprinted with permission from Canuto et al. (2010a). © NASA all rights reserved)

the ocean, are passively carried along by the flow without altering it. It follows that the diffusivity of a passive tracer must be different from that of heat and salt. Identifying the passive tracer diffusivity with that of salt is, however, a frequent assumption most likely because *no expression for a passive tracer diffusivity was available in the literature.* In C10a, the vertical diffusivities for momentum, heat and salt were derived as functions of the two large scale parameters Ri and R_ρ. The thermal expansion and haline contraction coefficients are computed using the non-linear UNESCO equation of state (UNESCO 1981a, b). (It was pointed out to the author that there are cases of lakes in which the concentrations of CO_2 and CH_4 contribute to the density stratification to give rise to double-diffusive convection). Since in C10a we did not derive an expression for the diffusivity of a passive tracer, here we present the derivation which is not trivial. There are two kinds of problems. *First*, since the starting dynamic equation is: valid for an arbitrary tracer, one must identify where and how a generic tracer is treated as passive or active. *The most difficult part is the determination of the dissipation time scales that characterize*

Fig. 2.14 The dependence of the heat flux on R_ρ. (Reprinted with permission from Kelley et al. (2003). © NASA all rights reserved)

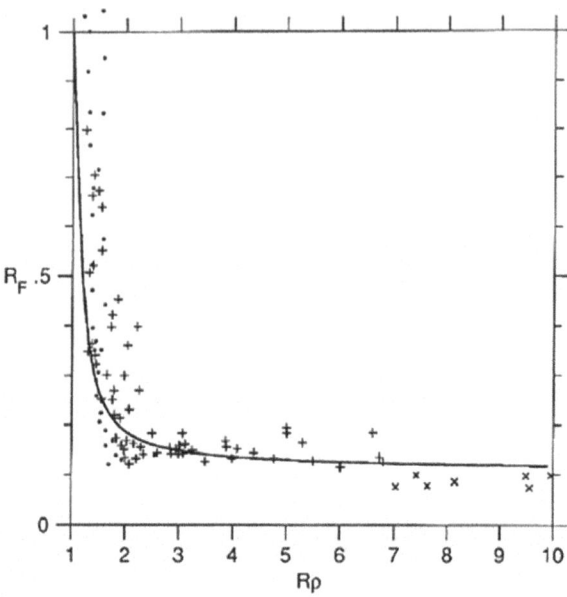

Fig. 2.15 Heat conductivity vs. R_ρ. (Reprinted with permission from Kelley et al. (2003) © NASA all rights reserved)

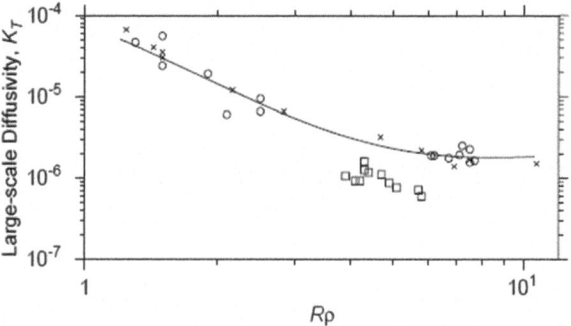

small scale processes in the passive tracer case. Assuming a constant ratio of the dissipation time scales to the dynamical time scale, as it was assumed for many years because of lack of data did not motivate other suggestions, yielded a rather poor fit to the NATRE data (North Atlantic Tracer Release Experiment, Ledwell et al. 1993, 1998; St. Laurent and Schmitt 1999). A better fit was achieved when the time scale ratios were made Ri, R_ρ dependent. Within the present context, the question is how to relate those dissipation times scales to the new ones that enter the passive tracer case. We do so by requiring that the diffusivity obtained as a solution of the turbulent equation for a passive scalar should coincide with those of either salinity and/or temperature when the latter are so diluted to not affect the density field thus behaving like passive tracers. Such a requirement allowed us to find a relation between the dissipation time scale of a passive tracer and those already determined in the temperature-salinity case.

2.10 Passive Tracers

Derivation of the fluxes. The dynamic equation for an arbitrary tracer reads as follows:

$$\partial_t \rho c + \partial_i(\rho c u_i) = \chi_c \partial_i^2 c \qquad (2.96)$$

where χ_c is the kinematic diffusivity. Following standard procedure, the total fields c and u_i are split into average and fluctuating components. We further employ the incompressibility condition, average the result and subtract it from the equation for the total c-field. These purely algebraic steps lead to the following equation for the fluctuating component of the generic field c':

$$D_t c' + u'_i \partial_i \overline{C} + \partial_i \left(u'_i c' - \overline{u'_i c'} \right) = \chi_c \partial_i^2 c' \qquad (2.97)$$

Using an analogous procedure for the velocity field described by the Navier–Stokes equations, one derives the equation for the field w' which reads as follows:

$$\partial_t w' + \partial_j \left(w' u'_j - \overline{w' u'_j} \right) = -\rho_0^{-1} \partial_z p' - g \rho_0^{-1} \rho' + \nu \partial_i^2 w' \qquad (2.98)$$

where ν is the kinematic viscosity of seawater. As expected, the averages of (2.97) and (2.98) yield $0 = 0$ since all fluctuating variables have zero averages. Next, we recall that if the field c is not affecting the density, which is the assumption of c being a passive tracer then, the density is only affected by the truly active scalars, temperature and salinity and thus the density field ρ' in (2.98) is taken to be $\rho'/\rho_0 = -\alpha\theta + \beta s$. The next steps are of purely algebraic nature: multiply Eq. (2.97) by w' and Eq. (2.98) by c', average the results and sum the two equations. The result is the dynamic equation for the *vertical flux of the passive tracer* $J_c = \overline{w'c'}$:

$$\partial_t J_c = -\overline{w'^2} \partial_z \overline{C} + g \left(\alpha \overline{\theta'c'} - \beta \overline{s'c'} \right) - \tau_{pc}^{-1} J_c \qquad (2.99)$$

where we have lumped together pressure correlations and dissipation terms in the dissipation-relaxation time scale τ_{pc}. Next, we must find an expression for the two covariances in Eq. (2.99). This is done by using the two equations for the fluctuating temperature and salinity fields which are obtained using a procedure analogous to the one used to arrive at (10.3). The results, that were derived in the literature in many textbooks (e.g., Leslie 1973), are as follows:

$$D_t \theta' + u'_i \partial_i T + \partial_i \left(\theta' u'_i - \overline{\theta' u'_i} \right) = \chi_\theta \partial_i^2 \theta' \qquad (2.100)$$

$$D_t s' + u'_i \partial_i S + \partial_i \left(s' u'_i - \overline{s' u'_i} \right) = \chi_s \partial_i^2 s' \qquad (2.101)$$

where $\chi_{\theta, s}$ are the heat and salt molecular diffusivities whose ratio is about 100 while $\nu/\chi_\theta \simeq 7$. Multiply (2.100) by c' and (2.101) by θ', sum the two after averaging them and carry out an analogous operation with (2.101). The result are the two equations:

$$\partial_t \overline{\theta' c'} = -\left(J_h \overline{C}_{,z} + J_c T_{,z}\right) - \tau_{c\theta}^{-1} \overline{\theta' c'}$$

$$\partial_t \overline{s' c'} = -\left(J_s \overline{C}_{,z} + J_c S_{,z}\right) - \tau_{cs}^{-1} \overline{s' c'} \quad (2.102)$$

where the time scales represent the relaxation–dissipation time scales of the corresponding correlations. Before we proceed with their derivation, we pause to point out an important difference between passive and active tracers. Notice that Eqs. (2.102) do not depend on density. Suppose, however, that c' is an active scalar: if c' had been either θ', s', Eqs. (2.102) would have given rise to the two terms:

$$\overline{\theta^2}, \overline{s^2} \quad (2.103)$$

that represent the potential energies stored in the temperature-salinity fields which are part of the energy budget; however, there is no $\overline{c'^2}$ in the above equations since a passive scalar field does not contribute to the energy balance. In the derivation presented in sec. 2.4, the dynamic equations for (2.103) are given in Eq. (2.71). In the $c' \to \theta'$ case, for example, Eq. (2.99) becomes the equation for the heat flux and the second term would be twice the potential energy in the temperature field, representing a source of the heat flux. In such a case, Eq. (2.99) would coincide with (2.64). Using (2.84), the c-flux is then given by:

$$J_c = \overline{w'c'} = -K_c \frac{\partial \overline{C}}{\partial z}, \quad K_c = \Gamma_c \frac{\varepsilon}{N^2}, \quad \Gamma_c = \frac{p_1(\tau N)^2}{1 + p_1 p_2 (\tau N)^2} \left(\frac{\overline{w^2}}{2K} - p_2 \Gamma_\rho\right) \quad (2.104)$$

Relation (2.104) requires several comments. *First*, Γ_ρ is the mixing efficiency of the buoyancy (density) field defined as follows:

$$\Gamma_\rho = \frac{\Gamma_h - \Gamma_s R_\rho}{1 - R_\rho} \quad (2.105)$$

where the heat–salt mixing efficiencies are given by the second relation in (2.84), the variable $(\tau N)^2$ is from the solution of (2.88) and the p's are defined as follows:

$$p_1 = \frac{\tau_{pc}}{\tau}, \quad p_2 = \frac{\tau_{c\theta}}{\tau} = \frac{\tau_{cs}}{\tau} \quad (2.106)$$

whose determination is the most difficult part of the problem for which we suggest the following procedure. Consider the active tracer salt in the limit when the haline contraction coefficient becomes very small:

$$\beta \to 0, \quad R_\rho \to 0, \quad K_c \to K_s \qquad (2.107)$$

which physically means that salinity is so diluted as to not affect the density, in which case salinity behaves like a passive scalar which justifies the last relation above. Furthermore, from (2.83) we have:

$$p \to \pi_4\pi_5 \text{-} \pi_4\pi_2, \quad q \to \pi_1\pi_2 \qquad (2.108)$$

which lead to the relations:

$$p_1 = \pi_1(R_\rho = 0), \quad p_2 = \pi_2(R_\rho = 0) \qquad (2.109)$$

Next, we treat the case when heat becomes passive which is represented by the relations:

$$\alpha \to 0, \quad R_\rho \to \infty, \quad K_c \to K_h \qquad (2.110)$$

In analogy to the previous case, this leads to the relations:

$$p_1 = \pi_4(R_\rho = \infty), \quad p_2 = \pi_2(R_\rho = \infty) \qquad (2.111)$$

Using (2.109) and (2.111), we suggested the following form for the variables $p_{1,\,2}$:

$$p_1 = \frac{R_\rho + R_\rho^{-1}}{\pi_4^{-1} R_\rho + \pi_1^{-1} R_\rho^{-1}}, \quad p_2 = \pi_2 \qquad (2.112)$$

In the figures below (Fig. 2.16, 2.17, 2.18, 2.19, 2.20, 2.21 and 2.22), we show the results that confirm the results (2.95).

2.11 Solar Convection, Rotation, Overshooting and Angular Momentum

Convection in stars. A reliable description of stellar heat transport by turbulent convection has been and continues to be a challenge. In 1958, Bohm-Vitense suggested the Mixing Length Theory MLT to describe turbulent convection in stars. The model, which has been widely used ever since, contains however the seed of its own limitations for the following reasons. First, in a highly turbulent flow,

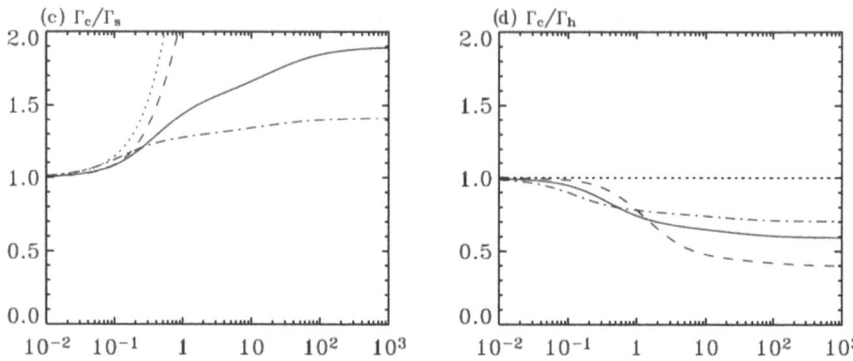

Fig. 2.16 . (Reprinted with permission from Canuto et al. (Reprinted with permission from Canuto et al. 2011. © NASA all rights reserved)

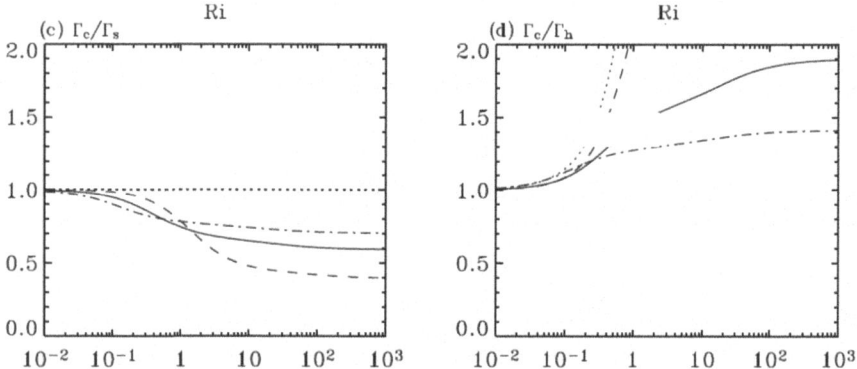

Fig. 2.17 The SF case $R_\rho = 0(....), 0.2(---), 0.6(-), 0.8(-,-,)$. (Reprinted with permission from Canuto et al. (2011). © NASA all rights reserved)

it is quite difficult to identify with any certainty a variable such as a mixing length whose description has given rise to the suggestions that it is proportional to the pressure scale height with a coefficient alpha, whose determination is equally difficult as was discussed in several papers. A further, deeper model independent argument is that the width of a turbulent spectrum is proportional to 3/4 power of the Reynolds number Re and therefore in a 3D case, the number of grid points or degrees of freedom to be accounted for, goes like the 9/4 power of Re. Considering that in a stellar context, Re may be considerably larger than say 10^{12}, one has a wide spectrum of eddies, see Fig. 2.23 in clear contrast with one-eddy model represented by the delta function of the MLT. The MLT heat flux is traditionally presented in the following form:

2.11 Solar Convection, Rotation, Overshooting and Angular Momentum

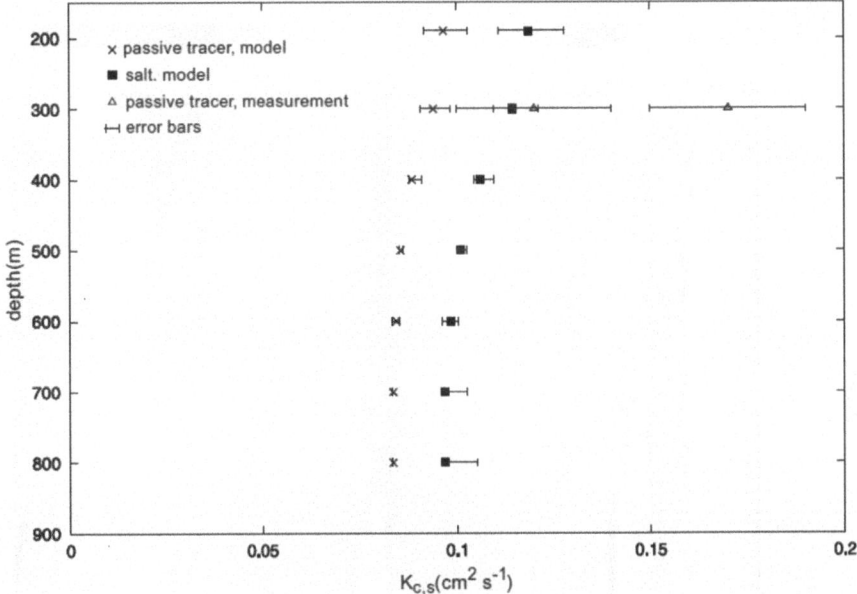

Fig. 2.18 z-profile of passive tracer and salinity diffusivities vs NATRE. (Reprinted with permission from Canuto et al. (2011). © NASA all rights reserved)

$$F_c = c_p\rho <w\theta> = -K_t\left[\frac{dT}{dz} - \left(\frac{dT}{dz}\right)_{ad}\right] = K_r TH_p^{-1}(\nabla - \nabla_{ad})\Phi \quad (2.113)$$

where $K_r = 4acT^3/3\kappa\rho$ is the radiative conductivity, K_t is the turbulent thermal conductivity and:

$$\Phi = \frac{K_t}{K_r} \quad (2.114)$$

The MLT form of the dimensionless function Φ is given by:

$$\Phi_{MLT} = \frac{1}{2}a_o\Sigma^{-1}\left[(1+\Sigma)^{1/2} - 1\right]^3 \quad (2.115)$$

where:

$$a_0 = 9/4, \quad \Sigma = 4A^2(\nabla - \nabla_{ad}), \quad A = \frac{\ell^2}{9\chi}\left(\frac{g}{2H_p}\right)^{1/2} \quad (2.116)$$

where ℓ is the mixing length and $\chi = K_r/c_p\rho$ is the thermometric conductivity.

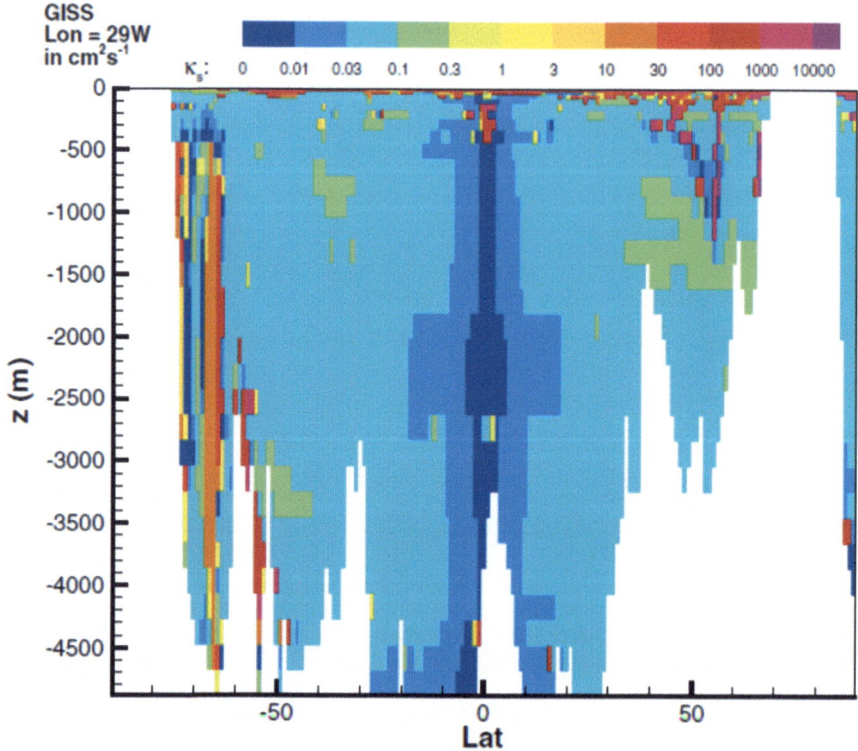

Fig. 2.19 Salt diffusivity

In 1991, the one-eddy MLT model was replaced by a *full spectrum model* (Canuto and Mazzitelli 1991). The results shown in that and subsequent work (e.g., Ventura et al. 1998), documented the quantitative advantages of using the full eddy spectrum instead of the MLT. In such a model, the new function Φ was represented as follows:

$$\Phi = a_1 \Sigma^m [(1 + a_2 \Sigma)^n - 1]^p \tag{2.117}$$

where $a_1 = 24.868$, $10^2 a_2 = 9.7666$, $m = 0.14972$, $n = 0.18931$, $p = 1.8503$. The ratio of (2.117) to (2.115) is shown in Fig. 2.24. Given the larger number of eddies to transport heat, it is clear that the new function Φ is larger than the one due to the MLT.

Effect of rotation on the convective flux. Helio-seismological data have shown the existence of a transition from rigid to differential rotation when reaching the upper convective zone. A relevant question then becomes, does rotation affect the convective flux and if so, by how much and in which direction? The problem was studied by Canuto and Hartke (1986) with the following result. The importance of rotation was found to be governed by the dimensionless ratio:

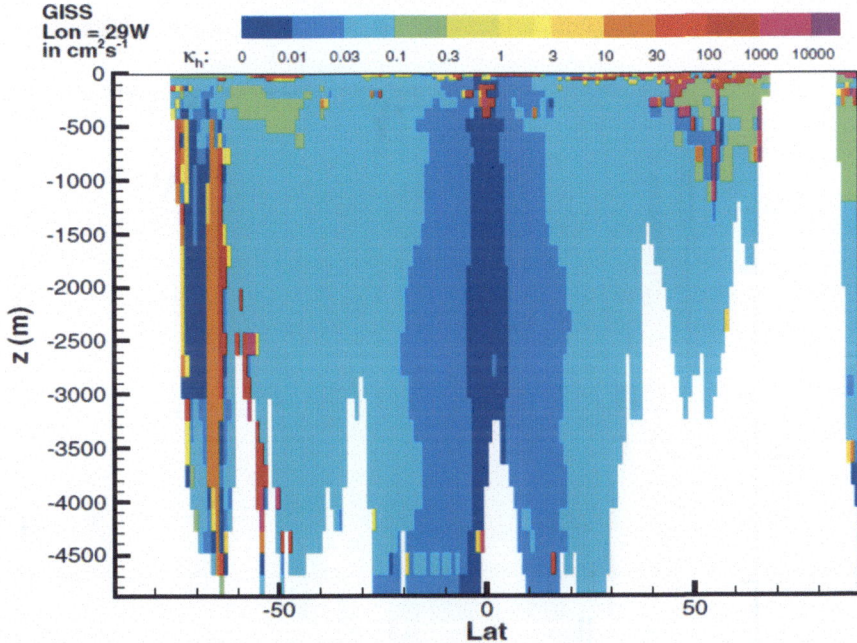

Fig. 2.20 Heat diffusivity

$$\omega^2 = \frac{4\Omega^2}{N^2} \tag{2.118}$$

where N is the Brunt-Vaisala frequency. Figure 2.25 shows that for values of (2.118) of order 100, the decrease in the convective flux, when gravity and rotation are parallel, can be significant. However, in the solar case $\Omega = 2 \times 10^{-6}$ s^{-1}, it seems unlikely that such value could be attained.

A further progress was made in Canuto (1992) when it was first suggested to employ the even more complete ***Reynolds Stress Model RSM*** that provides the time dependent, non-local equations of the heat flux, buoyancy variance, z-component of the kinetic energy and kinetic energy. It is instructive to recall that the 1992 proposal to adopt the RSM was made 47 years after the first application of the RSM to shear driven flows made in 1945 by P.Y. Chou. Between that interval there was an important step, the application of RSM to geophysical flows by Mellor and Yamada in 1982 that was then followed by several subsequent studies (e.g., Canuto 1997a, b, c; Canuto et al. 2001, 2002, Canuto et al. 2010a, b; Canuto 2011; Cheng et al. 2002, 2020). The RSM results are the fluxes of heat, salt and momentum which however contain adjustable parameters due to the well-known difficulties to treat non-linearity using the Navier-Stokes equations, NSE.

Overshooting. When the buoyancy force on a fluid particle vanishes, the acceleration vanishes but the velocity dos not which means that the fluid particle will keep

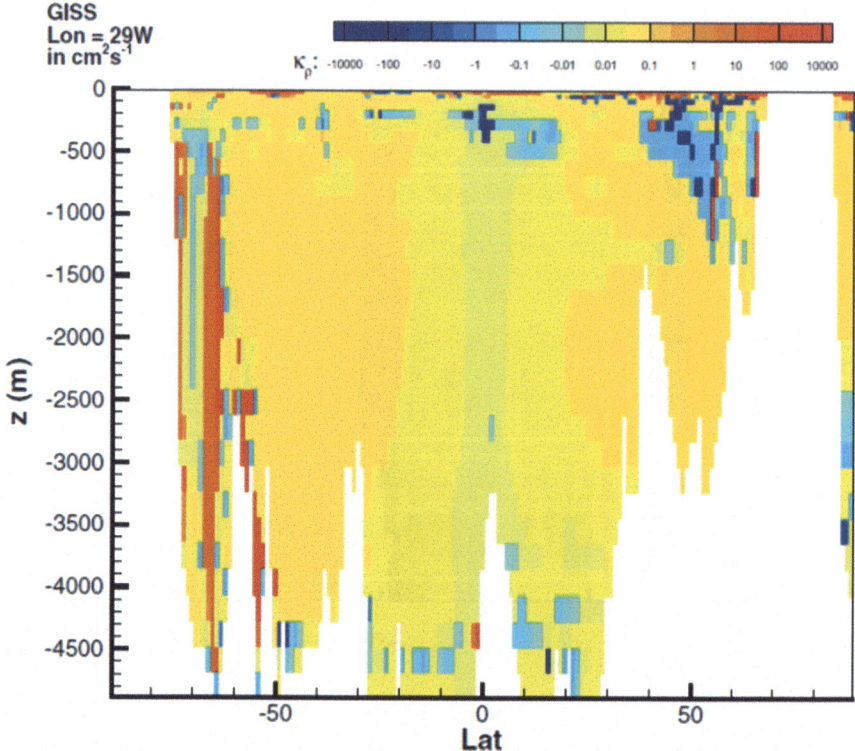

Fig. 2.21 Buoyancy diffusivity. (Reprinted with permission from Canuto et al. (2011). © NASA all rights reserved)

on moving into the adjacent region. The extent of that excursion, known as overshooting OV, carries flows with a given molecular weight into a region of different nature and given the strong dependence of the solar luminosity on the molecular weight, the effect of OV may be significant, which requires a reliable quantification of the OV extent. This problem has attracted considerable attention, but no final solution is available for which we suggest an alternative route. The key variable has been identified as the flux of EKE, $F_{K,i} = \overline{Ku_i}$ (Roxburgh 1978; Kupka and Muthsam 2008; Kupka 2008) that enters the EKE equation as follows:

$$D_t K + \nabla \cdot \mathbf{F}_K = -R_{ij}S_{ij} + g\lambda_i J_i^\rho - \varepsilon \qquad (2.119)$$

where $J_i^\rho = -\bar{\rho}^{-1}\overline{\rho' u'_i} = \alpha_T J_i^h - \alpha_\mu J_i^\mu$ is the buoyancy (density flux) and $g\bar{\rho}\lambda_i = -\partial \bar{p}/\partial x_i$. Thus far, the OV problem was treated without the shear term in (2.119) and using a parameterization of the third-order moment $\mathbf{F}_K$ e.g., a downgradient formulation as discussed in the last two references above. We suggest a procedure that avoids the need to parameterize $\mathbf{F}_K$ and instead, derives a differential

2.11 Solar Convection, Rotation, Overshooting and Angular Momentum

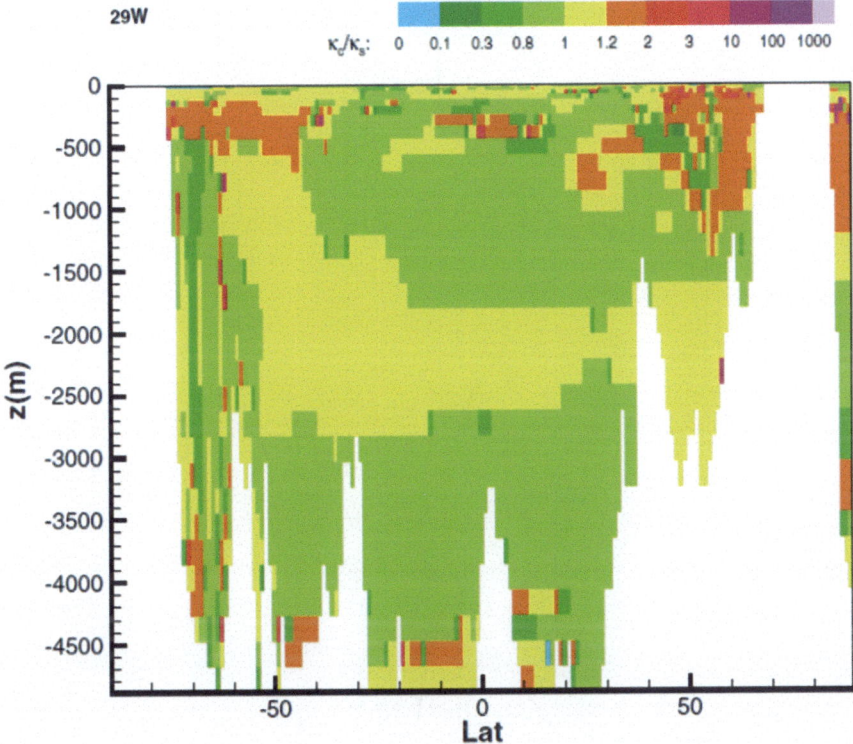

Fig. 2.22 Ratio of passive scalar to salt diffusivity. (Reprinted with permission from Canuto et al. (2011). © NASA all rights reserved)

equation for $\mathbf{F}_K$. This in turn will yield the OV extent as the radial distance where $\mathbf{F}_K$ vanishes. The novel feature is the *total energy E* satisfies the equation (Canuto 1999, Eq.115):

$$\bar{\rho}\frac{DE}{Dt} = -\frac{\partial \Phi_i}{\partial x_i} - \frac{\partial \bar{p}}{\partial t} \qquad (2.120)$$

where:

$$E = c_p T + K + K_M + G, \quad \rho K = \frac{1}{2}\overline{\rho u'_i u'_i} \quad K_M = \frac{1}{2}\bar{u}^2, \quad g_i u_i = \frac{DG}{Dt} \qquad (2.121)$$

which represents the sum of enthalpy, turbulent kinetic energy, mean kinetic energy and gravitational energy G; the *total flux* Φ_i is the sum of radiative, heat, turbulent kinetic energy and flux of Reynolds stresses:

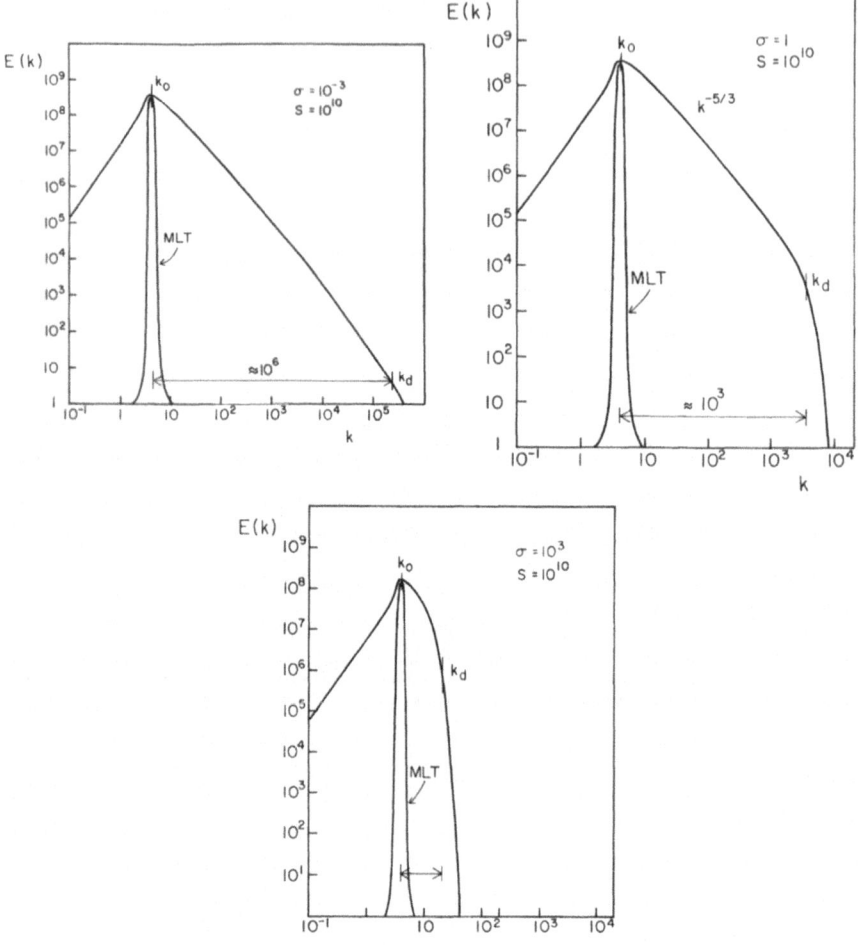

Fig. 2.23 The width of kinetic energy spectrum for different values of viscosity represented by the Prandtl number $\sigma = \nu/\chi$ as the ratio of the kinematic viscosity to the thermometric conductivity. The smaller the ratio, the more different is the spectrum from the MLT form. (Reprinted with permission from Canuto and Mazzitelli (1991). © NASA all rights reserved)

$$\Phi_i = F_i^r + F_i^h + F_i^{ke} + \overline{\rho} R_{ij} \overline{u}_j \qquad (2.122)$$

where:

$$F_i^r = -c_p \rho \chi_r \frac{\partial T}{\partial x_i}, \quad F_i^h = c_p \overline{\rho u_i' T'}, \quad \overline{\rho} R_{ij} = \overline{\rho u_i' u_j'} \qquad (2.123)$$

where χ_r is the thermometric conductivity. Eq (2.122) generalizes the commonly used equation:

2.11 Solar Convection, Rotation, Overshooting and Angular Momentum

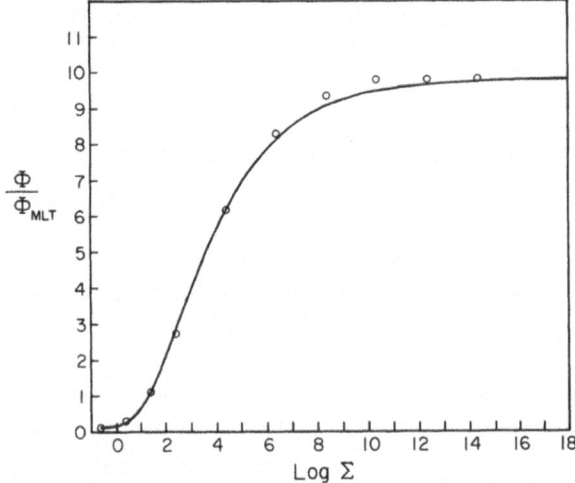

Fig. 2.24 Ratio of (2.117) to (2.115). (Reprinted with permission from Canuto and Mazzitelli (1991). © NASA all rights reserved)

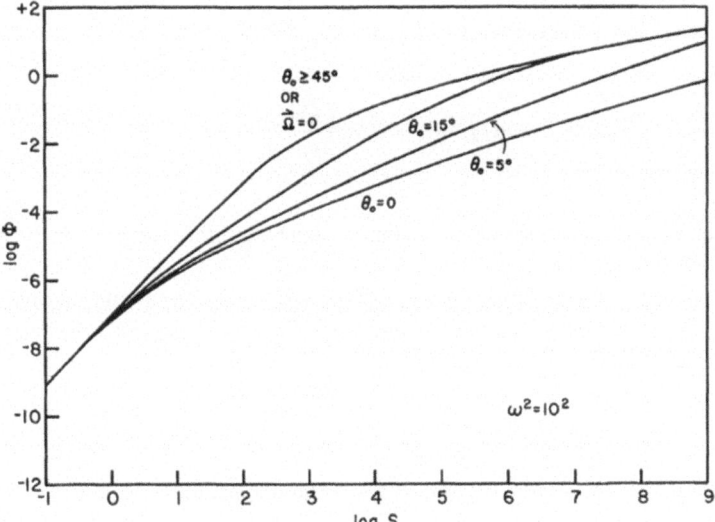

Fig. 2.25 The effect of rotation on the convective heat flux. (Reprinted with permission from Canuto and Hartke (1986). © NASA all rights reserved)

$$\bar{\rho} c_v \frac{DT}{Dt} = -\frac{\partial}{\partial x_i} \left(F_i^r + F_i^h \right) \quad (2.124)$$

which, in the stationary case, yields the flux conservation of radiative plus heat (convective) fluxes:

$$F_i^r + F_i^h = \text{const} \qquad (2.125)$$

Next, consider the equation for a mean scalar denoted by $c = \overline{C} + c'$ (e.g., molecular weight):

$$\overline{\rho}\frac{D\overline{C}}{Dt} = -\frac{\partial \overline{\rho}J_i^c}{\partial x_i} - \frac{\partial \overline{\rho}F_i^c}{\partial x_i}, \quad F_i^c = -\chi_c\frac{\partial \overline{C}}{\partial t}, J_i^c = \overline{u_i'c'} \qquad (2.126)$$

Next, we multiply (2.120) and (2.126) by α_T, $c_p\alpha_c$ and subtract the results:

$$\overline{\rho}\frac{DE_*}{Dt} = -\frac{\partial \Phi_i^*}{\partial x_i} + \frac{\partial \overline{p}_*}{\partial t} \qquad (2.127)$$

where:

$$E_* = c_p\alpha_T T(1+\xi) - c_p\alpha_c\overline{C}, \quad \xi = \frac{1}{c_pT}(K + K_M + G), \overline{p}_* = \alpha_T\overline{p} \qquad (2.128)$$

$$\overline{\rho}^{-1}\Phi_i^* = c_pJ_i^\rho + \alpha_T F_i^{ke} - c_pJ_i^{r,m} + \alpha_T\overline{u}_jR_{ij},$$

$$J_i^{r,m} \equiv \alpha_T\chi_r\frac{\partial T}{\partial x_i} - \alpha_c\chi_c\frac{\partial \overline{C}}{\partial x_i} \qquad (2.129)$$

where the notation (r,m) stands for radiative-molecular, the variable ξ is the ratio of the sum of turbulent kinetic energy, mean kinetic energy and gravitational energy to thermal energy (or enthalpy); finally and the term $\overline{u}_jR_{ij}$ represents the flux of the Reynolds stresses. In the stationary limit, (2.127), (2.128) and (2.129) become the *generalized conservation law*:

$$c_pJ_i^\rho + \alpha_T F_i^{ke} - c_pJ_i^{r,m} + \alpha_T\overline{u}_jR_{ij} = \text{const.} \qquad (2.130)$$

Eliminating J_i^ρ between (2.119) and (2.130), one obtains the equation for F_i^{ke}:

$$\partial_i F_i^{ke} + \left(\frac{g\alpha_T}{c_p}\right)\lambda_i F_i^{ke} = \Phi_{old} + \Phi_{new} \qquad (2.131)$$

where:

$$\Phi_{old} = \text{const.} - g\lambda_i J_i^{r,m} - \varepsilon, \quad \Phi_{new} = -R_{ij}S_{ij} - gc_p^{-1}\lambda_i\overline{u}_j\left(\delta_{ij}E_* + \alpha_T R_{ij}\right) - \overline{u}_j\partial_j K \qquad (2.132)$$

For a given value of the constant, differential equation (2.131) is linear, can be solved analytically to yield the radial dependence of F_i^{ke}. Once a stellar code provides the large-scale variables in the above equations, the location where the flux F_i^{ke} vanishes, yields the OV extent.

The solar angular momentum problem. Helio-seismological data have highlighted an interesting new feature, solar rotation changes from uniform in the radiative region RZ to differential in the convective zone CZ. The problem has been studied for many years by several authors, and in what follows we present a brief summary of the present situation:

(a) 3D numerical simulations by Brummell et al. (2002) and Brun and Toomre (2002) have made "promising contacts" with the convective zone but have not yet explained the uniform rotation in the radiative regime. The reason suggested by the authors of those studies is the limited range of physical parameters (compared to the true solar values) allowed by numerical simulations, e.g., the simulations are still too viscous,

(b) *use of the angular momentum equation with Reynolds stresses of the down gradient flux for shear, leads to extraction of angular momentum from the interior that is too weak to explain the helio data. Such models yield large rotation gradients (in the stellar interiors) that are not consistent with helio data,*

(c) several authors suggested that an additional process is needed, *internal gravity waves* (IGW) (Kumar and Quataert 1997; Zahn et al. 1997; Kumar et al. 1999; Charbonnel and Talon 2005, 2007; Talon and Charbonnel 2005). The time scale needed for the IGW mechanism to arrive at a solar rotational curve compatible with the RZ data was estimated to be $\sim 10^7$ yrs.

(d) we show that if, in addition to the shear generated down-gradient fluxes, one includes *vorticity and stable stratification*, the Reynolds stresses contain a new term that that helps the RZ rotation to become close to rigid. For small Ri, i.e., strong shear, as expected in the CZ-RZ transition zone, the time scale can be of the same order as the one provided by the IGW mechanism.

Next, we show why the angular momentum equation employed in all previous studies is an approximation of the complete one. We then derive and show how to include *vorticity* and *stratification*, which contribute with opposite signs in the CZ and in the RZ zones where they produce an up-gradient flux in the angular momentum equation. Finally, we show how the new Reynolds stress model includes unstable stratification, stable stratification, differential rotation, double-diffusion, arbitrary Peclet number (accounting for radiative losses), and meridional currents. Talon and Charbonnel (2005) documented successes and failures of the "hydrodynamically induced" processes in stellar structure studies. Here, we limit ourselves to discussing the "failures" or "incompleteness" of such models, the most notorious of which concerns the Sun since, as Thompson et al. (2003, sec. 7.3) stated, *"calculations based on the angular momentum equation:*

$$\frac{\partial r^2 \Omega}{\partial t} = r^{-2} \frac{\partial}{\partial r} \left(r^4 K_m \frac{\partial \Omega}{\partial r} \right) + \ldots \tag{2.133}$$

predict rotation of the solar interior at a rate several times higher than the surface rate, in stark disagreement with helio data of nearly uniform rotation. Equation (2.133) is the standard equation used in all studies we have consulted (e.g., Talon and Zahn 1998; Maeder and Maynet 2001; Talon and Charbonnel 2003; Mathis et al. 2004; Palacios et al. 2003, 2006). Equation (2.133) is a particular case of the general equation derived from the Navier-Stokes equations for the mean velocity:

$$\frac{\partial}{\partial t}(\overline{\rho u_i}) + \frac{\partial}{\partial x_j} \overline{\rho}(R_{ij} + \overline{u}_i \overline{u}_j) = -g_i \overline{\rho} - \frac{\partial \overline{p}}{\partial x_i} \tag{2.134}$$

where $R_{ij} = \overline{u'_i u'_j}$ are the Reynolds stresses representing the effect of turbulence on the mean flow (the total velocity field has a mean component denoted by an overbar and a fluctuating component denoted by a prime). The $\overline{u}_\varphi = \Gamma r \Omega$, $\Gamma = \sin\theta$ component of Eq. (2.134) reads as follows, with $J = \Gamma^2 r^2 \Omega$:

$$\frac{\partial J}{\partial t} = -\Gamma^{-1} r^{-2} \frac{\partial}{\partial r} \left(R_{r\varphi} + \overline{u}_r \overline{u}_\varphi \right) - \Gamma^{-3} \frac{\partial}{\partial \theta} \Gamma^2 \left(R_{\theta\varphi} + \overline{u}_\theta \overline{u}_\varphi \right) \tag{2.135}$$

in which the equation of the meridional currents is given by:

$$\frac{\partial}{\partial t}(\rho \overline{u}_r) = A_1 - \frac{\partial P}{\partial r} + \rho \frac{\partial \Phi}{\partial \theta}, \quad \frac{\partial}{\partial t}(\rho \overline{u}_\theta) = A_2 - r^{-1} \frac{\partial P}{\partial \theta} + \rho r^{-1} \frac{\partial \Phi}{\partial \theta} \tag{2.136}$$

where P is the mean pressure, Φ is gravitational potential and:

$$A_1 = -r^{-2} \frac{\partial}{\partial r} \left(r^3 \psi_{rr} \right) - (r\Gamma)^{-1} \frac{\partial}{\partial \theta} (\Gamma \psi_{r\theta}) + r^{-1} \left(\psi_{\theta\theta} + \psi_{\varphi\varphi} \right)$$

$$A_2 = -r^{-2} \frac{\partial}{\partial r} \left(r^3 \psi_{rr} \right) - (r\Gamma)^{-1} \frac{\partial}{\partial \theta} (\Gamma \psi_{\theta\theta}) + (r \operatorname{tg}\theta)^{-1} \psi_{\varphi\varphi}$$

$$\psi_{ij} \equiv \overline{\rho}(R_{ij} + \overline{u}_i \overline{u}_j) \tag{2.137}$$

Equations (2.137) show that the meridional currents also depend on the Reynolds stresses, which we consider next. Thus far, all models have assumed that R_{ij} is contributed only by shear S_{ij}, that is, they adopted the down-gradient relation:

$$R_{ij} = -2K_m S_{ij}, \quad 2S_{ij} = \overline{u}_{i,j} + \overline{u}_{j,i}, \quad S_{r\varphi} = \frac{1}{2} r \Gamma \frac{\partial \Omega}{\partial r} \tag{2.138}$$

2.11 Solar Convection, Rotation, Overshooting and Angular Momentum

where K_m is the momentum diffusivity. Use of this model in (2.135), yields (2.133) which is often but incorrectly, referred to as a "diffusion equation". It can be shown that including *vorticity and stratification*, one obtains a true diffusive equation:

$$\frac{\partial J}{\partial t} = -r^{-2}\frac{\partial}{\partial r}\left(r^2 x K_m \frac{\partial J}{\partial r}\right) + \dots \qquad (2.139)$$

where $x = (\tau N)^2$, N is the Brunt-Väisäla frequency, $\tau = 2K/\varepsilon$ is the dynamical time scale, K is the eddy kinetic energy and ε its rate of dissipation. Clearly, regimes with opposite signs of x alter the down-up gradient nature of the contribution. This means that the generally assumed down-gradient shear instability is incomplete since the transition from unstable to stable stratification changes the nature of the instabilities that operate since in the CZ $x < 0$ while in the RZ $x > 0$ and the sign of the second term in (2.139) changes sign.

There are several problems with (2.133). The first is the need to justify why only shear enters, and second is the determination of the momentum diffusivity. The first item is almost never discussed and the determination of K_m is handled with heuristic models that is, most of the attention was devoted to the determination of K_m and very little, if any, to the completeness of the "shear alone" model (2.138). Since (2.138) is an arbitrary choice, a reliable model must include shear S, vorticity V, buoyancy B, radiative losses, Peclet number:

$$R_{ij}(S, V, B, Pe), \quad Pe = \frac{\pi^2}{125} Re (\tau N)^2, \quad Re = \frac{\varepsilon}{\chi N^2} \qquad (2.140)$$

where Pe is large in the CZ and small in the RZ; furthermore, χ is the thermometric diffusivity and $Pr = \nu/\chi \approx 10^{-8}$ is the Prandtl number. It is instructive to point out that the combinations:

$$\text{shear alone, shear + buoyancy, shear + vorticity} \qquad (2.141)$$

failed to reproduce the Reynolds stresses:

$$R_{\theta\varphi} = \overline{u_\theta' u_\varphi'} \quad \text{vs.} \theta \qquad (2.142)$$

measured at the solar surface (Ulrich et al. 1988; Virtanen 1989; Pulkkinen et al. 1993). On the other hand, the combination:

$$\text{vorticity + buoyancy} \qquad (2.143)$$

was able to reproduce the above data (Canuto et al. 1994a, b). The general form of the Reynolds stresses is given in Eqs. (2.76), (2.77), (2.78) and (2.79).

Solution with no meridional currents *(Canuto and Minotti 2001).* The solution presented in that paper highlighted a key feature of the model, *the existence of a* *counter-gradient angular momentum flux within the hydrodynamic instability framework.* We begin by introducing the following dimensionless variables:

$$r_{ij} = \frac{R_{ij}}{K}, \quad s_{ij}, v_{ij} = \tau(S_{ij}, V_{ij}), \quad z_{ij} = \frac{\tau Z_{ij}}{K},$$
$$\widehat{b}_{ij} = \frac{\tau B_{ij}}{K}, \quad n_i = g\alpha\tau^2\beta_i, \quad j_i = g\alpha\tau\frac{J_i}{K} \tag{2.144}$$

Equations (2.76), (2.77), (2.78) and (2.79) then become:

$$r_{ij} = \frac{2}{3}\delta_{ij} - \frac{8}{75}s_{ij} - \frac{1}{10}z_{ij} + \frac{1}{10}\widehat{b}_{ij}, z_{ij} = \left(r_{ik} - \frac{2}{3}\delta_{ik}\right)v_{jk} + \left(r_{jk} - \frac{2}{3}\delta_{jk}\right)v_{ik}, \widehat{b}_{ij} = \lambda_i j_j + \lambda_j j_i - \frac{2}{3}\delta_{ij}\lambda_k j_k \tag{2.145}$$

while the dimensionless form of the first of (2.68) reads as follows:

$$(\delta_{ik} + \mu_{ik})j_k = \pi_4 r_{ik} n_k, \quad \mu_{ij} = \pi_4\left(s_{ij} + v_{ij} - \pi_5 \lambda_i n_j\right) \tag{2.146}$$

Equations (2.145) and (2.146) form a system of linear, coupled, algebraic equations that may be solved using a method of symbolic algebra; additionally, one must solve the relation $P = \varepsilon$ which, using the first of (2.145) reads as follows:

$$-\left(r_{\varphi r}s_{\varphi r} + r_{\theta\varphi}s_{\theta\varphi}\right) + \frac{1}{2}j_r = 1 \tag{2.147}$$

For illustrative purposes, consider the case:

$$\lambda_r = 1, \quad \lambda_{\theta,j} = 0, \quad n_i = (n_r, 0, 0), \quad x = \tau^2 N^2, \quad N^2 = -g\rho_0^{-1}\partial\overline{p}/\partial z \tag{2.148}$$

Solving the mentioned equations, the resulting Reynolds stress and heat fluxes have the forms:

$$R_{r\varphi} = -2K_m\left(S_{r\varphi} + xV_{r\varphi}\right), \quad J_r = -K_h\frac{\partial T}{\partial r} \tag{2.149}$$

together with Eq. (2.147). Using the relations:

$$u_\phi = r\Gamma\Omega(r,\theta), \quad u_r = u_\theta = 0, \Gamma = \sin\theta \tag{2.150}$$

we further have:

$$S_{\varphi r} = \frac{1}{2}r\Gamma\frac{\partial\Omega}{\partial r} = -\frac{1}{2}p\frac{\omega}{\tau}, \quad V_{r\varphi} = -\left(1-\frac{1}{2}p\right)\frac{\omega}{\tau}\Gamma \tag{2.151}$$

with $S_{\phi r} + V_{\phi r} = -\Gamma\Omega$ and where we introduced the dimensionless variables:

2.11 Solar Convection, Rotation, Overshooting and Angular Momentum

$$p = -\frac{r}{\Omega}\frac{\partial \Omega}{\partial r}, \quad \omega = \tau\Omega, \quad Ri = \frac{1}{p^2\Gamma^2}\frac{N^2}{\Omega^2} \tag{2.152}$$

Solving (2.147), one obtains the functions:

$$\omega(Ri, p), \quad K_{m,h}(Ri, p) \tag{2.153}$$

and the angular momentum then becomes:

$$\frac{\partial J}{\partial t} = -r^{-2}\frac{\partial}{\partial r}\left(r^4 K_m \frac{\partial \Omega}{\partial r}\right) - r^{-2}\frac{\partial}{\partial r}\left(xr^2 K_m \frac{\partial \Omega}{\partial r}\right) \tag{2.154}$$

The first term has the same form as in (2.139), while the second term includes the contribution of vorticity and stratification x. Several comments are in order:

(a) Vorticity and buoyancy appear together which is somewhat surprising since such a combination was not obvious in the starting Eq. (2.145).
(b) Since x represents stratification, we have x < 0 in the CZ (convective zone) and x > 0 in the RZ (radiative zone).
(c) The first term in the rhs of (2.154) is independent of stratification and does not have the form of a diffusion of J, as we have already pointed out. The second term in (2.154) is of the diffusion type, and its sign depends on whether one is in the CZ or RZ zone.

In the CZ, $x = \tau^2 N^2 < 0$, the second term in (2.154) and thus both terms are positive. Due to the p dependence exhibited in relation (2.151), p = 2 corresponds to pure shear and no vorticity.

In the RZ, $x = \tau^2 N^2 > 0$, the second term is now negative helping achieve a rigid body rotation.

Comparison with previous models. It is instructive to compare the momentum diffusivity K_m in this work with the one used in the literature (Charbonnel and Talon 2005; Talon and Charbonnel 2005, Eq. (5); Zahn 2008):

$$Ri < Ri(cr) = \frac{1}{4}, \quad \frac{K_m}{\chi} = \frac{1}{2Ri} \tag{2.155}$$

where χ is thermometric conductivity that enters the radiative diffusivity $K_r = c_p \rho \chi$. Several comments about (2.155) are in order.

The *first* point concerns the absence of a factor representing the amount of energy (or power) that generates mixing. In fact, turbulence is a process that does not generate or destroy energy, rather, it distributes the available energy among a wide variety of scales. Without such an energy, there would be no turbulent motion or, alternatively, turning off such source of energy would lead to a decaying turbulent mixing. Kolmogorov law $E(k) = Ko\epsilon^{2/3} k^{-5/3}$ gives the eddy energy spectrum generated by the nonlinear interaction and contains the rate of dissipation or energy input ε which is considered an external input.

The *second* comment is that in the absence of DD processes, the momentum diffusivity is a decreasing function of Ri, see Fig. 2.3a, but in the DD case, the opposite occurs which is not accounted for in (2.155).

The *third* comment is that (1.155) is limited to values Ri < 1/4 while most recent data show that there is no critical value of Ri.

Internal gravity waves, IGW. As discussed by several authors (Kumar and Quataert 1997; Zahn et al. 1997; Kumar et al. 1999; Charbonnel and Talon 2005, 2007), the most recently employed form of the angular momentum equation (Talon and Charbonnel 2005, Eq. (9)), has an additional term due to IGW in the right-hand side of Eq. (2.133):

$$-\frac{3}{8\pi\rho}r^{-2}\frac{\partial}{\partial r}L_{IGW}(r) \tag{2.156}$$

where the form of the IGW luminosity is given the references just cited, its magnitude is approximately 10^{29} erg s^{-1} which is to 0.004% of the solar luminosity at the base of the CZ, 2.5×10^{33} ergs^{-1} (Talon and Charbonnel 2005, sec. 3.3). From the new equation for the angular momentum, one obtains a time scale of:

$$\tau(IGW) = \frac{8\pi}{3}\frac{\rho r^5 \Omega}{L_{IGW}} \simeq \text{a few } 10^7 \text{yrs} \tag{2.157}$$

where we used $r = 3.5 \times 10^9$ cm, $\rho = 125$ gr cm^{-3}, $\Omega = 2 \times 10^{-6}$ s^{-1}, $L_{IGW} = 10^{29}$ erg s^{-1}. By comparison, if we use the standard Eq. (2.133) down-gradient model (DG), we obtain:

$$\tau(DG) = \frac{r^2}{K_m} \simeq \frac{4 \times 10^4}{K_m} \ 10^7 \text{yrs} \tag{2.158}$$

Since the values of K_m given in Fig. 14 of Talon and Charbonnel (2005) at the bottom of the CZ, are of the order of 50 cm^{2}s^{-1}, the τ from (2.158) is two-to-three orders of magnitude larger than (2.157). On the other hand, if we consider the last term in (2.154), we have:

$$\tau(new) = F(Ri)10^7 \text{yrs}, \quad F(Ri) = \frac{4}{3}\frac{10^3}{G_m \Gamma_m Ri} \tag{2.159}$$

For Ri=1, $G_m = 10^2$, F(Ri)=O(1), the time scale (2.159) is compatible with the IGW result (2.157).

Concluding remarks. Given the challenge of trying to reproduce the data of the solar rotation in the RZ, we have examined the ingredients of the angular momentum equation. It contains two terms, Reynolds stresses and meridional currents which, in a large Re regime such as the one that characterizes stellar interiors and in a steady state, must balance each other. It is worth noticing that this balance has not yet been exhibited by the numerical simulations published thus far (see Figs. 11 of Brun and Toomre 2002), most likely because they are still too viscous. Since the equations for the meridional currents also depend on the Reynolds stresses, the latter constitute a key ingredient and we have therefore concentrated on how they have been modeled thus far and what the missing terms are that must be included. Since the final new formula for the Reynolds stress Eq.(2.76) is relatively simple, the hope is that it will be tested and assessed to ascertain which angular momentum profiles it produces and what improvement it brings with respect to the expression used thus far, Eq. (2.133).

A key feature of the new RSM is that all the relevant equations governing Reynolds stresses and heat fluxes are obtained by solving linear algebraic equations. This is a welcome feature if one considers the large amount of information that the new model contains: stable stratification, unstable stratification, double-diffusion, differential rotation, shear, radiative losses (arbitrary Peclet number) and meridional currents.

As an illustrative example, we have worked out the case of no double diffusion and no meridional currents in order to highlight a key feature of the model. The standard RSM model based on shear only is of the down-gradient type and fails to reproduce the helio data that point to a rigid body rotation below the solar convective zone. It was then suggested that IGW, which operate on an up-gradient flux, may be responsible for such a rotational state, and quantitative computations by several authors confirmed that on a time scale of the order of 10^7 yrs, a rigid body rotation can be achieved. Here, we have presented an alternative possibility that is not ad hoc but part of the RSM model. It is unavoidable and it turns out that the combination *vorticity+stable stratification* yields an up-gradient term in the angular momentum equation, yielding a time scale comparable to the one by the IGW model. As Talon and Charbonnel (2005) pointed out, a mixing model must do more than reproduce the solar rotation curve and the model presented here must await those tests before a final judgment can be made.

2.12 Overflows. Gravity Currents. New Turbulence Models

Density driven currents down topography (overflows) are a key oceanic process in the North Atlantic, Demark Straits and Faroe Bank Channel where they supply most of the North Atlantic Deep water and ultimately the thermohaline circulation deep branch. Since overflows cannot be resolved numerically in coarse resolution OGCMs, they must be parameterized in terms of resolved fields. Without them, the locations cited in Table 1 of the NSF sponsored *"The gravity current entrainment climate process team"* (Legg et al. 2009), may exhibit deep sites in the wrong places as well as incorrect meridional overturning circulation vertical structure heat transport distribution. The determination of the *entrainment function E* for gravity currents defined below, was studied by many authors (Morton et al. 1956; Ellison and Turner 1959, E-T; Turner 1973, 1986; Price and Baringer 1994; Price and Yang 1998; Fernando 1992; Girton and Sanford 2003; Arneborg et al. 2007; Legg and Klymak 2008; Legg et al., 2009; Jackson et al. 2008; Cenedese and Adduce 2010). In 1959, E-T used laboratory data to suggest the following form of the entrainment E:

$$E = \frac{w_E}{\bar{u}}, \quad E_{ET} = \frac{0.08 Fr^2 - 0.1}{5 + Fr^2}, \quad Fr^2 \geq 1.25, Fr^2 = Ri^{-1} \quad (2.160)$$

Here, $\bar{u}$ is its men velocity, w_E is the entrainment velocity, Fr and Ri are the Froude and Richardson numbers, respectively. Since the E-T formula has been used in almost all the studies of overflows, we discuss some of its features. *First,* its validity is restricted to $Fr^2 \geq 1.25$ implying that below that value, there is no entrainment. Since such a limit is not supported by the data, Cenedese and Adduce (2010) introduced a new empirical relation whose novelty, compared to ET, is the presence of the Reynolds number Re. In the model discussed in this work, the range of validity is improved to $Fr^2 \geq 1$. *Second,* bottom friction has been found to be important in overflows (Girton and Sanford 2003) but is not accounted for in the E-T formula. *Third,* ET represents *shear instability,* but in the ocean other instabilities can also contribute, e.g., baroclinic instabilities in the diabatic regime, baroclinic instabilities in the BBL, internal gravity waves, internal baroclinic tides and bottom boundary layer whose contributions to E have not yet been quantified which we do here. Given such different features, it seems unlikely that one could account for them using heuristic arguments. Rather, one needs a formulation of *the entrainment function E in terms of an arbitrary source of mixing.* Such relation was derived in Canuto et al. (2005) with the following, model independent form of E in terms of the shear production P_s:

$$E = \frac{2h}{\bar{u}^3} P_s, \quad P_s = -h^{-1} \int_0^\infty dz \overline{w'\mathbf{u}'} \cdot \frac{\partial \bar{\mathbf{u}}}{\partial z} \quad (2.161)$$

Here, h is the thickness of the descending plume, P_s is the shear production of eddy kinetic energy, and $\overline{w'\mathbf{u}'}$ is the vertical momentum flux. The overbar on the turbulence variables denoted by primes, represents a time or ensemble average which, under ergodic conditions, yield the same result (Bradshaw 1976). *Relations (2.161) are the central ingredients used to construct new parameterizations of E.* The momentum flux $\overline{w'\mathbf{u}'}$ is obtained by solving a turbulence model, e.g., the local, second-order closure SOC models developed by Mellor and Yamada (1982), Canuto et al. (2001, 2002), Canuto et al. (2010a, b) and Cheng et al. (2002, 2020). The last three references represent state-of-the art turbulence models and were assessed on a variety of data. Since in the MY models the is a critical Richardson number of 0.19 corresponding to $Fr^2 \geq 5.26$ which is *more restrictive* than the E-T value; the model of Cheng et al. (2002) has a critical Richardson number of one corresponding to

2.12 Overflows. Gravity Currents. New Turbulence Models

$Fr^2 \geq 1$ which is *less restrictive* than the E-T value and in agreement with data and results of non-linear stability analysis (Abarbanel et al. 1984).

Entrainment based on a turbulence model Since the goal is to derive the form of E due to a variety of mixing processes, e.g., shear instabilities, baroclinic instabilities, bottom boundary layer instabilities, internal gravity wave instabilities etc., the priority was to find a model independent expression for the shear production P_s that determines the function E in (2.161). Turbulence models further provide the momentum, heat, diffusivities and eddy turnover time as already discussed. Using the relation $P = P_s(1 - R_f)$ and $P = \varepsilon$, the first of (2.161) yields:

$$E = \frac{2h}{\overline{u}^3}(1-R_f)^{-1}\varepsilon \qquad (2.162)$$

which yields E in terms of the kinetic energy dissipation ε.[1] In this work, we proceed as follows:

(a) in the case of baroclinic instabilities, e.g., sub-mesoscales, we employ (2.161) and compute the shear production,
(b) in the BBL (bottom boundary layer) and IGW (internal gravity waves), we employ (2.162) and employ existing parameterizations of the dissipation.
(c) in the mixed layer, we follow the methodology of sec. 11 in Canuto et al. (2001), sec. 9a in Canuto et al. (2005) and Mellor and Yamada (1982), where the dissipation is given by Eq. (62) of C2010b:

$$\varepsilon = \eta_0(\tau\Sigma)^{-3}\ell^2\Sigma^3 \qquad (2.163)$$

where ℓ is the dissipation length scale and $\eta_0 = (21.6)^2$ Cheng et al. (2002). Using (2.163) in (2.162), one obtains:

$$E = 2\eta_0\Delta(1-R_f)^{-1}(\tau\Sigma)^{-3}.\Delta = \frac{h\ell^2\Sigma^3}{\overline{u}^3} \qquad (2.164)$$

Using $\tau = 2K/\varepsilon$, the $P = \varepsilon$ relation becomes:

$$\frac{1}{2}(\tau\Sigma)^2(1-R_f)S_m = 1 \qquad (2.165)$$

which we use to eliminate $\tau\Sigma$, yielding the final form of E:

---

[1] As discussed in Canuto et al. (2010a, b), ε is a notoriously difficult variable to model. Though the exact equation was derived by Davydov (1961), is unworkable because of the large number of terms requiring closure while the forms suggested by other authors (e.g., Jackson et al. 2008), are heavily parameterized. There is only one attempt to derive an ε-equation from a two-point closure model (Canuto et al. 2010b).

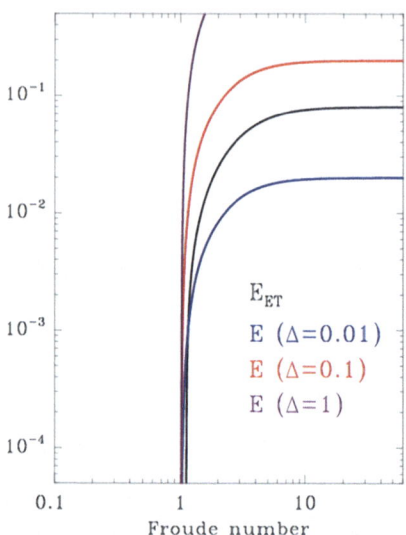

Fig. 2.26 E vs. Fr from relation (2.166)

$$E = E_0 \Delta (1 - R_f)^{1/2} S_m^{3/2}, \quad E_0 = 2^{-1/2} \eta_0 \qquad (2.166)$$

Using the structure functions $S_{m,h}$ and the constants in Table 2 of Cheng et al. (2002), Eq. (2.166) yields the function E(Fr) shown in Fig. 2.26.

The results in Fig. 2.26 show that the E-T *heuristic relation is: derivable from a second-order closure turbulence model*. Should one search for a value of Δ to reproduce the magnitude of the E-T? We suggest one should not because Δ contains physical features not present in the E-T formula, e.g., thickness of the plume, mean velocity, mean shear and dissipation length scale. Given the different nature of overflow regimes, it would be restrictive to assume that they can all be represented by a single Δ. Finally, the E-T restriction Ri < 0.8 is now extended to Ri < 1.

Double counting. Some overflows parameterizations, e.g., Legg and Klymak (2008), Papadakis et al. (2003), employ an entrainment function E(Fr) usually of the E-T form, plus a small-scale mixing model. However, since it has now been shown that *small-scale mixing is the basis of the E-T formula, such a procedure may constitute double counting*. In this context, Ezer (2005) employed *only* a small-scale parameterization (MY model) and no *E(Fr)*. Since the MY model solves the prognostic equations of the model and length scale, Ezer (2005) concluded there was no need for an additional model of E.

Bottom Boundary Layer BBL Consider the bottom drag cited in Legg et al. (2006):

$$\tau_b = C_D \rho \bar{u} |\bar{u}|, \quad W = C_D \rho |\bar{u}| \bar{u}^2, \quad u_* = (\tau_b/\rho)^{1/2}, \quad C_D = 0.003 \qquad (2.167)$$

where W is the work done by bottom friction. In the deep ocean where $\bar{u} = 2 \, \text{cms}^{-1}$:

2.12 Overflows. Gravity Currents. New Turbulence Models

$$W = 2*10^{-5} Wm^{-2} \quad (2.168)$$

Multiplying (2.168) by the ocean's surface of $5*10^{14} m^2$, the result amounts to 1% of the T/P altimetry data of 1TW (Egbert and Ray 2000). Relations (2.167) *are therefore not a viable parameterization of the BBL.*

Barotropic and baroclinic tides. Bottom mixing is generated by the *conversion of barotropic to baroclinic tides through the interaction of the former with rough bottom topography,* a process that was studied by Kantha and Tierney (1997), Llewellyn Smith and Young (2002), St. Laurent and Garrett (2002), Legg (2004), Simmons et al. (2004) and Saenko and Merryfield (2005). Jayne and St. Laurent (2001) and Jayne (2009), derived the barotropic tidal velocity $\mathbf{u}_t$ by solving the 2D Laplace shallow water equations (see their Eq.3–4), at a resolution of ½ of a degree; the drag contained in such equations depending on bottom topographic roughness h was taken from Smith and Sandwell (1997) data at a 1/32 of a degree resolution and then binned into the ½ degree resolution of the 2D code that provides $\mathbf{u}_t$. The map of $\mathbf{u}_t$ in Fig. 2.28 was kindly provided by S. Jayne. A form of the rate of energy conversion E (x, y) from barotropic to baroclinic tides was suggested by Jayne and St. Laurent (2001) to be:

$$E(x,y) = \frac{1}{2}\rho N \kappa h^2 \overline{\mathbf{u}_t^2} \quad (2.169)$$

were h was taken from high resolution bathymetry while $\kappa = 2\pi/10$ km was chosen to provide the best fit to observations. The global maps of h and $\mathbf{u}_t$ are shown in Figs. 2.27, 2.28 and 2.29 courtesy of S. Jayne.

Entrainment in the BBL. We employ the model of St. Laurent and Garrett (2002):

$$\varepsilon_{tides} = q\rho^{-1}\zeta^{-1} E(x,y) F(z), \quad F(z) = A \exp\left(-\frac{z+H_b}{\zeta}\right) \quad (2.170)$$

where $A^{-1} = 1 - \exp(-H_b/\zeta)$, $\zeta = 500$m (based on fitting an exponential profile to the Brazilian Basin microstructure survey), and H_b is the ocean depth; though the fraction q of the baroclinic energy that creates mixing is not known, a value of 1/3 has been suggested; the 1-q fraction is radiated to the ocean interior to enhance the background diffusivity. Using (2.162), the form of E becomes:

$$E = \frac{2h}{\overline{u}^3}(1-R_f)^{-1}\varepsilon_{tides} \quad (2.171)$$

Because of the strong location dependence of E(x,y) shown in Fig. 2.29 it is not possible to present a "characteristic E". However, if one employs $E=10^{-x} Wm^{-2}$, h = 100 m, H_b = 3km, H_b/ζ = 6, q = 1/3, $\overline{u} = 2$ cm s^{-1}, $2h\overline{u}^{-3} = 1/4*10^8 s^3 m^{-2}$, $F(0) = 5/2*10^{-3}$, we obtain:

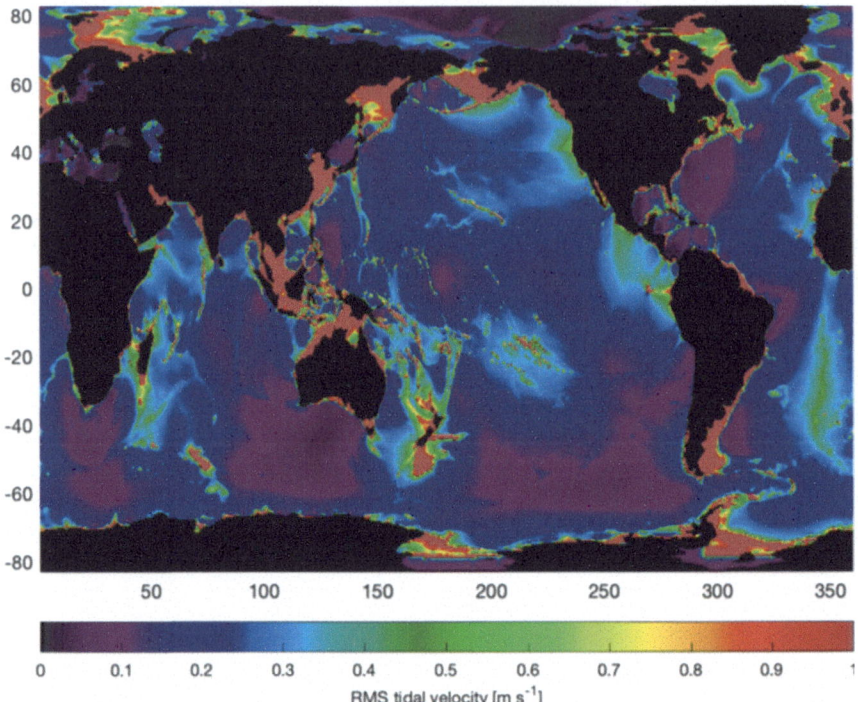

Fig. 2.27 Tidal velocity

$$\varepsilon_{\text{tides}} = \frac{5}{3} 10^{-9-x} \text{Wkg}^{-1}, E = 0.8 \, 10^{-2-x} \quad (2.172)$$

where we assumed $(1 - R_f)^{-1}$ to be of order unity as seen from Fig. 2.30. In the regimes where x is 1–2, relation (12.12) puts E in the lower left corner of Fig. 2.32 where Fr $\simeq 1$.

Baroclinic instabilities, BI. Since the ocean is globally *baroclinically unstable*, it is necessary to inquire if such an instability contributes to entrainment. Several authors (Allen and Newbeger 1998; Wenegrat et al. 2018; Wenegrat and Thomas 2020; Yankovsky and Legg 2019) studied baroclinic instabilities in the BBL. Most of the papers dealt with stability problems while to compute E in (12.2) one needs the momentum flux. J. Wenegrat kindly pointed out to us that Fig.18 of Wenegrat et al. (2018) showed the results of their non-linear analysis from which one derives that:

$$P_{\text{BBL}} \simeq 2^* 10^{-5} \text{cm}^2\text{s}^{-3} \quad (2.173)$$

With h = 100 m, $\bar{u} = 2\,\text{cms}^{-1}$, the entrainment function from the first of (2.161) becomes:

2.12 Overflows. Gravity Currents. New Turbulence Models

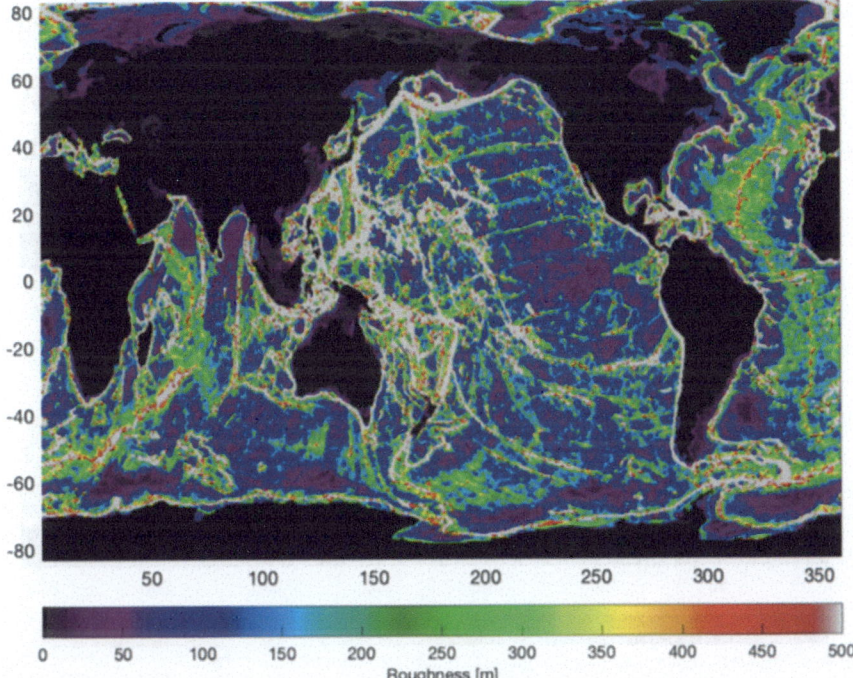

Fig. 2.28 Roughness in m

$$E \simeq \frac{1}{20} \quad (2.174)$$

which in Fig. 2.32 corresponds to the Fr = O(1) regime.

Sub-mesoscales. The model presented in Appendix B was used to derive the following expression for the shear production.

$$P_s(z) = \underbrace{p_0(z)h^2|f|^{-1}|\nabla_H \overline{b}|^2}_{GSP} + \underbrace{p_1(z)EBF}_{ASP1} + \underbrace{p_2(z)(h^2|f|)^{-1}(\tau_w/\rho)^2}_{ASP2} \quad (2.175)$$

where the dimensionless functions p(z)'s are given in Eq. (2.254). The three terms in (2.175) are called GSP (geostrophic production) and ASP1,2 (a-geostrophic productions).

The results in Fig. 2.31 compare favorably with those of numerical simulations in Fig. 1 in Thomas and Taylor (2010) and Fig. 9 of Thomas et al. (2013) neither of which however included the ASP2 term. The first and third terms in (2.175) correspond to the products of the (g, g) and (ag, ag) components of the mean velocity $\overline{u}$. Finally, EBF is the Ekman buoyancy flux:

$$\rho EBF = f^{-1}(\nabla_H \overline{b} \times \tau_w) \cdot e_z \quad (2.176)$$

Depth averaging <...> (2.175) using relations (2.259), one obtains:

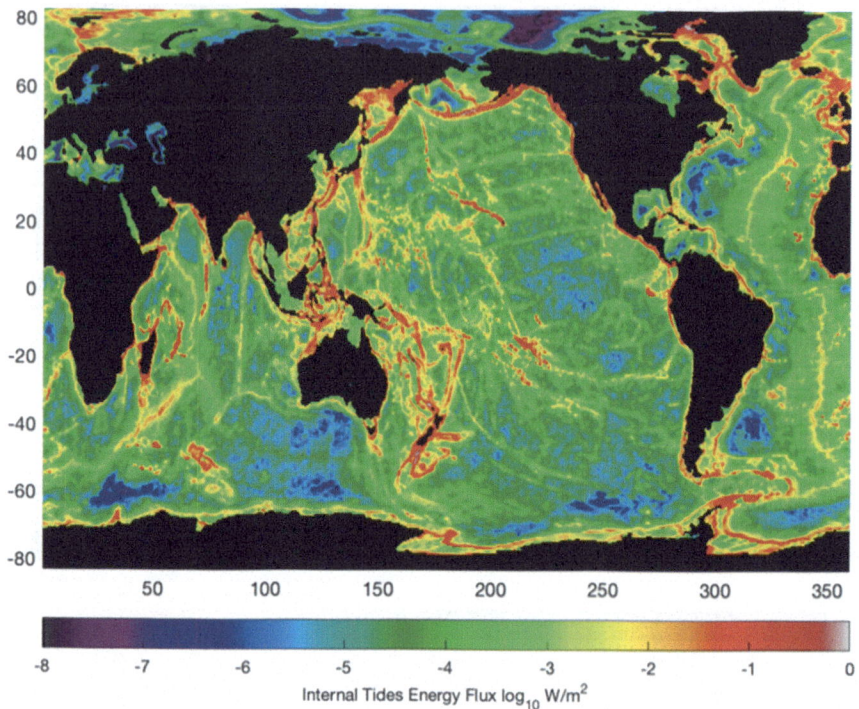

Fig. 2.29 Internal tidal energy in W/m^2

Fig. 2.30 The function $(1 - R_f)^{-1}$ vs Fr

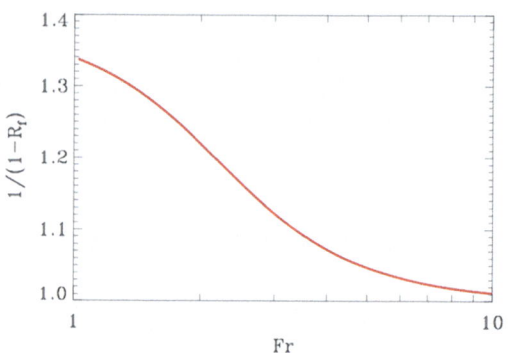

$$\frac{<P_s(z)>}{|f|h^2N^2} = \frac{A(Ro)^2}{6Ro}\left(Ri_g^{-1} + 12\,\omega_0\right), \qquad \omega_0 = E_k^{1/2}e_* + E_k^{-1/2}\left(1 - E_k^{1/2}\right)\tau_*$$

(2.177)

where the geostrophic Richardson number $Ri_g = f^2 N^2 |\nabla_H \bar{b}|^{-2}$ is related to Ri:

Fig. 2.31 Buoyancy flux, shear production and their sum

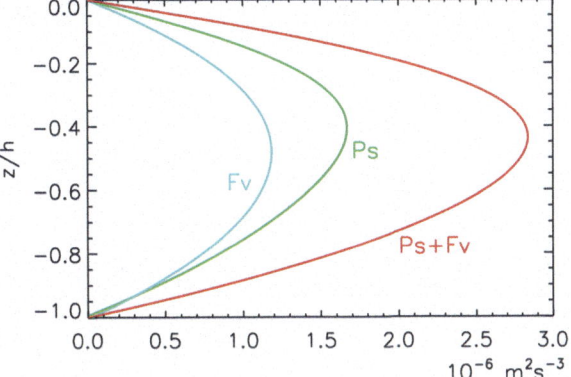

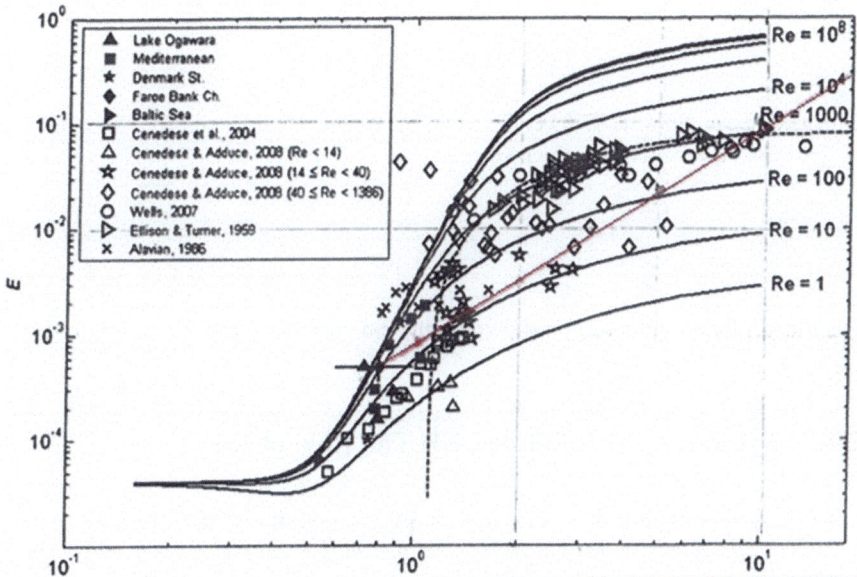

Fig. 2.32 Red curve, Eq. (2.182) due to SM superimposed on the data from Cenedese and Adduce (2010)

$$\mathrm{Ri}_g^{-1} = \mathrm{Ri}^{-1} - \Delta, \quad \Delta = 2E^{-1/2}(e_* + E^{-1}\tau_*), \quad e_* \equiv \frac{\mathrm{EBF}}{|f|h^2 N^2}, \quad \tau_* \equiv \left(\frac{\tau_w/\rho}{h^2 N f}\right)^2 \tag{2.178}$$

showing that in the absence of wind stresses, Ri and Rig coincide. Using the shear production (2.177) in the first of (2.161), one obtains after some algebra:

$$E_{SM} = E_* Fr^2 \left(1 - \omega Fr^{-2}\right) \tag{2.179}$$

where:

$$E_* = \frac{A(Ro)^2}{3Ro} \frac{h^3 N^2 |f|}{\bar{u}^3}, \quad \omega = 2e_*(1\text{-}6E)E^{-1/2} + 2\tau_* E^{-3/2} \left[1\text{-}6E\left(1 - E^{1/2}\right)\right] \tag{2.180}$$

Since the Ekman number E<1 (see Appendix B), the second of (2.180) becomes:

$$\omega = 2(1\text{-}6E)\left(e_* E^{-1/2} + \tau_* E^{-3/2}\right) \tag{2.181}$$

which vanishes for 1 - 6E = 0, corresponding to an Ekman depth of 0.4 h, and thus:

$$E_{SM} = E_* Fr^2 \tag{2.182}$$

which is the sought-after E dependence on Fr. As an example, consider h = 100 m, $N^2 = 10^{-6} s^{-2}$, $\bar{u} = 2 cms^{-1}$ in which case:

$$E_* = 4 \frac{Ro^3}{(1 + Ro^2)^2} \tag{2.183}$$

For Ro = 1, relation (2.182) is shown by the red curve in Fig. 2.32.

2.13 Symmetric Instabilities, SI. The Role of Ro

Yankovsky and Legg (2019, YL) were the first to point out that *"the role of SM symmetric instability SI has not been examined in the context of overflows"*. They presented a numerical study of SI whose inception is when the Ertel potential vorticity Q is opposite in sign to the Coriolis parameter f, where:

$$Q = \boldsymbol{\omega} \cdot \nabla_H \bar{b}, \quad \boldsymbol{\omega} = \mathbf{e}_z f + \boldsymbol{\zeta} \tag{2.184}$$

where $\boldsymbol{\omega}$ is the *absolute vorticity*, $\boldsymbol{\zeta} = \nabla \times \mathbf{u}$ is the *relative vorticity* and the other symbols have the usual meaning. Haynes and McIntyre (1987) derived the following conservation equation for Q:

$$\frac{\partial Q}{\partial t} + \nabla \cdot \mathbf{J} = 0 \tag{2.185}$$

where the **J** flux is defined as follows:

2.13 Symmetric Instabilities, SI. The Role of Ro

$$\mathbf{J} = Q\mathbf{u} - \omega\frac{D\overline{b}}{Dt} + \nabla\overline{b} \times \mathbf{F} \equiv \mathbf{J}_a + \mathbf{J}_d + \mathbf{J}_f, \mathbf{F} = \frac{D\mathbf{u}}{Dt} + f\mathbf{e}_z \times \mathbf{u} + \rho_0^{-1}\nabla p \quad (2.186)$$

The vectors **J** are called *advective, diabatic and frictional* which we consider next. Before we do so, it must be pointed out that the treatment of SM requires special considerations, they are a-geostrophic and their Rossby number is >1. The first feature means that the mean velocity contains geostrophic and a-geostrophic components:

$$\mathbf{u} = \mathbf{u}_g + \mathbf{u}_{ag}, \quad f\mathbf{e}_z \times \partial_z \mathbf{u}_g = -\nabla_H \overline{b}, \quad \rho f\mathbf{e}_z \times \mathbf{u}_{ag} = \partial_z \boldsymbol{\tau} \quad (2.187)$$

where the form of the wind stress $\boldsymbol{\tau}$ must be specified, for example, the Ekman form:

$$\boldsymbol{\tau}(z) = \alpha(\zeta)\boldsymbol{\tau}_w + \beta(\zeta)\frac{f}{|f|}\mathbf{e}_z \times \boldsymbol{\tau}_w \quad (2.188)$$

where $\alpha(\zeta), \beta(\zeta) = e^\zeta(\cos\zeta, \sin\zeta)$, $\zeta = z/\delta_e$, $\delta_e = (2\nu_t/|f|)^{1/2}$, see Appendix B. For an observational confirmation of (2.188), see Chereskin (1995). As shown in Canuto (2015), the presence of the a-geostrophic velocity enters as a third term in the new function Q:

$$Q = \left(1 + \frac{\zeta}{f}\right)fN^2 - f^{-1}|\nabla_H\overline{b}|^2 - \left(\frac{\alpha-\beta}{h^2|f|}\right)fEBF \quad (2.189)$$

and therefore, the standard Hoskins' relation for $Q < 0$, changes to:

$$Q < 0: 1 + \frac{\zeta}{f} < \frac{1}{Ri_g} + (\alpha-\beta)\frac{EBF}{h^2N^2|f|} \quad (2.190)$$

where $Ri_g = f^2N^2|\nabla_H\overline{b}|^{-2}$ is the geostrophic Richardson number. Relation (2.190) implies that for EBF > 0, *the a-geostrophic component of the velocity makes Hoskins relation easier to satisfy and helps the onset of SI*. The second feature characterizing is that the SM Rossby number is >1 which must appear in the SM parameterization, while in the case of the geostrophic mesoscales M, the corresponding Ro < 1 which therefore does not appear in M parameterizations.

Next, consider the *diabatic and frictional* components in (2.186). From Canuto and Cheng (2021, Eq.6.1), the geostrophic and a-geostrophic contributions to the vertical buoyancy flux F_v^{SM} given by Eq.(A.7) give rise to:

$$J_d = \mathbf{J}_d \cdot \mathbf{e}_z = f\frac{\partial F_v^{SM}}{\partial z} = a(z)\frac{hf}{|f|}|\nabla_H\overline{b}|^2 + b(z)\frac{f}{h}EBF \quad (2.191)$$

$$a(z) = -\frac{A(Ro)}{2Ro}\left(1 + \frac{2z}{h}\right), \quad b(z) = A(Ro)\left[1 - \frac{\alpha-\beta}{E^{1/2}} - \frac{\alpha+\beta}{RoE^{1/2}}\right]$$

Figure 2.33 shows that J_d is negative for $|z| < h/2$ and positive for $|z| > h/2$ In their fig. 7c, YL give the divergence of $\mathbf{J}_d$ which is negative and so is (2.191).

Next, we consider the frictional component:

$$J_f = \left(\frac{\partial}{\partial z}\overline{\mathbf{u}'w'} \times \nabla_H \overline{b}\right) \cdot \mathbf{e}_z + \left(\frac{\partial \boldsymbol{\tau}}{\partial z} \times \mathbf{e}_z\right) \cdot \nabla_H \overline{b} \qquad (2.192)$$

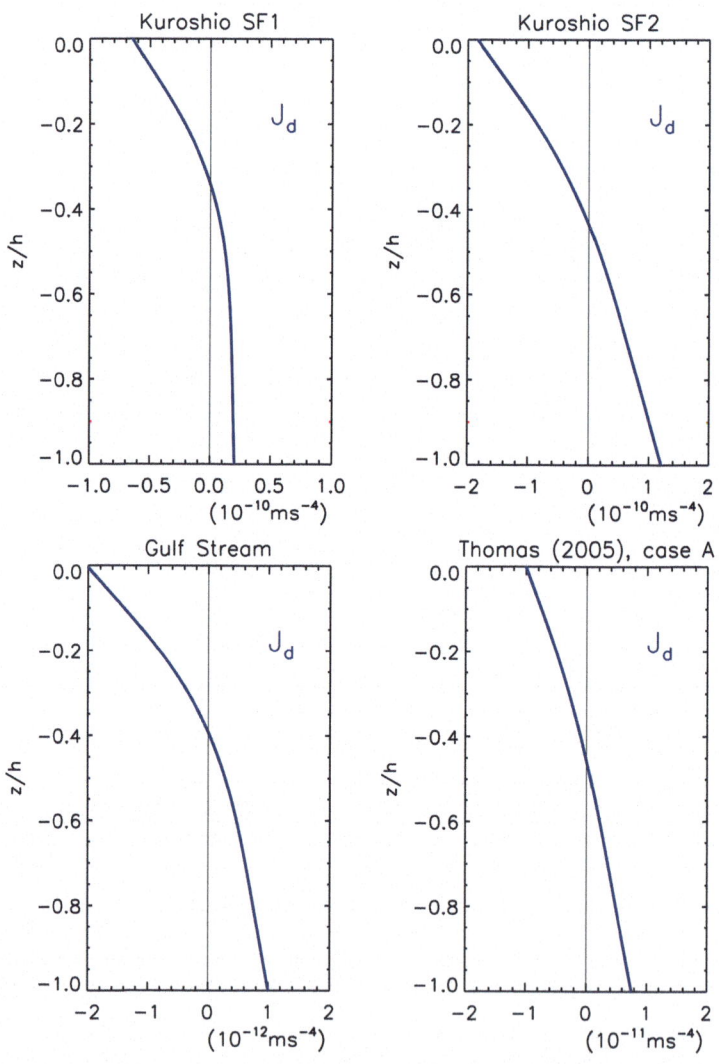

Fig. 2.33 Vertical profiles of J_d. (Reprinted with permission from Canuto and Cheng (2021) © NASA all rights reserved)

2.13 Symmetric Instabilities, SI. The Role of Ro

which exhibits the momentum flux $\overline{u'w'}$ given in (1.131) and the (wind stress) Ekman contribution given in B.11. The two components were computed in Canuto and Cheng (2021) to be:

$$J_f = -\Gamma_2(z)|\nabla_H \bar{b}|^2 - \Gamma_3(z)EBF$$

$$\Gamma_2(z) = \frac{A(Ro)^2}{Ro}\left(1 + \frac{2z}{h}\right), \quad \Gamma_3(z) = 2A(Ro)\frac{f}{h}\left[1 - \frac{\alpha-\beta}{E^{1/2}} + \frac{A(Ro)}{Ro}\frac{\alpha+\beta}{E^{1/2}}\right]$$
(2.193)

Figure 2.34 shows the vertical profiles of the first component of (2.193) called BI (baroclinic instability, blue), the second component called SW (wind stress, green) and their sum (red) together with $J_f(Ek)$ (black) given by:

$$J_f(Ek) = (\alpha-\beta)\frac{f}{hE^{1/2}}EBF$$
(2.194)

At $z = 0$, the total frictional given by the sum of (2.193) and (2.194) is shown in Fig. 2.35 which exhibits a clear effect of Ro: *only values between 1–10 yield a negative frictional component favoring SI.* We cannot, however, assess this result against YL's work since there was no Ro.

Internal Gravity Waves IGW. The ocean thermocline is the seat of internal gravity wave (IGW) described by the Gregg-Henyey-Polzin model (Polzin et al. 1995) with the result:

$$\varepsilon_{IGW} = 0.288 A N^2 (\text{cgs units})$$
(2.195)

where $A = O(2)$ is the deviation from the Garrett-Munk background internal gravity spectrum. At NATRE, where $N^2 = 1.7^*10^{-5} s^{-2}$, (2.195) yields:

$$\varepsilon_{IGW} = 0.5 A^* 10^{-9} W kg^{-1}$$
(2.196)

which agrees with the data in Fig.5 of Canuto et al. (2010a, b). The ratio of (2.196) and (2.182) is:

$$\frac{E_{tides}}{E_{IGW}} = \frac{\varepsilon_{tides}}{\varepsilon_{IGW}} = \frac{10}{3A} 10^{-x}$$
(2.197)

In the regime where $x = 1–2$, the two contributions may be of the same order.
Summary of the entrainment functions.

- *shear instability:* $E = $ Eq. (2.166)
- *baroclinic instability, sub-mesoscales in the diabatic regime:* $E = E_* Fr^2$:

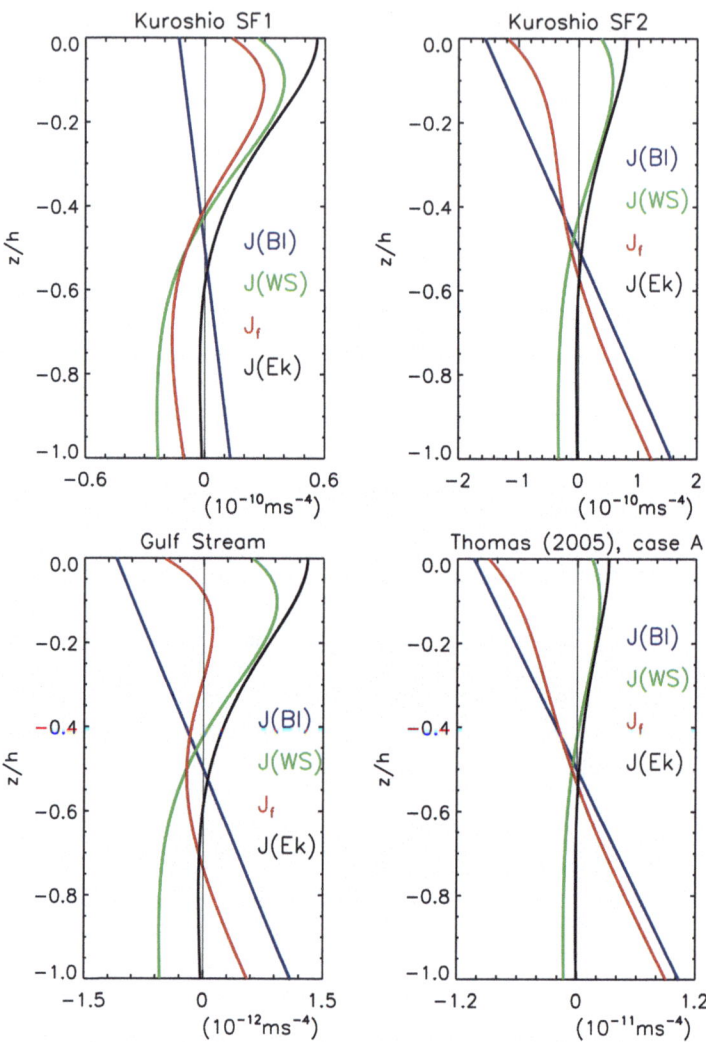

Fig. 2.34 Vertical profiles of J_d. (Reprinted with permission from Canuto and Cheng (2021) © NASA all rights reserved)

$$E_* = \frac{1}{3}\frac{h^3 N^2 |f|}{\bar{u}^3}\frac{Ro^3}{(1+Ro^2)^2}$$

- *baroclinic instabilities in the BBL:* $P_s \simeq 2*10^{-5} \text{cm}^2\text{s}^{-3}$,
- *internal gravity waves:* $E = \frac{2h}{\bar{u}^3}(1-R_f)^{-1}\varepsilon_{IGW}$, $\varepsilon_{IGW} = 0.288 AN^2$ (cgs units),
- *internal baroclinic tides:* $E = \frac{2h}{\bar{u}^3}(1-R_f)^{-1}\varepsilon_{tides}$, $\varepsilon_{tides} = \rho^{-1}qE(x,y)\zeta^{-1}F(z)$.

2.13 Symmetric Instabilities, SI. The Role of Ro

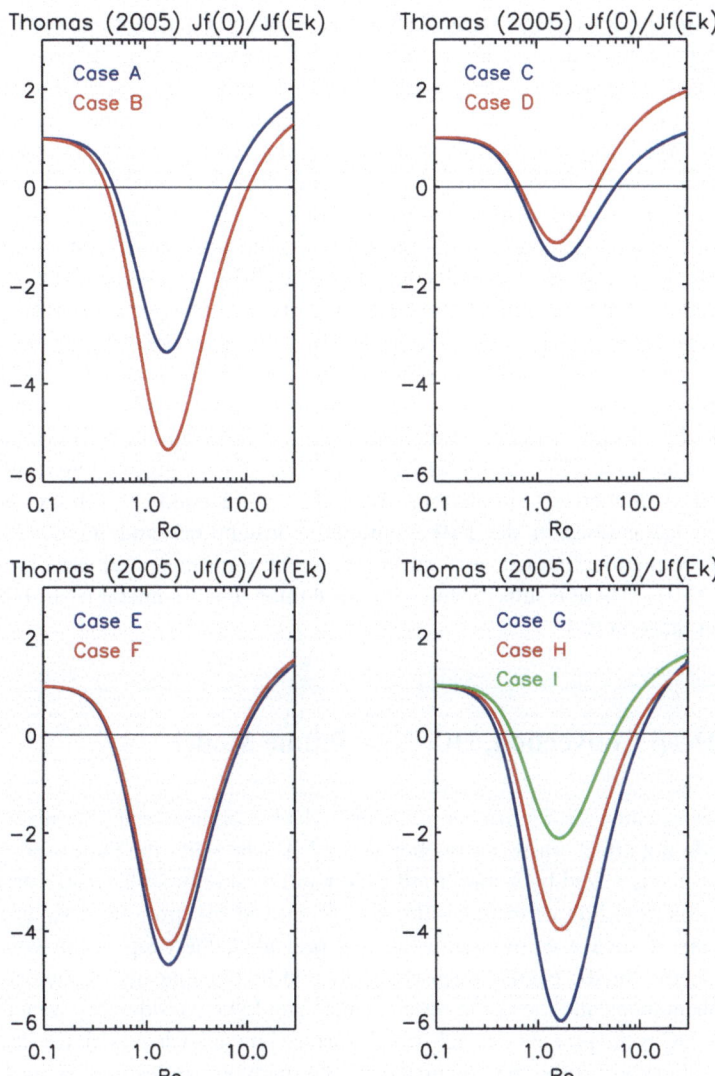

Fig. 2.35 Plot of J_f in units of $J_f(Ek)$ at $z = 0$ for different Ro. (Reprinted with permission from Canuto and Cheng (2021). © NASA all rights reserved)

Conclusions. For many years, (2.160) was the only formula available to represent entrainment by shear instability at the interface between a descending plume and the environment. The goal was to extend the contributions to E by other instabilities. This was possible because relation (2.161) expresses E in terms of the *shear production* generated by the turbulent momentum flux $\overline{w'u'}$. Using a second-order closure turbulence model SOC, a further possibility was to represent E in terms of the rate of dissipation ε, Eq. (2.162). The first goal was to show that the E-T relation could be derived from a SOC model as shown by Eq. (2.166) that brings into the problem features that are not present in the E-T formula, e.g., thickness of the plume, mean velocity, mean shear and dissipation length scale. A second possibility is to employ (2.162) that gives E in terms of dissipation of eddy kinetic energy. To obtain the latter, we employed the barotropic-to-baroclinic conversion model that requires the tidal velocity obtained by solving the 2D Laplace shallow waters equations. The resulting map of the tidal velocity is shown Fig. 2.27. It must be pointed out, however, that this is not the only possible parameterization, an alternative having been proposed by Decloedt and Luther (2012) who presented global and basin (Pacific, Atlantic and Indian oceans) comparisons of the two approaches. Their general conclusion was that both approaches "significantly underestimate mixing in the abyssal ocean", the discrepancy being likely due to the constant scale height $\zeta=500$ m adopted in (2.170) that, it was argued, should depend on z. In addition, estimates of ζ vary from smaller (Polzin 2004) to larger (Kunze et al. 2006) than the 500 m assumed in (2.170). As for *baroclinic instabilities*, we present two results, the first from a numerical simulation, the other corresponding sub-mesoscales. Baroclinic instabilities in the BBL, symmetric instabilities and internal gravity waves in the thermocline whose contribution to entrainment is compared to that of tides in (2.197). These results extend the contribution to entrainment by instabilities other than those in E-T.

2.14 Deep Convection, DC. New Plume Model

Wind stresses cause strong mixing in the first ~100 m of the ocean (the mixed layer ML) but do not affect water masses below the ML where lies the largest portion of the ocean characterized by stable stratification and weaker mixing. Stable stratification acts as a rigid lid that insulates the strongly mixed ML from the weakly mixed deep ocean. If such a configuration always prevailed, the deep ocean would be shielded from climatic events, deep waters would be dynamically decoupled from surface phenomena and the ocean deep currents would be considerably weaker than observed. DC is the process through which surface buoyancy losses pierce the lid of strong stable stratification that characterizes the thermocline, leading to the formation of *deep waters* (Chu and Gascard 1991; Maxworthy 1997; Marshall and Schott 1999; The Labrador Sea Deep Convection Experiment, 2002). Open ocean deep convection occurs in few locations, Labrador, Greenland, Weddell Sea and Western Mediterranean Sea and yet, it is the dominant mechanism for the production of the North Atlantic Deep Water (NADW) and the Antarctic Bottom Water (AABW, Weaver et al. 1999), both of which play major roles in earth's climate. Loss of surface buoyancy leads to a top-heavy, unstable configuration which acts as precursor of turbulent motion that leads to deep convection that upwells warmer waters that can melt ice and reduce the albedo resulting in a "negative feedback' that affects climate (Killworth 1983). Regrettably, however, DC is still poorly represented in coarse resolution OGCMs: though laboratory and numerical simulations (Sander

2.14 Deep Convection, DC. New Plume Model

et al. 1995; Denbo and Skyllingstad 1996; Maxworthy 1997; Marshall and Schott 1999) have stressed key features of DC though translating them into an OGCMs has not yet been achieved.

DC is a highly turbulent process. This is illustrated by the large vertical diffusivity:

$$K_v \approx \ell w \simeq (1\text{-}10)10^5 \text{cm}^2\text{s}^{-1} \quad (2.198)$$

where $\ell \sim (1-2)$km, $w \sim (1-5)$cms^{-1} from Marshall and Schott (1999).

Effect of rotation. Consider the characteristic length scale (Golystin 1980):

$$\ell(\text{rot}) \simeq \left(B_* f^{-3}\right)^{1/2} \simeq (0.15 - 0.56)\text{km} \quad (2.199)$$

where B_* is the surface buoyancy and the estimate corresponds to the Greenland and Mediterranean Seas. While in the atmosphere $\ell(\text{rot})$ > height of the planetary boundary layer $\simeq 1$km, in the ocean the reverse is true since, $\ell(\text{rot}) <<$ smaller than the ocean depth. This yields a small Rossby number Ro = $\ell(\text{rot})/H$ = 0.1 – 0.3 showing the importance of rotation.

Local mixing schemes. The NCAR-CSM global ocean model (Large et al. 1997) with a 3°x 3° resolution was used to simulate DC using both KPP and GISS mixing schemes. The measurements by Lavender et al. (2002, LDO) in Fig. 2.36 show that the *deepest convection DC, defined as the one with mixed-layers deeper than 0.8 km,* is confined to a small region in the western Labrador Sea (Fig. 2.36). Figures 2.37 and 2.38 show results from GISS and KPP models.

Within the GISS model, DC is further evident in both the flat vertical temperature profile and large heat diffusivity (see Fig. 2.39) down to 1.2 km. The GISS model heat diffusivity increases from 0.04 m^2 s^{-1} near the surface to a maximum of 6 m^2 s^{-1} at 400 m. In other columns, maximum diffusivities up to 10 m^2 s^{-1}. The maximum

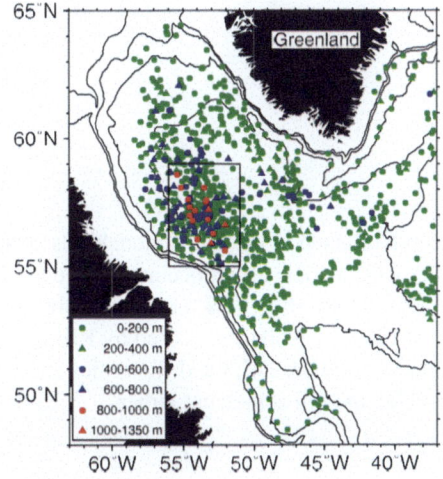

Fig. 2.36 Mixed layer depths as from Lavender et al. (2002). (Reprinted with permission from Canuto et al. (2004). © NASA all rights reserved)

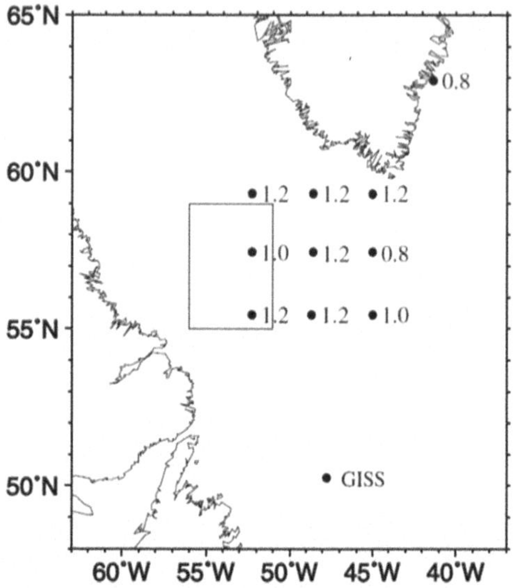

Fig. 2.37 DC (in km): GISS mixing model. (Reprinted with permission from Canuto et al. (2004) © NASA all rights reserved)

Fig. 2.38 DC (in km): KPP mixing model. (Reprinted with permission from Canuto et al. (2004) © NASA all rights reserved)

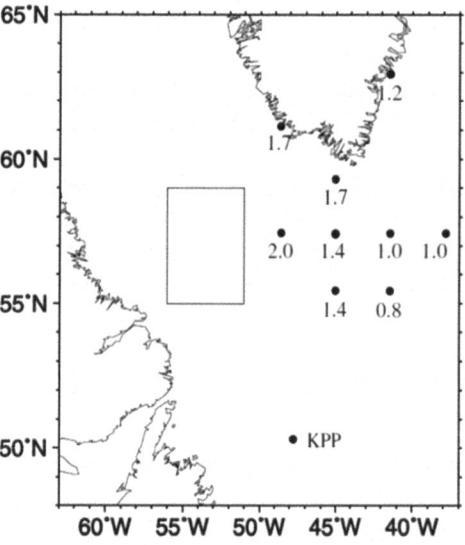

depth attained in the GISS model falls within the observational range while that for the KPP model, which occurs outside the LDO box, is larger than observed. In the LDO region, the GISS model diffusivity lies between $1 m^2 s^{-1}$ used by Marotzke (1991) and $10–50 m^2 s^{-1}$ used by Klinger et al. (1996). At the points where the KPP model produces DC, the diffusivities are of the same order of magnitude as the GISS model, though somewhat larger.

2.14 Deep Convection, DC. New Plume Model

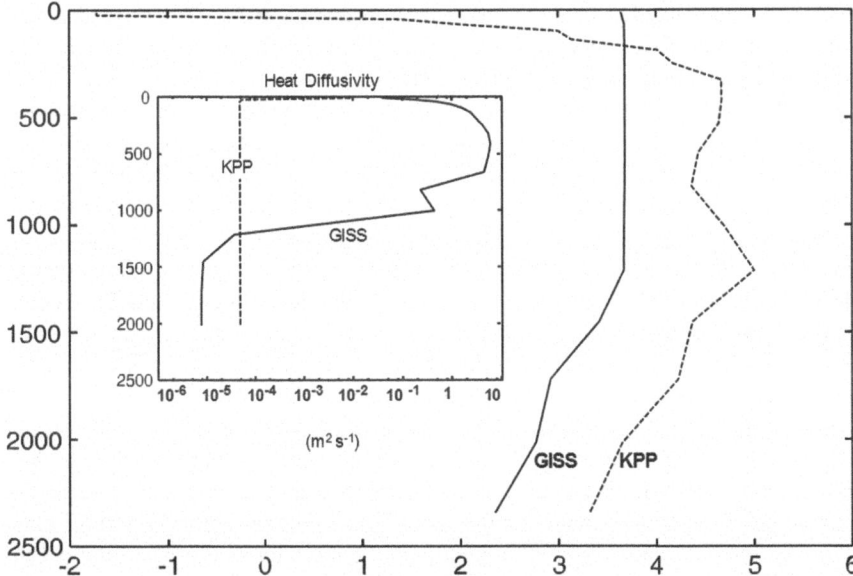

Fig. 2.39 Temperature and heat diffusivity) at 52.2 W, 57.5 N. (Reprinted with permission from Canuto et al. (2004) © NASA all rights reserved)

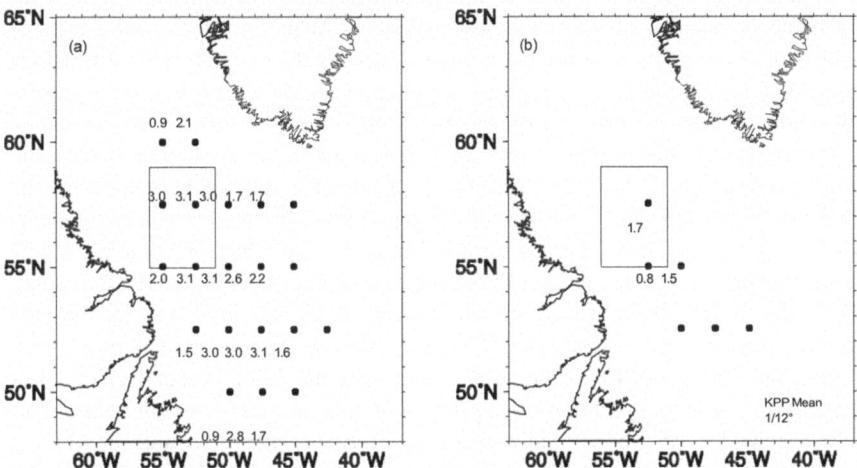

Fig. 2.40 Results from a1/12°· 1/12° resolution. (Reprinted with permission from Canuto et al. (2004) © NASA all rights reserved)

A HYCOM model with a horizontal resolution of 1/12° was also run. The maximum and mean mixed layer depths in the 2.5° bins are shown in Fig. 2.40, respectively. As with the 1/3° results, there are more maximum mixed layer depths than mean mixed layer depths, for the reasons described above. However, in the

1/12° model, the mixed layer depths are much deeper than in the 1/3° case. The range of the mean mixed layer depths) is 0.8–1.7 km. These depths are consistent with the observations, although most of the deep convection sites.

Conclusions. The KPP model was constructed to alleviate the "too little mixing" problem of the Mellor-Yamada models. It is interesting to compare its performance vs. the MY models: KPP yields correct mixed layer depths in stable situations, but it produces too little mixing in the LDO region. Since the GISS model predicts Ri_{cr} of $O(1)$, it gives rise to mixed layer depths in stable situations in agreement with the data. In unstable convective situations, the predictions of the GISS model for the MLD are closer to the LDO data than those of the KPP model.

Non-local mixing schemes. Turbulent fluxes can be represented by a diffusivity tensor, the symmetric part of which describes "turbulent diffusion" while the anti-symmetric part describes "advection". Diffusion is a local process in the sense that it depends only on the local gradients of the mean fields while advection is non-local for it is represented by an integral over all length scales (all eddies) that can "fit" from say the bottom of the physical domain to the z where the fluxes are computed. In the ocean, there are two main regimes where non-local transport is important. One regime is where storms release a sudden burst of mechanical energy to the ocean surface that is then transported downward by energetic eddies that deepen the mixed layer. Even relatively simple non-local models yield results considerably more realistic than those of local models. The second regime is deep convection (DC) caused by loss of surface buoyancy, the description of which is required for a reliable assessment of water masses formation. At present, there is no reliable model for either of these non-local regimes individually or much less a formalism capable of accounting for both regimes simultaneously. In what follows, we present a formalism that provides the expressions for the *non-local fluxes for momentum, heat and salinity encompassing both cases*. Since the resulting number of dynamic equations involves is however large, we consider the case of buoyancy. In the Reynolds Stress formalism, non-locality is represented by the third-order moments which in turn depend on the fourth-order moments for which we employ a new model that has been tested against LES data, aircraft data and a full PBL simulation. We further rewrite the non-local model in terms of *Plumes* since thus far the only non-local model used to treat oceanic DC has been the "plume model" of Morton, Taylor and Turner (MTT model). We show that the MTT model has two key limitations, (1) an important physical process such as the rate of entrainment cannot be determined by the model and remains an adjustable parameter and (2) MTT is purely advective and thus only applicable to the initial stages of DC but not to the whole process which is both advective and diffusive. The model we derive bypasses these limitations, is a generalization of the MTT model and is applicable to the entire development of deep convection. The non-local version of Eq. (2.6) requires modelling the following diffusion terms:

2.14 Deep Convection, DC. New Plume Model

$$D_f(J) = \frac{\partial \overline{w^2\theta}}{\partial z}; \quad D_f(\overline{\theta^2}) = \frac{\partial \overline{w\theta^2}}{\partial z}$$

$$D_f(\overline{w^2}) = \frac{\partial \overline{w^3}}{\partial z}, \quad D_f(K) = \frac{3}{4}\frac{\partial \overline{w^3}}{\partial z}$$
(2.200)

Canuto et al. (2007) suggested the following closures:

$$\overline{w^3} = -0.06 g\alpha\tau^2 \overline{w^2}\frac{\partial J}{\partial z}, \quad \overline{w\theta^2} = -\tau J\frac{\partial J}{\partial z}, \quad \overline{w^2\theta} = -0.3\tau\overline{w^2}\frac{\partial J}{\partial z}$$
(2.201)

which in Figs. 2.41, 2.42 and 2.43 are compared with the LES data of Mironov et al. (2000) and with aircraft data of Hartmann et al. (1999). The data are reproduced satisfactorily.

The first of (2.201) yields the correct negative skewness below the cooling surface where $dB/dz > 0$ while it yields a positive skewness near a surface heated from below where $dB/dz < 0$. By contrast, a down-gradient type model:

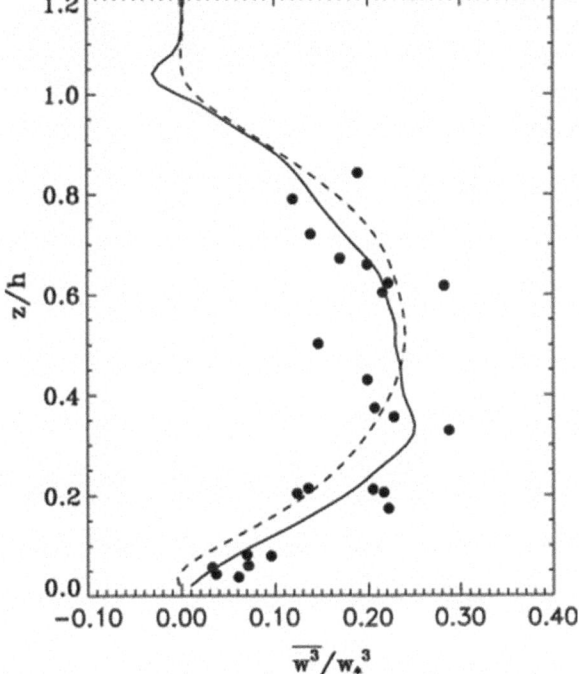

Fig. 2.41 The third moment vs. height normalized by the PBL depth, h. The filled circles represent the aircraft data of Hartmann et al. (1999). The dashed line represents the LES data of Mironov et al. (2000). The solid line represents the result of the new model based on a second-order moments from the same LES data as input. The normalization is the Deardorff's standard normalization, $w_* = (g\alpha h \overline{w\theta}|_{surf})^{1/3}, \theta_* = w_*^{-1}\overline{w\theta}|_{surf} w_* = (g\alpha h \overline{w\theta}|_{surf})^{1/3}, \theta_* = w_*^{-1}\overline{w\theta}|_{surf}$

Fig. 2.42 Same as Fig. 2.41 for $\overline{w^2\theta}$

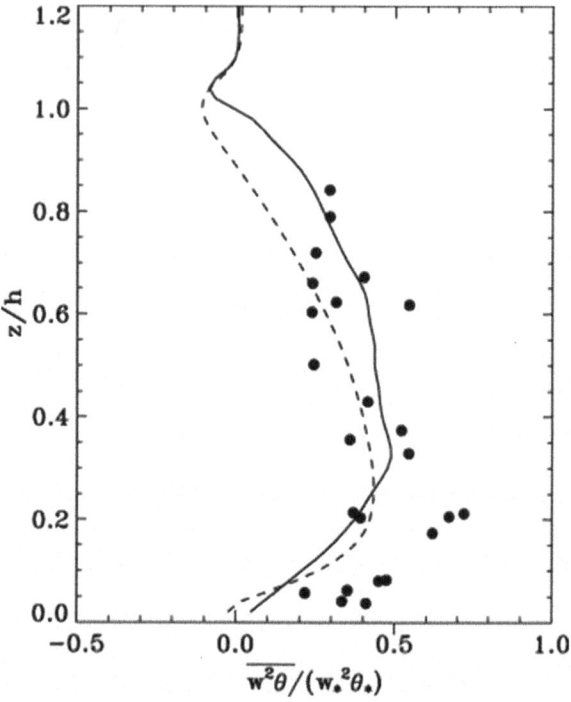

Fig. 2.43 Same as Fig. 2.41 for $\overline{w\theta^2}$

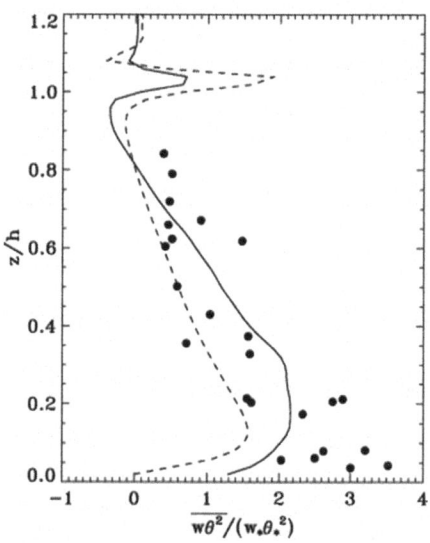

2.14 Deep Convection, DC. New Plume Model

$$\overline{w^3} \approx -\tau \overline{w^2} \frac{\partial \overline{w^2}}{\partial z} \tag{2.202}$$

would yield the wrong sign of the skewness in both cases discussed above. To further highlight the role of skewness, we can re-write the last two relations in (2.201) as follows:

$$g\alpha \overline{w\theta^2} = 17\tau^{-1} J \left(\overline{w^2}\right)^{1/2} S_w, \quad g\alpha \overline{w^2\theta} = 5\tau^{-1} \left(\overline{w^2}\right)^{3/2} S_w, \quad S_w = \overline{w^3}/\left(\overline{w^2}\right)^{3/2} \tag{2.203}$$

as it was emphasized by Wyngaard and Weil (1991) and Hamba (1995).

Plume models, old and new. Due to its importance in the formation of Atlantic deep waters, the Deep Convection regime has been extensively studied with a non-local model, specifically a plume model, that, at first sight is different from the turbulence-based model just presented above. It is therefore important to discuss the relationship between the turbulence-based model and the plume model. Paluszkiewicz et al. (1994), Alves (1995), Paluszkiewicz and Romea (1997) studied Deep Convection using the only plume model available, the MTT plume model (Morton et al. 1956; Turner 1973; Turner 1986) which was based on an original suggestion by Taylor (1945).

Plume models are attractive since they provide an intuitive visualization of the DC process exhibited by LES results showing the existence of narrow descending plumes and wide ascending plumes (the environment). Contrary to a non-local turbulence-based model that, as we have seen, is both advective and diffusive, the MTT model is only *advective*. It contains two equations representing conservation of momentum and buoyancy. However, since there are three unknowns, the third being the fraction of space occupied by the plumes that varies with z (or the plume's radius), Taylor suggested a phenomenological *"entrainment equation"* which contains an entrainment parameter α that the MTT is unable to determine. The parameter α has therefore been treated as an adjustable parameter but in reality, is a function of the large scale features of the flow. Ellison and Turner (1959) used laboratory data to determine $\alpha = \alpha(Ri)$, but the model provides a poor fit to the Mediterranean outflow data (Price and Baringer 1994; Price and Yang 1998). A second problem with the MTT is that it assumes that σ, the fractional area occupied by a plume, is much smaller than unity:

$$\sigma << 1, \quad \sigma = \frac{A^p}{A^p + A^e} \tag{2.204}$$

where Λ is the total cross-section of the *plumes and environment* at a given depth. However, since during the plume's evolution, the plumes entrain fluid from the environment, σ is bound to increase with depth to the point where (2.204) becomes invalid. Specifically, *entrainment* causes the plume's mass flux $w_p A^p \propto \sigma w_p$ to increase while stable stratification decreases w_p, the net result being an increase of σ to the point when (2.204) is no longer valid. In addition, a small σ model cannot satisfy the *zero- mass flux relation:*

$$\sigma w_d + (1-\sigma)w_u = 0 \qquad (2.205)$$

where $w_{u,d}$ are the velocities of the up and down drafts and z is considered upward. In addition, in the limit of small σ, Eq. (2.205) implies that:

$$w_d \mid \gg w_u \qquad (2.206)$$

On the other hand, for the argument given above, when $\sigma = 1/2$, Eq. (2.205) implies that:

$$w_u = |w_d| \qquad (2.207)$$

which is not consistent with (2.204). Finally, the mass conservation (2.205) is invariant under the transformation:

$$w_u \to w_d, \quad \sigma \to 1-\sigma \qquad (2.208)$$

and so should be a plume model. The MTT model is not invariant under (2.208), and is valid only in the plumes' early development stages when the fraction of space occupied by the plumes is still small. In summary, standard MTT plume model has the advantage of simplicity but at present:

(a) is restricted by Eq. (2.204),
(b) it depends on the undetermined rate of entrainment α,
(c) it is only advective and it leaves out diffusion.

In the Table below, we sketch the key features of the initial ($t=0$) and final stages ($t=\infty$) of DC:

Table

t=0: $S_w<0$, $\sigma=$ small, Diffusive: No, Downflows, Advective: Yes

t=∞ : $S_w \simeq 0$ $\sigma=1/2$, Diffusive: Yes, Downflows$\simeq$upflows, Advective: No

In the early stages, the downdrafts dominate over updrafts, while in the final stages, updrafts and downdrafts are equally important; the initial stages are governed by advection, the final stages are governed by diffusion. To correct the limitations of the MTT model, we proceed as follows:

(a) We use the turbulence mixing model and limit the non-locality to the heat and salinity fluxes.

2.14 Deep Convection, DC. New Plume Model

(b) The new plume model is such that all relations are invariant under (2.208) and is valid throughout the entire plume's development. As a basic simplifying assumption, we write the non-local TOMs in the "plume approximation" which assumes a top hat profile that consists of two delta functions for the pdf of each state variable, corresponding to ascending and descending plumes. This implies (Lappen and Randall 2001) that such a profile has 100% probability of having one of just two possible values, the two allowed states being up-drafts and down-drafts. This introduces a considerable simplification to the problem since it reduces substantially the number of higher-order moments that are needed, it assures the realizability condition of the higher order moments and requires fewer prognostic equations.

New plume model. To "*plumenize*" the TOMs using the up-down draft notation, we employ the following relations (Canuto and Dubovikov 1998):

$$\overline{w^2} = \sigma(1-\sigma)(w_u-w_d)^2 = \beta_\sigma w^2, \quad \beta_\sigma = \sigma(1-\sigma)^{-1}$$

$$\overline{\theta^2} = \sigma(1-\sigma)(\theta_u-\theta_d)^2 = \beta_\sigma^{-1} w^{-2} J_h^2 \qquad (2.209)$$

$$\overline{w\theta} = \sigma(1-\sigma)(\theta_u-\theta_d)(w_u-w_d)$$

where $w \equiv w_d$. Analogous relations hold for the salinity field. These relations are invariant under (2.208). The plumenized TOMs then become:

$$\overline{w^3} = -\sigma(1-\sigma)(1-2\sigma)(w_u-w_d)^3$$

$$\overline{w^2\theta} = -\sigma(1-\sigma)(1-2\sigma)(\theta_u-\theta_d)(w_u-w_d)^2 \qquad (2.210)$$

$$\overline{w\theta^2} = -\sigma(1-\sigma)(1-2\sigma)(\theta_u-\theta_d)^2(w_u-w_d)$$

Small σ limit, MTT model. In this section we prove that the plume model just derived, which is valid for any σ, contains the MTT in the limit for small σ. The MTT model contains three equations representing the plume's kinetic energy $1/2w^2$, the fractional area σ occupied by the plume and the buoyancy flux $B(\text{cm}^2\ \text{s}^{-3})$. They are (Turner 1973):

$$\frac{\partial w^2}{\partial z} = \frac{2B}{w\sigma} - \frac{4\alpha}{\ell} \frac{w^2}{\sigma^{1/2}} \frac{\partial \sigma}{\partial z}$$

$$\frac{\partial B}{\partial z} = -\sigma w N^2, \quad N^2 = -g\rho_{\text{ref}}^{-1} \partial_z \rho_{\text{env}} \qquad (2.211)$$

Equations (2.211) are not invariant under (2.208) as they are only applicable in the regime (2.204). If one substitutes the buoyancy equation into the mean temperature equation, one obtains:

$$\frac{\partial T}{\partial t} + (\overline{w} + w_{adv})\frac{\partial T}{\partial z} = 0, \qquad w_{adv} = -\sigma w \qquad (2.212)$$

Since the rhs of the first equation is: zero, there is no diffusion which proves what stated earlier that the MTT is only advective with an advection velocity w_{adv} that is σ times the plume's velocity. An additional interesting variable is the plume's "mass flux" defined as $M=\sigma w$. Using (2.211), one obtains the relation:

$$M^{-1}\frac{\partial M}{\partial z} = E\text{-}D = \frac{2\alpha}{\ell\sigma^{1/2}} > 0 \qquad (2.213)$$

where E and D are the rates of entrainment-detrainment respectively. Since the rhs is positive, MTT *accounts only for entrainment but not detrainment* because detrainment requires a dynamical environment which is excluded in the MTT model which assumes the environment to be quiescent. In the ocean case (w < 0) and the small σ limit, we have from Eqs. (2.209) and (2.210):

$$\beta_\sigma = \sigma, \quad S_w = -\sigma^{-1/2}, \quad \overline{w^2} = \sigma w^2 \qquad (2.214)$$

and thus from Eq. (2.210) it follows that:

$$\overline{w^3} = \sigma w^3, \quad \overline{w^2\theta} = wJ, \quad \overline{w\theta^2} = \sigma^{-1/2}J\overline{\theta^2}^{1/2} \qquad (2.215)$$

Using (2.215), the first of Eq. (2.201) becomes:

$$\frac{\partial J}{\partial z} = -\frac{w}{Cg\alpha\tau^2} \qquad (2.216)$$

Next, Eq. (2.9), (2.11), together with (2.200), become:

$$w^3\frac{\partial \sigma}{\partial z} + \frac{3}{2}\sigma w\frac{\partial w^2}{\partial z} = \frac{2}{3}(1+2\beta_5)B - 4\sigma w^2\tau^{-1} \qquad (2.217)$$

$$w\frac{\partial J}{\partial z} + \frac{J}{2w}\frac{\partial w^2}{\partial z} = -\sigma w^2 T_{,z} = (1-\gamma_1)g\alpha\sigma^{-1}w^{-2}J^2 - \pi_4^{-1}\tau^{-1}J \qquad (2.218)$$

Solving Eqs. (2.215), (2.217) and (2.218), and using:

$$\tau = -C_0 \sigma^{1/2} \ell w^{-1} \tag{2.219}$$

and using the values $C = 0.06$, $\beta_5 = 1/2$, $\gamma_1 = 1/3$, $\pi_4 = 0.08$, one obtains:

$$\frac{\partial w^2}{\partial z} = 1.3 B w^{-1} \sigma^{-1} + 24 C_0^{-1} w^2 \ell^{-1} \sigma^{-1/2} - 2 B^{-1} \sigma w^3 \Delta_1 \tag{2.220}$$

$$\frac{\partial \sigma}{\partial z} = -0.7 B w^{-3} - 32 C_0^{-1} \ell \sigma^{1/2} + 3 w B^{-1} \sigma^2 \Delta_1 \tag{2.221}$$

where:

$$\Delta_1 = N_h^2 - 17 C_0^{-2} \sigma^{-2} \ell^{-2} w^2 \tag{2.222}$$

Since in our system z is positive upward, and the descending plume is small near the surface getting progressively larger at depth, $d\sigma/dz. < 0$, the second term in the rhs of (2.221) that represents entrainment, must be negative. By the same token, the second term in (2.220) is positive since $dw^2/dz. > 0$. If $C_0 = 6$ (which comes from relating the dissipation length scale in (2.219) to the ℓ in Eqs. (2.211), Eqs. (2.220) and (2.221) compare well with Eq. (17a) of the MTT model and Eqs. (10)–(11) of Paluszkiewicz and Romea (1997).

Conclusions. We have developed a non-local mixing model based on the Reynolds Stress formalism. We have presented the general non-local model that today's knowledge in turbulent closure allowed us to formulate. Clearly, the full model in which all the fields of temperature, salinity and 3D velocities are treated non-locally, is fairly complex. To assess the validity of non-locality, we have worked out the case of Deep Convection (DC) which in the past was treated with the MTT plume model of Morton et al. (1956). Such a model has two substantial limitations, it is only advective while in the final stages of DC diffusion dominates and the rate of entrainment is treated as an adjustable quantity. By rewriting the turbulence-based mixing model developed here in the plume formalism, a new Plume Model emerges which no longer suffers from the limitations of the MTT model and is valid for the entire development of a convective regime.

2.15 Derivation of the ε Equation from a Two-Point Closure

In 1961, Davydov used a one-point closure model to derive the ε equation from first principles but the final result contained undetermined terms and thus lacked predictive power. Schiestel (1987) and Rubinstein and Zhou (2001) attempted to derive the ε equation from first principles using a two-point closure, but their methods relied on a phenomenological assumption. The standard practice has been to employ a heuristic form of the ε equation that contains three empirical ingredients, two constants, $c_{1,2}$ and a diffusion term D_ε. Here, we employ a two-point closure yielding the following results: (1) the empirical constants get replaced by $c_{1,2}$ that are no longer constant but functions of K and ε, (2) $c_{1,2}$ are not independent because a general relation is: derived that links the two; (3) $c_{1,2}$ become constant with values close to the empirical values (Pope 2000) in the case of homogenous flows and 4) the empirical form of the diffusion term D_ε is no longer needed because it gets substituted by the K–εdependence of $c_{1,2}$, which plays the role of the diffusion, together with the diffusion of the turbulent kinetic energy which now enters the new ε equation (i.e., inhomogeneous flows). The empirical $c_{1,2}, D_\varepsilon$ are now replaced by a single function $c_1(K, \varepsilon)$ or $c_2(K, \varepsilon)$, plus a D_K term. Tests of the new εequation are presented.

The dynamic equations for the kinetic energy K and the dissipation ε are two basic relations in turbulence. While the first equation is: exact, the second is "entirely empirical" Pope (2000) and the commonly used form is (P stands for production and D for diffusion):

$$\frac{\partial \varepsilon}{\partial t} + D_\varepsilon = \frac{\varepsilon}{K}(c_1 P - c_2 \varepsilon), \quad c_1 = 1.44, \quad c_2 = 1.92 \quad (2.223)$$

Since the approach by Davydov (1961) using a one-point closure yielded a result containing many unknown terms, the procedure had no predictive power. An attempt by Canuto et al. (2010b) to use a two-point closure using $c_{1,2} \rightarrow c_{1,2}(\varepsilon, K)$, was relatively successful. We begin by considering the following spectra corresponding to a shear driven flow. Following Schiestel (1987) and Rubinstein and Zhou (2001), we consider the following spectra:

$$E(k) = Ak^s, \quad k_c \leq k \leq k_m, \quad E(k) = Ko\varepsilon^{2/3} k^{-5/3}, \quad k \geq k_m \quad (2.224)$$

and:

$$m(k) = Bk^s, \quad k_c \leq k \leq k_m, \quad m(k) = CS\varepsilon^{1/3} k^{-7/3}, \quad k \geq k_m \quad (2.225)$$

where m(k) is the Reynolds stress spectrum $\overline{uw} = -\int_{k_c}^{\infty} dk\, m(k), S = \partial \overline{u}/\partial z$ is the shear and $P = -\overline{uw} S$ is the production in Eq. (2.223). In addition, k_m is the wavenumber of the energy containing eddies where the functions E(k) and m(k) exhibit their maxima. Continuity of the spectra at k_m requires that the following relations be satisfied:

$$A = Ko\varepsilon^{2/3} k_m^{-s-5/3}, \quad B = CS\varepsilon^{1/3} k_m^{-s-7/3} \quad (2.226)$$

Both kinetic energy and energy production are easily found to have the following forms:

2.15 Derivation of the ε Equation from a Two-Point Closure

$$\psi = \left(\frac{k_c}{k_m}\right)^{1+s} \quad : K = \frac{3s+5\text{-}2\psi}{2(s+1)} K_o \varepsilon^{2/3} k_m^{-2/3}, \quad P = \frac{3s+7\text{-}4\psi}{4(s+1)} C S^2 \varepsilon^{1/3} k_m^{-4/3} \quad (2.227)$$

Taking the time derivative of the second relation and using $\partial K/\partial t = P - \varepsilon$, one obtains:

$$\frac{1}{\varepsilon}\frac{\partial \varepsilon}{\partial t} = \frac{3}{2K}(-D_K + P\text{-}\varepsilon) + \Psi \frac{1}{k_m}\frac{\partial k_m}{\partial t}, \quad \Psi \equiv \frac{1\text{-}\psi}{1\text{-}2\psi(3s+5)^{-1}} \quad (2.228)$$

CD10a further derived the new relation:

$$\frac{1}{2}(3s+5)\frac{1}{k_m}\frac{\partial k_m}{\partial t} = \frac{1}{\varepsilon}\frac{\partial \varepsilon}{\partial t} - \frac{3(3s+5\text{-}2\psi)}{3s+7\text{-}4\psi}\frac{P}{K} \quad (2.229)$$

Using (2.229) into (2.228), yields the new equation for ε:

$$\frac{1}{\varepsilon}\frac{\partial \varepsilon}{\partial t} = \frac{\varepsilon}{K}(-c_2 D_K + c_1 P\text{-}c_2\varepsilon) \quad (2.230)$$

where:

$$c_1 = \frac{3}{2} - \frac{3(1\text{-}\psi)}{3(s+1)+4(1\text{-}\psi)}, \quad c_2 = \frac{3}{2} + \frac{1\text{-}\psi}{s+1} \quad (2.231)$$

Thus, for any ψ, we have the relation:

$$\frac{1}{c_1} + \frac{1}{c_2} = \frac{4}{3} \quad (2.232)$$

showing that the harmonic mean of $c_{1,2}$ is 3/2. Finally, we need a relation linking ψ to the key variables K,ε which is done by combining relations (2.227) with the result:

$$\frac{K}{K_c} = \psi^p\left(1 + \frac{2}{3}\frac{1\text{-}\psi}{1+s}\right), \quad K_c = \frac{3}{2} K_o \varepsilon^{2/3} k_c^{-2/3}, \quad p = \frac{2}{3(1+s)} \quad (2.233)$$

We have the following cases:

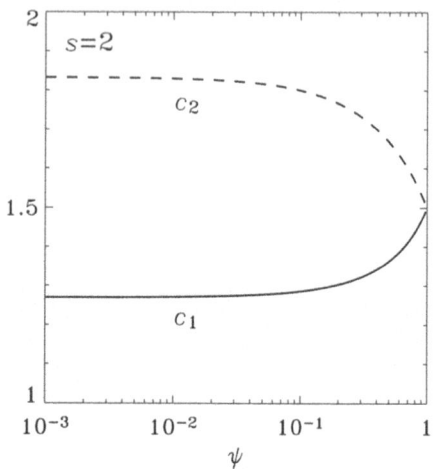

Fig. 2.44 The c_1 (solid line) and c_2 (dashed line) obtained from the new split-spectrum model vs $\psi = \left(\frac{k_c}{k_m}\right)^{s+1}$ for $s = 2$. For $\psi \ll 1$, c_1 and c_2 are constants and $c_1 < c_2$; as $\psi \to 1$, both c_1 and c_2 approach 1.5, their harmonic mean. (Reprinted with permission from Canuto et al. (2010b) © NASA all rights reserved)

(a) Homogeneous case $\psi \ll 1$:

$$c_1 \to 1.27, \quad 1.34 \quad for \quad s = 2, 4$$
$$c_2 \to 1.83, \quad 1.70 \quad for \quad s = 2, 4 \qquad (2.234)$$

which are close to the so-called "compromised values" of $c_{1,2} = 1.44, 1.92$ (Pope 2000). On the other hand, in the inhomogeneous case when eddy size approaches the maximum value, we have:

$$\psi \to 1, \quad c_{1,2} \to \frac{3}{2} \qquad (2.235)$$

which interprets the harmonic mean 3/2 as limit $\psi \to 1$. In Fig. 2.44, for $\psi \ll 1$, $c_{1,2}$ are constant and close to the empirical values; as $\psi \to 1$, $c_{1,2}$ exhibit opposite behavior reaching 3/2 at $\psi = 1$ (Fig. 2.45).

Conclusions. We have employed a two-point closure spectrum to derive a new ε equation, Eqs. (2.230) and (2.231). With respect to the previous attempts by Schiestel (1987) and Rubinstein and Zhou (2001) who used a similar methodology, we substituted their empirical closure for the time derivative of k_m suggested in Schiestel (1987) with a new relation (2.229) based on a physical model of the transfer function T(k, t) in the infrared region. In the homogeneous case, theεequation contains only production and dissipation terms with constant coefficients and zero diffusion D_ε. The present model predicts such a form with constants $c_{1,2}$ close to the "compromise" values (Pope 2000).

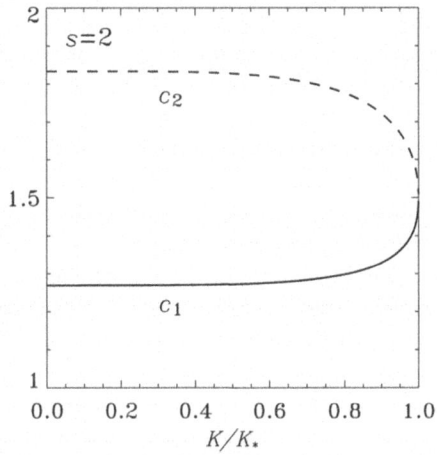

Fig. 2.45 vs K/K^*. For $K/K^* < <1$, c_1, c_2 are constants and $c_1 < c_2$; as $K/K^* \to 1$, c_1, c_2 approach 1.5. (Reprinted with permission from Canuto et al. (2010b) © NASA all rights reserved)

2.16 Planetary Boundary Layer, PBL

Since the early 1970s, higher-order turbulence closure models in geophysics were pioneered and developed by Mellor and Yamada and others (e.g., Mellor and Yamada 1974, 1982, see references in Cheng et al. 2020, C20), and were broadly successful. These models were derived from the basic dynamic equations that need to be closed and there is room for improvement of the turbulence closures. The Galperin et al. (1988, G8) formulation that significantly simplified the original MY82 model while preserving its important physics, was welcomed by the geophysics community and became widely used. Both MY82 and G8 have limitations, one of which is the presence of a finite critical Richardson number beyond which turbulence ceases to exist. The predicted value of $Ri(cr) = 0.19$, used by both formulations, is also far too small since various data show turbulence does exist for $Ri \gg 1$ (see a detailed discussion in C20). Cheng et al. (2002), Canuto et al. (2008) and others (e.g., Zilitinkevich et al. 2013) addressed deficiencies in the MY model and its variants while the Canuto et al. (2008) model enhanced $Ri(cr)$ to infinity. The changes resulted in a better agreement with available observational, experimental, and high-resolution numerical data. A drawback was that these models were more complicated than MY82 and not as efficient and convenient as G8. In the Canuto et al. (2008) model, while $Ri(cr)$ is infinity and no longer appears in the model, some of the originally constant coefficients become flow dependent, leading to an increase in the complexity of the model. Duran et al. (2014) demonstrated that an alternative relation between the pressure–temperature time scale and the dissipation time scale derived by Canuto et al. (2008), can be used to make the model coefficients flow independent without a $Ri(cr)$, a feature that deserves further exploration. Despite these advances, the cause of a finite $Ri(cr)$ was not yet evident. Since turbulence exists at high Ri in geophysical flows, there is a need for clearer understanding of this behavior and for an efficient model that reproduces it. C20 turbulence closure model leads to new horizontal and vertical heat flux eqs. A careful comparison between the old and new heat flux equations reveals the cause of a finite (and unphysical) critical Richardson number Ri(cr), while the new equations are without a finite Ri(cr), consistently with a variety of meteorological observation, laboratory experimental, direct numerical simulation (DNS) and large-eddy simulation (LES) data. Furthermore, the new model's stability functions are structurally simpler. The new model consists of a hierarchy of levels based on MY82 and G8's terminology: nonequilibrium (levels 3 and 2.5), quasi equilibrium (levels 2.75 and 2.25) and equilibrium (level 2) with level 2.75 being our addition to MY82 and G8 hierarchy (see Appendix B of C20). All levels higher than 2 can be combined with a nonlocal treatment of the turbulent kinetic energy and/or the turbulent potential energy to enhance transport processes. Since the levels 3 and 2.75 are not the main topic, they were present in Appendix B of C20. In sec. 2 of C20 we derived the new heat flux equations using the new turbulence closure, leading to a nonequilibrium model (level 2.5) that improves upon the MY82 level 2.5 model while in sec. 3 we modified the G8 procedure and derived a new quasi-equilibrium model (level 2.25). The realizability conditions of the new models are derived in sec. 4, and in sec. 5 we derived the new equilibrium model (level 2) that improves upon the MY82 and G8 level 2 models. In sec. 6 we compared the model results with various data. In sec. 7, we discussed the length scale parameterization appropriate to the new models. In sec. 8 we showed improved predictions of PBL height and other variables in the single-column model (SCM) of the GISS general circulation model (GCM), with additional discussion and overall conclusions provided in sec. 9. Appendix A provided the derivation of the new heat flux equations, Appendix B presented the new model at levels 3 and 2.75, and Appendix C summarized the MY82 and G8 models for easy reference and comparison (Figs. 2.46, 2.47, 2.48 and 2.49).

Conclusions. The results in Figs. 2.46, 2.47, 2.48 and 2.49 show that the Cheng et al. (2020) is a demonstrable improvement over MY82 and G8.

Appendices

Appendix A: Derivation of the Structure Function $S_b(Ri)$

In order to find λ defined in (2.56) in terms of resolved fields, we use the fact that the dissipations of EKE and of that of half of the buoyancy variance ε_b are defined by the integrals over the whole spectra of the works of the external forces:

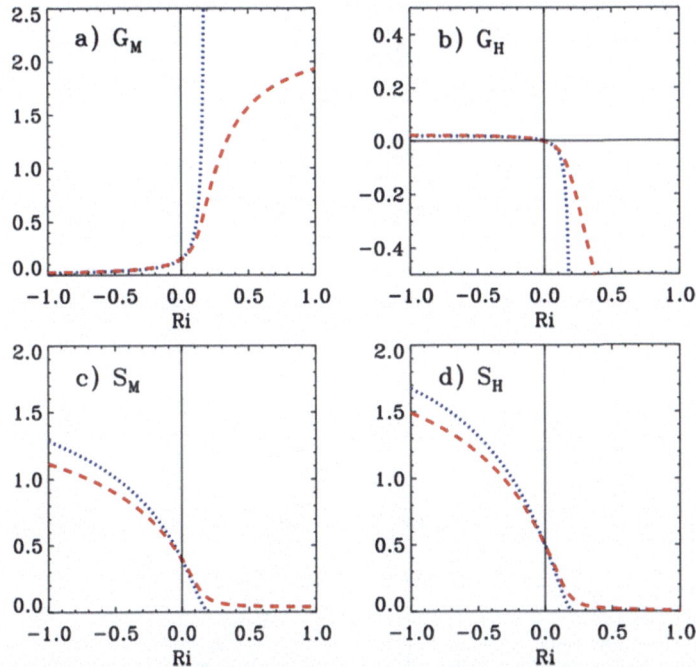

Fig. 2.46 Turbulent Prandtl number σ_t and flux Richardson number R_f versus Richardson number Ri in the equilibrium version of the new model (red dashed line), the MY82 or G8 model (blue dotted line, overlapping the red dashed line for Ri < 0.19), meteorological observations (Kondo et al. 1978, slanting black triangles; Bertin et al. 1997, snowflakes), laboratory experiments (Strang and Fernando 2001, black circles; Rehmann and Koseff 2004, slanting crosses; Ohya 2001, diamonds), LES (Zilitinkevich et al. 2007, 2008, triangles), and DNS (Stretch et al. 2001, five-pointed stars). (Reprinted with permission from Cheng et al. (2020) © NASA all rights reserved)

$$\varepsilon = \int dk A^{ext}(k) = A^{ext}_{sh} + A^{ext}_{b}, \quad A^{ext}_{sh} = \int dk\, A^{ext}_{sh}(k), \quad A^{ext}_{b} = \int dk\, A^{ext}_{b}(k) \tag{2.236}$$

Using (2.30) and the definition of the buoyancy flux in a homogeneous flux, together with the incompressibility condition, we have:

$$\overline{b'^{*}(\mathbf{k'})u'_i(\mathbf{k})} = J_i(\mathbf{k})\delta(\mathbf{k}-\mathbf{k'}), \quad k_i J_i(\mathbf{k}) = 0 \tag{2.237}$$

from the last in (2.236) we deduce:

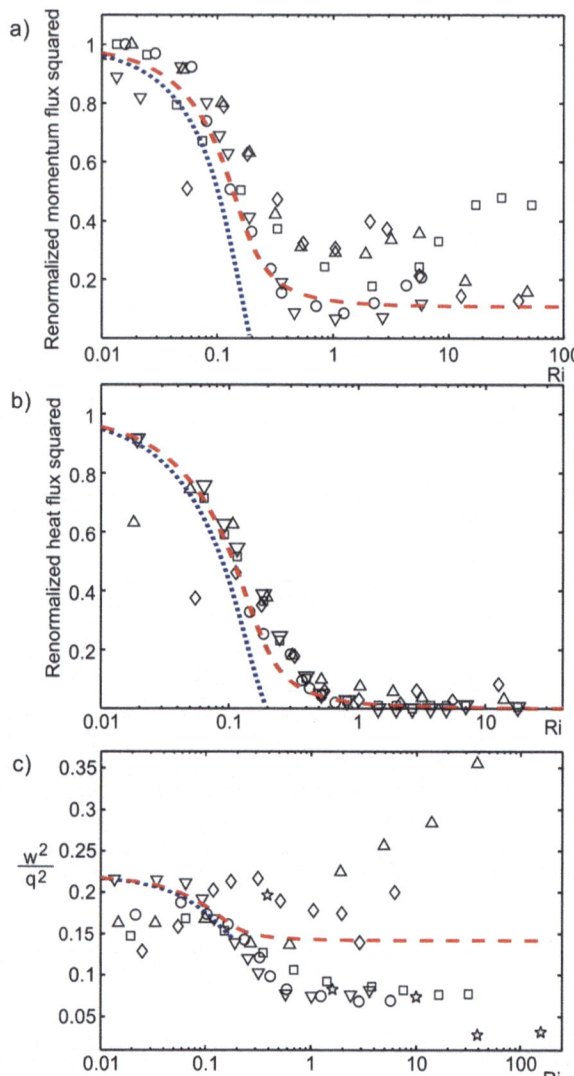

Fig. 2.47 (a) As in the previous figure but for the squared dimensionless turbulent momentum flux (normalized to its value at Ri = 0) defined in Eq.(11a) of C20), compared with lab experiments (Ohya 2001, diamonds), LES (Zilitinkevich et al. 2007, 2008, triangles), and meteorological observations (Mahrt and Vickers 2005, squares; Uttal et al. 2002, circles; Poulos et al. 2002, Banta et al. 2002, overturned triangles). (b) As in a) but for the squared dimensionless heat fluxes; (c) a in a) but for the dimensionless w-variance with the DNS data of Stretch et al. (2001) shown by five-pointed stars. (Reprinted with permission from Cheng et al. (2020). © NASA all rights reserved)

$$A_b^{ext} = \text{Re} \int dk dk' \overline{f_i^b(\mathbf{k'}) u_i'^*(\mathbf{k})} = -\text{Re} \int dk dk' g^{-1} g_i^\perp(\mathbf{k'}) \overline{b'(\mathbf{k'}) u_i'^*(\mathbf{k})} =$$

$$-\int d\mathbf{k} \left(\delta_{ij} - k^{-2} k_i k_j \right) g^{-1} g_j J_i(\mathbf{k}) = -g^{-1} g_i \int d\mathbf{k} J_i(\mathbf{k}) = \int d\mathbf{k} j(\mathbf{k}) = j$$

(2.238)

Using the shear part of (2.236) and (2.238), the external work of the mean shear is:

Appendices

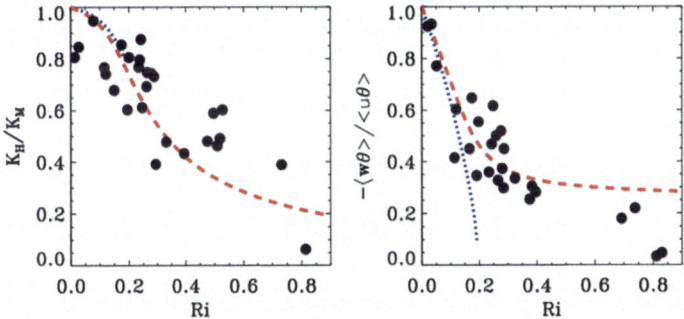

Fig. 2.48 The inverse turbulent Prandtl number (normalized by its value for neutral stratification) and the ratio of the vertical heat flux and the horizontal heat flux (normalized by its value for neutral stratification) as a function of Ri from the new model at level 2 (red dashed line), the MY82 or G8 models (blue dotted line), and the Webster (1964) experimental data (filled circles). (Reprinted with permission from Canuto and Cheng et al. (2020) © NASA all rights reserved)

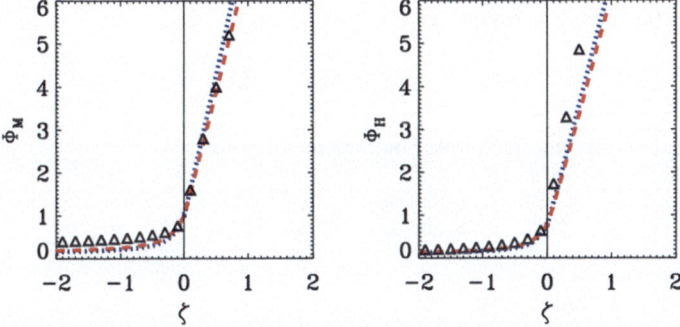

Fig. 2.49 The non-dimensional shear Φ_M and the potential temperature gradient Φ_H as a function of $\zeta = z/L$ from the MY82 or G8 model (blue dotted line), the new model (red dashed line), Businger et al. (1971)'s formula as modified by Högström (1988) (triangles). (Reprinted with permission from Cheng et al. (2020) © NASA all rights reserved)

$$A_{sh}^{ext} = \mathrm{Re} \int dk dk' \overline{f_i^{sh}(k) u_i'^*(k')} = \mathrm{Re} \int dk dk' \overline{u_i'^*(k') \left[-S_{ij} u_j'(k) + k_j S_{jm} P_{i\ell} \partial u_\ell'(k)/\partial k_m \right]}$$
(2.239)

Using (2.238) and the incompressibility condition $k_i u_i'(k) = 0$, we obtain:

$$A_{sh}^{ext} = \mathrm{Re} \int dk dk' \overline{u_i'^*(k') \left[-S_{ij} u_j'(k) + k_j S_{jm} \partial u_i'(k)/\partial k_m) \right]}$$
(2.240)

In the case of homogeneous turbulence, we have:

$$R_{ij}(\mathbf{k}) = \operatorname{Re}\overline{u_i{}'(\mathbf{k}')u'_j{}^*(\mathbf{k})} = R_{ij}(\mathbf{k})\delta(\mathbf{k}-\mathbf{k}') \quad (2.241)$$

Separating (2.240) into the two terms and integrating, use of (2.241), gives:

$$A_{sh}^{ext} = A_{sh}^{ext1} + A_{sh}^{ext2}, \quad A_{sh}^{ext1} = -S_{ij}R_{ij}, \quad (2.242)$$

$$A_{sh}^{ext2} = \frac{1}{2}\int dk k_j S_{jm} \partial R_{ii}(\mathbf{k})/\partial k_m \quad (2.243)$$

In the case of isotropic turbulent spectra which is satisfied by the UV part of the spectra, Eq. (2.243) yields:

$$A_{sh}^{ext2} = S_{ii}\int_0^\infty k dk\, \partial E(k)/\partial k = 0 \quad (2.244)$$

since $S_{ii} = 0$. Thus, we obtain:

$$\varepsilon \simeq A^{ext} = A_{sh}^{ext} + A_{b}^{ext} = -S_{ij}R_{ij} + j \quad (2.245)$$

In the ML, the only not negligible components of S_{ij} is:

$$S_{i3} = \frac{\partial \overline{u}_i}{\partial z}, \quad i = 1, 2 \quad (2.246)$$

in which case, we have:

$$P_k = \varepsilon = j + \frac{1}{45}K\tau|\partial_z \overline{u}|^2, \quad P_b = \varepsilon_b = -N^2 j \quad (2.247)$$

in which case the quadratic equation follows:

$$\lambda^2 + aN^2\lambda - (2\sigma_t)^{-1}N^4 = 0, \quad a = 1 - \frac{1}{2\sigma_t} + \frac{1}{12Ri}\sigma_t^{-1}(1+\sigma_t^{-1}) \quad (2.248)$$

which has only one positive root.

Appendix B: SM Momentum, Buoyancy Fluxes and Shear Production

momentum fluxes:

$$\frac{\overline{w'\mathbf{u}'}}{2|f|A(Ro)} = \frac{A(Ro)}{Ro}\boldsymbol{\kappa}(z) + \frac{f}{|f|}\boldsymbol{\kappa}(z)\times\mathbf{e}_z \qquad (2.249)$$

where the diffusivity $\boldsymbol{\kappa}(z)$ is defined in terms of the 2D mean velocity $\overline{\mathbf{u}}$:

$$\boldsymbol{\kappa}(z) = \int_0^z dz'[\overline{\mathbf{u}}(z') - <\overline{\mathbf{u}}(z')>] \quad , \quad <\overline{\mathbf{u}}(z)> \equiv h^{-1}\int_{-h}^0 \overline{\mathbf{u}}(z)dz \qquad (2.250)$$

$$\overline{\mathbf{u}} = \mathbf{u}_g + \mathbf{u}_{ag}, \partial_z \mathbf{u}_g = f^{-1}\mathbf{e}_z \times \nabla_H \overline{b}, \overline{\mathbf{u}}_{ag} = -f^{-1}\rho^{-1}\mathbf{e}_z \times \partial_z \boldsymbol{\tau} \qquad (2.251)$$

where $\boldsymbol{\tau}$ is the wind stress discussed below. The momentum flux vanishes at the surface $z = 0$ and bottom $z = -h$ of the vertical extent of the SM; A(Ro) and Ro are defined as follows:

$$A(Ro) = \frac{Ro^2}{1+Ro^2}, \qquad \frac{1}{2}Ro^2 = \frac{K_s}{f^2 r_s^2} \qquad (2.252)$$

where Ro is the SM Rossby number, K_s is the SM kinetic energy and r_s = (0.1–10) km are the estimated SM horizontal sizes. If one employs (2.251), it is a matter of algebra to derive the following momentum fluxes:

$$geostrophy: \overline{\mathbf{u}'w'} = -A(Ro)|\sigma(z)|h^2\left(\nabla_H \overline{b} + \frac{A(Ro)}{Ro}\frac{f}{|f|}\mathbf{e}_z \times \nabla_H \overline{b}\right)$$

$$a-geostrophy: \overline{\mathbf{u}'w'} = 2A(Ro)\left(c_1\boldsymbol{\tau}_w + c_2\frac{f}{|f|}\mathbf{e}_z \times \boldsymbol{\tau}_w\right)\rho^{-1} \qquad (2.253)$$

where $\sigma(z) = z/h(1 + z/h)$, $c_1 = 1 - \alpha + z/h + Ro^{-1}A(Ro)\beta$, $c_2 = -\beta + Ro^{-1}A(Ro)(1 + z/h - \alpha)$ and the dimensionless functions p(z) are given by:

$$\frac{Ro}{A(Ro)^2}p_0(z) = -\sigma(z), \qquad \frac{Ep_2(z)}{4A(Ro)} = \alpha^2 + \beta^2 - \left(\alpha + \frac{A(Ro)}{Ro}\beta\right)(1+z/h),$$

$$\frac{p_1(z)}{2A(Ro)} = -\left(1 + z/h - \alpha + \frac{A(Ro)}{Ro}\beta\right) - E^{-1}\sigma(z)\left(\frac{A(Ro)}{Ro}\alpha - \beta\right)$$

$$(2.254)$$

In Fig. 2.31 we show the vertical profiles of (2.175) and of the vertical buoyancy flux:

$$F_v(z) = \frac{A(Ro)}{2Ro}|\sigma(z)|h^2|f|^{-1}\ |\nabla_H \overline{b}|^2 + A(Ro)\lambda(z)EBF \qquad (2.255)$$

where $\lambda(z) = 1 + z/h - \alpha - Ro^{-1}\beta$. Using the depth averages (2.259), one has:

$$\frac{<P_s(z)>}{|f|h^2N^2} = \frac{A(Ro)^2}{6Ro}\left(Ri_g^{-1} + 12\,\omega_0\right), \qquad \omega_0 = E^{1/2}e_* + E^{-1/2}\left(1 - E^{1/2}\right)\tau_* \qquad (2.256)$$

while the depth average of the buoyancy flux (2.255) is given by:

$$\frac{<F_b(z)>}{|f|h^2N^2} = \frac{A(Ro)}{12Ro}Ri_g^{-1} + \frac{1}{2}A(Ro)\left[1-E^{1/2}(1-Ro^{-1})\right]e_* \qquad (2.257)$$

where:

$$Ri_g = \frac{f^2N^2}{|\nabla_H\overline{b}|^2}, \qquad e_* \equiv \frac{EBF}{|f|h^2N^2}, \qquad \tau_* \equiv \left(\frac{\tau_w/\rho}{h^2Nf}\right)^2 \qquad (2.258)$$

Ekman form of the wind shear. The Ekman form of $\tau(z)$ is (Chereskin 1995) $\tau(z) = \alpha(\zeta)\tau_w + f|f|^{-1}\beta(\zeta)e_z \times \tau_w$, where $\alpha(z)$, $\beta(z) \equiv e^\zeta(\cos\zeta, \sin\zeta)$, $\zeta = z/\delta_E$, $\delta_E = (2\nu_t|f|^{-1})^{1/2}$, ν_t is the turbulent viscosity, δ_E is the Ekman depth and $E_k = (\delta_E/h)^2$ is the Ekman number. The following depth averages are needed:

$$2<\alpha,\beta> = \left(E^{1/2}, -E^{1/2}\right), \quad 8<\alpha^2,$$
$$<\beta^2> = (3,1)E^{1/2}, \quad <\alpha z/h> = 0, \quad 2<\beta z/h> = E,$$
$$2<(\alpha,\beta)z^2/h^2> = -E^{3/2}$$
$$6<p_0> = a(Ro), \quad <p_1> = 2E^{1/2}a(Ro), \quad <p_2> = 2E^{-1/2}\left(1 - E^{1/2}\right)a(Ro),$$
$$a(Ro) = Ro^{-1}A(Ro)^2$$

$$(2.259)$$

References

Abarbanel, H.D., et al.: Richardson number criterion for the nonlinear stability of 3D stratified flow. Phys. Rev. Lett. **52**, 2352–2355 (1984)

Allen, J.S., Newbeger, P.A.: On symmetric instabilities in the ocean BBL. J. Phys. Oceanogr. **28**, 1131–1151 (1998)

Alves, J.O.S.: Open-Ocean Deep Convection: Understanding and Parameterization., Ph.D. thesis. University of Reading, Reading (1995)

References

Arneborg, L., Fiekas, V., Umlauf, L., Burchard, H.: Gravity current dynamics and entrainment. A process study based on observations in the Arkona Basin. J. Phys. Oceanogr. **37**, 2094–2117 (2007)

Banta, R.M., et al.: Nocturnal low-level jet characteristics over Kansas during CASES- 99. Bound-Layer Meteor. **105**, 221–252 (2002)

Bertin, F., Barat, J., Wilson, R.: Energy dissipation rates, eddy diffusivity, and the Prandtl number: an in situ experimental approach and its consequences on radar estimate of turbulent parameters. Radio Sci. **32**, 791–804 (1997)

Bradshaw, P. (ed.): Turbulence, Topics in Applied Physics Volume 12. Springer-Verlag, Berlin Heidelberg GmbH (1976)., 336pp

Brandt, A., Fernando, H.J.S.: Double Diffusive Convection, American Geophysical Union Monograph, p. 94. American Geophysical Union (1995)

Brummell, N.H., Clune, T.L., Toomre, J.: Penetration and overshooting in turbulent compressible convection. Ap. J. **570**, 825 (2002)

Brun, A.S., Toomre, J.: Turbulent convection under the influence of rotation: sustaining a strong differential rotation. Ap. J. **570**, 865 (2002)

Businger, J.A., et al.: Flux-profile relationships in the atmospheric surface layer. J. Atmos. Sci. **28**, 181–189 (1971)

Canuto, V.M.: Turbulent convection with overshooting: Reynolds Stress Approach. Ap. J. **392**, 218–232 (1992)

Canuto, V.M.: Compressible turbulence. Ap. J. **482**, 827–851 (1997a)

Canuto, V.M.: Overshooting in stars: five old fallacies and a new model. Ap. J. **489**, L71–L74 (1997b)

Canuto, V.M.: Sub-grid scale modeling. Correct models and others. Ap. J. **478**, 322–325 (1997c)

Canuto, V.M.: Turbulence in stars: unified treatment of diffusion convection, semi- convection, salt fingers and differential rotation. Ap. J. **524**, 311–340 (1999)

Canuto, V.M.: Stellar mixing. I–V. A&A. **528**, A76 (2011)

Canuto, V.M.: PV dynamics. The role of small scale turbulence, sub-mesoscales and mesoscales. J. Geophys. Res. **120**, 6971–6986 (2015)

Canuto, V.M., Cheng, Y.: Do sub-mesoscales inhibit destruction of PV? J. Geophys. Res. Oceans. **126**(11), e2020JC016991 (2021). https://doi.org/10.1029/2020JC016991

Canuto, V.M., Dubovikov, M.S.: Stellar turbulent convection. I. Theory, Ap. J. **493**, 834 (1998)

Canuto, V.M., Dubovikov, M.S.: Reply to the comment by McDougall et al. Geophys. Astrophys. Fluid Dyn. **102**, 257–264 (2008)

Canuto, V.M., Hartke, G.J.: Convective turbulence with rotation and magnetic fields. A&A. **168**, 89–104 (1986)

Canuto, V.M., Mazzitelli, I.: Stellar turbulent convection: a new model and applications. Ap. J. **370**, 295–311 (1991)

Canuto, V.M., Minotti, F.: Mixing and transport in stars—I. Formalism: momentum, heat and mean molecular weight. Mon. Not. R. Astron. Soc. **328**, 829 (2001)

Canuto, V.M., Minotti, F., Shilling, O.: Differential rotation and turbulent convection: a new Reynolds stress model and comparison with solar data. Ap. J. **425**, 303 (1994a)

Canuto, V.M., Minotti, F., Ronchi, C., Ypma, R.M., Zeman, O.: Second-order closure PBL model with new third-order moments: comparison with LES data. J. Atmos. Sci. **51**, 1605–1618 (1994b)

Canuto, V.M., Howard, A., Cheng, Y., Dubovikov, M.S.: Ocean turbulence. Part 1: one –point closure model. Momentum and heat vertical diffusivities. J. Phys Oceanogr. **31**, 1413–1426 (2001)

Canuto, V.M., Howard, A., Cheng, Y., Dubovikov, M.S.: Ocean turbulence. Part II: vertical diffusivities of momentum, hat, salt, mas and passive tracers. J. Phys. Oceanogr. **32**, 240–264 (2002)

Canuto, V.M., et al.: Modeling Ocean deep convection. Ocean Model. **7**, 75–95 (2004)

Canuto, V.M., Dubovikov, M.S., Cheng, Y.: Entrainment: local and non-local turbulence models Is with double diffusion. Geophys. Res. Lett. **32**, L2264 (2005)

Canuto, V.M., Cheng, Y., Howard, A.M.: Non-local ocean mixing model and a new plume model for deep convection. Ocean Model. **16**, 28–46 (2007)

Canuto, V.M., et al.: Stably stratified flows: a model with no Ri(cr). J. Atmos. Sci. **65**, 2437–2447 (2008)

Canuto, V.M., et al.: Ocean turbulence III: new GISS vertical mixing scheme. Ocean Model. **34**, 70–91 (2010a)

Canuto, V.M., Cheng, Y., Howard, A.M.: An attempt to derive the dissipation equation from a two-point closure. J. Atmos. Sci. **67**, 1678–1685 (2010b)

Canuto, V.M., Cheng, Y., Howard, A.M.: Vertical diffusivities of active and passive tracers. Ocean Model. **36**, 198–207 (2011)

Cenedese, C., Adduce, C.: A new parameterization for entrainment in overflows. J. Phys. Oceanogr. **40**, 1835–1850 (2010). https://doi.org/10.1175/2010JPO4374.1

Charbonnel, C., Talon, S.: Influence of gravity waves on the internal rotation and li abundance of solar-type stars. Science. **309**, 2189 (2005)

Charbonnel, C., Talon, S.: Mixing a stellar cocktail. Science. **318**, 922 (2007)

Cheng, Y., et al.: An improved model for the turbulent PBL. J. Amos. Sci. **59**, 1550–1565 (2002)

Cheng, Y., et al.: A second-order closure turbulence model: new heat flux equations and no critical Richardson number. J. Atmos. Sci. **71**, 2743–2759 (2020)

Chereskin, T.K.: Direct evidence for an Ekman balance in the California current. J. Geophys. Res. **100**(C9), 18261–18269 (1995)

Chu, P.C., Gascard, J.C.: Deep Convection and Deep Water Formation in the Oceans, p. 397. Elsevier, New York. p (1991)

Davydov, B.I.: On statistical dynamics of an incompressible turbulent fluid. Sov. Phys. Dokl. **6**, 10–12 (1961)

Decloedt, T., Luther, D.S.: Spatially heterogeneous diapycnal mixing in the abyssal ocean: a comparison of two parameterizations to observations. J. Geophys. Res. **117**, C11025 (2012). https://doi.org/10.1029/2012JC008304

Denbo, D.W., Skyllingstad, E.D.: An ocean large-eddy simulation model with application to deep convection in the Greenland Sea. J. Gephys. Res. **101**, 1095–1110 (1996)

Duran, B., et al.: A compact model for the stability dependence of TKE production, destruction conversion term valid for the whole range of Richardson numbers. J. Atmos. Sci. **71**, 3004–3026 (2014)

Egbert, G.D., Ray, R.D.: Significant dissipation of tidal energy in the deep ocean inferred from satellite data. Nature. **405**, 775–778 (2000)

Ellison, T.H., Turner, J.S.: Turbulent entrainment in stratified flows. J. Fluid Mech. **6**, 423–448 (1959)

Ezer, T.: Entrainment, diapycnal mixing and three-dimensional bottom gravity current simulation using the Mellor-Yamada turbulence scheme. Ocean Model. **9**, 151–168 (2005)

Fernando, H.J.S.: Turbulent mixing in stratified flows. Ann. Rev. Fluid Mech. **23**, 455–493 (1992)

Galperin, B., et al.: A quasi-equilibrium turbulent energy model for geophysical flows. J. Atmos. Sci. **45**, 55 (1988)

Girton, J.B., Sanford, T.B.: Descent and modification of the overflow plume in the Denmark Strait. J. Phys. Oceanogr. **33**, 1351–1364 (2003)

Golystin, G.S.: Dokl. Akad. Nauk USSR. **251**, 1356 (1980)

Hamba, F.: An analysis of non-local scalar transport in the convective boundary layer using the Green function. J. Atmos. Sci. **52**, 1084–1095 (1995)

Hartmann, J., et al.: Arctic radiation and turbulence interaction study (ARTIST). In: Reports in Polar Research, Alfred Wegener Institute for Polar and Marine Research, Bremerhaven, vol. 305, p. 81 (1999)

Haynes, P.H., McIntyre, M.E.: On the evolution of vorticity and potential vorticity in the presence of diabatic heating and frictional or other force. J. Atmosph. Sci. **44**(5), 828–841 (1987)

References

Högström, U.: Non-dimensional wind and temperature profiles in the atmospheric surface layer: a re-evaluation. Bound.-Layer Meteor. **42**, 55–78 (1988)

Jackson, L., Hallberg, R., Legg, S.: A parameterization of shear-driven turbulence for ocean climate models. J. Phys. Oceanogr. **38**, 1033–1063 (2008)

Jayne, S.R.: The impact of abyssal mixing parameterizations in an ocean general circulation model. J. Phys. Oceanogr. **39**, 1756–1775 (2009)

Jayne, S.R., St. Laurent, L.C.: Parameterizing tidal dissipation over rough topography. Geophys. Res. Lett. **28**, 811–814 (2001)

Kantha, L.H., Tierney, C.G.: Global baroclinic tides. Prog. Oceanogr. **40**, 163–178 (1997)

Kelley, D.E.: Effective diffusivities within oceanic thermohaline staircases. J. Geophys. Res. **89**, 10484–10488 (1984)

Kelley, D.E.: Fluxes through diffusive staircases; a new formulation. J. Geophys. Res. **95**, 3365–3371 (1990)

Kelley, D.E.: Six questions about double diffusive convection. In: Muller, P., Garrett, C. (eds.) From Stirring to Mixing in Stratified Ocean 12th Aha Huliko, Hawaiian Winter Workshop. University of Hawaii (2001)

Kelley, D.E., et al.: The diffusive regime of double-diffusion convection. Prog. Oceanogr. **56**, 461–481 (2003)

Killworth, P.D.: Deep convection in the world ocean. Rev. Geophys. Space Phys. **21**, 1–26 (1983)

Klinger, B., Marshall, J., Send, U.: Representation of convective plumes by vertical adjustment. J. Geoph. Res. **101**(C8), 18,175–18,182 (1996)

Kondo, J., Kanechika, O., Yasuda, N.: Heat and momentum transfer under strong stability in the atmospheric surface layer. J. Atmos. Sci. **35**, 1012–1021 (1978)

Kumar, P., Quataert, E.J.: Angular momentum transport by gravity waves and its effect on the rotation of the solar interior. Ap. J. **475**, L143 (1997)

Kumar, P., Talon, S., Zahn, J.P.: Angular momentum redistribution by waves in the Sun. Ap. J. **520**, 859 (1999)

Kunze, E., et al.: Global abyssal mixing inferred from lowered ADCP shear and CTD strain profiles. J. Phys. Oceanogr. **36**, 1553–1576 (2006)

Kupka, F.: Shear driven turbulence and coherent structures in solar surface simulations. In: Deng, L., Chan, K.L. (eds.) The Art of Modelling Stars in the 21st Century IAU Symp. No. 252, pp. 451–462 (2008)

Kupka, F., Muthsam, R.J.: Analysis of contributions in moment equations of Reynolds stress models of convection with numerical simulations. In: Deng, L., Chan, K.L. (eds.) The Art of Modelling Stars in the 21st Century IAU Symp. No. 252, pp. 463–465 (2008)

Lappen, C.L., Randall, D.: Toward a unified parameterization of the boundary layer and moist convection. Part I: a new type of mass flux model. J. Atmos. Sci. **58**, 2021–2036 (2001)

Large, W.G., et al.: J. Phys Ocean. **27**, 2418–2447 (1997)

Lavender, K.L., et al.: Observations of open deep ocean convection in the Labrador Sea from subsurface floats. J. Phys. Ocean. **32**, 511–526 (2002)

Ledwell, J.R., et al.: Evidence for slow mixing across the pycnocline from open ocean tracer release experiment. Nature. **364**, 701–703 (1993)

Ledwell, J.R., et al.: Mixing of a tracer released in the pycnocline of a subtropical gyre. J. Geophys. Res. **103**, 21499–21529 (1998)

Legg, S.: Internal tides generated on a corrugated continental slope. Part cross-slope barotropic forcing. J. Phys. Ocean. **34**, 156–173 (2004)

Legg, S., and co-authors: Improving oceanic overflow representation in climate models. BAMS, 657–670 (2009)

Legg, S., Klymak, J.: Internal hydraulic jumps and overturning generated by tidal flow over a tall steep ridge. J. Phy. Oceanogr. **38**, 1949–1964 (2008)

Legg, S., Hallberg, R.W., Girton, J.B.: Comparison of entrainment in overflows simulated by z-coordinate, isopycnal and non-hydrostatic models. Ocean Model. **11**, 69–97 (2006)

Leslie, D.C.: Developments in the theory of turbulence. Reports Progr. Phys. **36**(11), 1365 (1973)

Llewellyn Smith, S.G., Young, W.R.: Conversion of a barotropic tide. J. Phys. Ocean. **32**, 1554–1566 (2002)

Maeder, A., Maynet, G.: Stellar evolution with rotation. VII. Low metallicity models and the blue to red supergiant ratio in the SMC. A&A. **373**, 555 (2001)

Mahrt, L., Vickers, D.: Boundary layer adjustment over small-scale changes of surface heat flux. Bound.-Layer Meteor. **116**, 313–330 (2005)

Marotzke, J.: Influence of convective adjustment on the stability of the thermohaline circulation. J. Phys. Ocean. **21**, 903–907 (1991)

Marshall, J., Schott, F.: Open-ocean convection, observations, theory and models *rev*. Geophys. **37**, 1–64 (1999)

Mathis, S., Palacios, A., Zahn, J.P.: On shear-induced turbulence in rotating stars. A&A. **425**, 243 (2004)

Maxworthy, T.: Convection domains with open boundaries. Ann. Rev. Fluid Dyn. **29**, 327–371 (1997)

Mellor, G.I., Yamada, T.: A hierarchy of turbulent closure models for planetary boundary layers, J. Amos. *Sci.* **31**, 1791–1806 (1974)

Mellor, G.I., Yamada, T.: Development of a turbulent closure model for geophysical fluid problems. Rev. Geophys. Space Phys. **20**(851–875), MY82 (1982)

Middleton, L., et al.: Estimating dissipation rates associated with double diffusion. Geophys. Res. Lett. **48**, e2021GL092779 (2021)

Middleton, L., et al.: Double diffusion, as driver of turbulence in the stratified boundary layer beneath George VI ice shelf. Geophys. Res Lett. **49**, e2021GL096119 (2022)

Mironov, D.V., et al.: Vertical turbulence structure and second-moment budgets in convection with rotation: a large-eddy simulation study. Quart. J. Roy. Meteor. Soc. **126**, 477–516 (2000)

Morton, B.R., Taylor, G.I., Turner, J.S.: Turbulent gravitational convection from maintained and isolated sources. Proc. Roy. Soc. London. **A234**, 1–23 (1956)

Ohya, Y.: Wind-tunnel study of atmospheric stable boundary layers over a rough surface. Bound.-Layer Meteor. **98**, 57–82 (2001)

Osborn, T.: Estimates of the local rate of vertical diffusion from dissipation measurements. J. Phys. Oceanogr. **10**, 83–89 (1980)

Osborn, T., Cox, C.S.: Oceanic fine structure. Geophy. Fluid Dyn. **3**, 321–345 (1972)

Palacios, A., Talon, S., Charbonnel, C., Forestini, M.: Rotational mixing in low-mass stars. I. Effect of the -gradients in main sequence and subgiant Pop I stars. A&A. **399**, 603 (2003)

Palacios, A., Charbonnel, C., Talon, S., Seiss, L.: Rotational mixing in low-mass stars II. Self-consistent models of Pop II RGB stars. A&A. **453**, 261 (2006)

Paluszkiewicz, T., Romea, R.D.: A one-dimensional model for the parameterization of deep convection in the ocean. Dyn. Atmos. and Oceans. **26**(2), 95–130 (1997)

Paluszkiewicz, T., Garwood, R.W., Denbo, D.W.: Deep convective plumes in the ocean. Oceanography. **7**, 37–44 (1994)

Papadakis, M.P., Chassignet, E.P., Hallberg, R.W.: Numerical simulations of the Mediterranean Sea outflow: impact of the entrainment parameterizations in an isopycnal coordinate model. Ocean Model. **5**, 325–356 (2003)

Polzin, K.: Idealized solutions for the energy balance of the finescale internal wave field. J. Phys. Oceanogr. **34**, 231–246 (2004)

Polzin, K., Toole, J.M., Schmitt, R.W.: Finescale parameterization of turbulent dissipation. J. Phys. Oceanogr. **25**, 306–328 (1995)

Pope, S.B.: Turbulent Flows. Cambridge University Press, Cambridge (2000)

Poulos, G.S., et al.: CASES-99: a comprehensive investigation of the stable nocturnal boundary layer. Bull. Amer. Meteor. Soc. **83**, 555–581 (2002)

Price, J.F., Baringer, M.O.: Outflows and deep-water production marginal seas. Progr. Oceanogr. **33**, 161–200 (1994)

Price, J.F., Yang, J.: Marginal seas overflows for climate simulations. In: Chassignet, E.P., Verron, J. (eds.) Ocean Modeling and Parameterizations, pp. 55–170. Kluwer (1998)

Pulkkinen, P., Tuominen, I., Branderburgh, A.: Rotational effects on convection simulated at different latitudes. A&A. **267**, 265 (1993)

Rehmann, C.R., Koseff, J.R.: Mean potential energy change in stratified grid turbulence. Dyn. Atmos. Oceans. **37**, 271–294 (2004)

Roxburgh, I.W.: Convection and stellar structure. A&A. **65**, 281–290 (1978)

Rubinstein, R., Zhou, Y.: Schiestel derivation of the epsilon equation and two equation modeling of rotating turbulence, NASA/CR-2001-211060, ICASE Rep. 2001–24, 6pp (2001)

Saenko, O.A., Merryfield, W.J.: On the of topographically enhanced mixing on the global ocean. J. Phys. Oceanogr. **35**, 826–834 (2005)

Sander, J., Wolf-Gladrow, D., Olbers, D.: Numerical studies of open ocean deep convection. J. Geophys. Res. **100**, 20579–20600 (1995)

Schiestel, R.: Multiple time scale modeling of turbulent flows in one-point closures. J. Fluid Mech. **37**, 581–593 (1987)

Schmitt, W.: Double diffusion in oceanography. Ann. Rev. Fluid Mech. **26**, 255–285 (1994)

Simmons, H.L., et al.: Internal wave generation in a global baroclinic tide model. J. Phys. Oceanogr. **34**(6), 1515–1543 (2004)

Smith, W.H.F., Sandwell, D.T.: Global Sea floor topography from satellite altimetry and ship data soundings. Science. **277**, 1956–1962 (1997)

St. Laurent, L., Garrett, C.: The role of internal tides in mixing the deep ocean. J. Oceanogr. **32**, 2882–2899 (2002)

St. Laurent, L., Schmitt, R.W.: The contribution of salt fingers to vertical mixing in the NATRE experiment. J. Phys. Oceanogr. **29**, 1404–1424 (1999)

Stern, M.E.: The 'salt fountain' and thermohaline convection. Tellus. **12**, 172–175 (1960)

Strang, E.J., Fernando, H.J.S.: Vertical mixing and transports through a stratified shear layer. J. Phys. Oceanogr. **31**, 2026–2048 (2001)

Stretch, D.D., et al.: Transient mixing events in stably stratified turbulence. In: 14th Australasian Fluid Mechanics Conference, pp. 625–628. Adelaide University, Adelaide (2001)

Talon, S., Charbonnel, C.: Angular momentum transport by internal gravity waves, I- Pop1 main sequence stars. A&A. **405**, 1025 (2003)

Talon, S., Charbonnel, C.: Hydrodynamical stellar models including rotation, internal gravity waves and atomic diffusion, I. Formalism and tests on Pop I dwarfs. A&A. **440**, 981 (2005)

Talon, S., Zahn, J.P.: Toward a hydrodynamical model predicting the observed solar rotation profile. *A&A.* **329**, 315 (1998)

Taylor, G.I.: Dynamics of a mass of hot gas rising in the air, USAEC Rep. MDDC-919 The Labrador Sea Deep Convection Experiment, 2002. J. Phys. Oceanogr. **32** (1945)

Thomas, L.N., Taylor, J.R.: Reduction of usable wind-work on the general circulation by forced symmetric instability. Geophys. Res. Lett. **37**, L18606 (2010). https://doi.org/10.1029/2010G044680

Thomas, L.N., et al.: Symmetric instability in the Gulf Stream. Deep Sea Res. II. **91**, 96–10 (2013)

Thompson, M.J., Christensen-Dalsgaard, J., Miesh, M.S., Toomre, J.: ARA&A. **41**, 17 (2003)

Turner, J.S.: Buoyancy Effects in Fluids. Cambridge University Press, Cambridge (1973)

Turner, J.S.: Turbulent entrainment: the development of the entrainment assumption and its application to geophysical flows. J. Fluid Mech. **173**, 431–471 (1986)

Ulrich, R.K., Boyden, J.E., Webster, L., et al.: Solar rotation measurements at Mount Wilson. V. Reanalysis of 21 years of data. Sol. Phys. **117**, 291 (1988)

UNESCO: The practical salinity scale 1978 and the international equation of state of seawater 1980. UNESCO Tech. Papers Marine Sci. **36**, 25 (1981a)

UNESCO: Background papers and supporting data on the international equation of state of seawater, 1980. UNESCO Tech. Papers Marine Sci. **37**, 144 (1981b)

Uttal, T., and Coauthors: Surface heat budget of the Arctic Ocean. Bull. Amer. Meteor. Soc. **83**, 255–276 (2002)

Ventura, P., D'Antona, F., Mazzitelli, I.: Full spectrum of turbulence convective mixing: I. Theoretical main sequences and turn-off for 0.6 to 15 M$\odot$. Ap. J. **497**(2), 737–750 (1998)

Virtanen, H.: Ph.D. thesis, University of Helsinki (1989)

Walsh, D., Ruddick, B.: Double-Diffusive interleaving in the presence of turbulence. The effect of a non- constant flux ratio. J.Phys. Oceanogr. **30**, 2231–2245 (2000)

Wang, Y., et al.: Seasonal variations in Circumpolar Deep Water intrusions into the Ross Sea continental shelf. Front. Marine Sci. **10**, 1020791 (2023)

Weaver, A.J., Bitz, C.M., Fanning, A.F., Holland, M.M.: Thermohaline circulation: high- latitude phenomena and the difference between the Pacific and Atlantic. Annu. Rev. Earth Planet. Sci. **27**, 231–285 (1999). https://doi.org/10.1146/annurev.earth.27.1.231

Webster, C.A.G.: An experimental study of turbulence in a density stratified shear flow. J. Fluid Mech. **19**, 221–245 (1964)

Wenegrat, J.O., Thomas, L.N.: Centrifugal and symmetric instability during Ekman adjustment at the BBL. J. Phys. Oceanogr. **50**, 1793–1812 (2020)

Wenegrat, J.O., Callies, J., Thomas, L.N.: Sub-mesoscale baroclinic instability in the bottom boundary layer. J. Phys. Oceanogr. **48**, 2571–2592 (2018)

Wyngaard, J.C., Weil, J.C.: Parameterizing turbulent diffusion in the atmospheric boundary layer. Bound.-Layer Meteorol. **56**, 231–246 (1991)

Yankovsky, E., Legg, S.: Symmetric and baroclinic instabilities in dense shelf overflows. J. Phys. Oceanogr. **49**, 39–60 (2019)

Zahn, J.P.: Instabilities and mixing in stellar radiation zone. In: Deng, L., Chan, K.L. (eds.) The Art of Modelling Stars in the 21st Century IAU Symp. 252, p. 83. Cambridge University Press (2008)

Zahn, J.P., Talon, S., Matias, J.: Angular momentum transport by internal waves in the solar interior. A&A. **322**, 320 (1997)

Zeman, O., Lumley, J.L.: Modeling salt-fingering structure. J. Mar. Res., Yale Univ. Peabody Museum, 315–330 (1982)

Zilitinkevich, S., et al.: Energy and flux budget (EFB) turbulence closure model for stably stratified flows. Part I, Steady-state, homogeneous regimes. Boundary Layer Meteor. **125**, 167–191 (2007)

Zilitinkevich, S., et al.: Turbulence energetics in steady stratified geophysical flows: strong and weak mixing regimes. Quart. J. Roy. Meteor. Soc. **134**, 793–799 (2008)

Zilitinkevich, S., et al.: A hierarchy of energy and flux-budget (EFB) turbulence closure models for stably stratified geophysical flows. Boundary Layer Meteor. **146**, 341–373 (2013)

The manufacturer's authorised representative in the EU is Springer Nature Customer Service Centre GmbH, Europaplatz 3, 69115 Heidelberg, Germany. If you have any concerns regarding our products, please contact ProductSafety@springernature.com

Printed and bound by CPI Group (UK) Ltd, Croydon, CR0 4YY

26/03/2026

02078983-0006

Why Does Everybody Hate Me?

Living and Loving with
Rejection Sensitivity Dysphoria

Alex Partridge

sheldon^PRESS

To anyone who's ever been called 'too sensitive'.

Contents

Foreword	vi
Introduction	xi
1 20,000 tiny cuts: What is rejection sensitive dysphoria?	1
2 Approval addicts: RSD and people-pleasing	29
3 'Are you mad at me?' Reading rejection in every pause	62
4 Taking the sting out of RSD: Some practical strategies	100
5 When every email feels personal: Navigating RSD at work	145
6 Holding a heart: Supporting a person with RSD	184
7 Dark places: RSD and the worst of times	203
8 The best and only way to truly beat RSD: A five-step plan	215
Final thoughts	225
Acknowledgements	231
Index	232

Foreword

I have been waiting for this book for 45 years. For at least that length of time it has been clear to everyone that difficulty controlling emotions has been a core feature of ADHD, and that it is a major source of impairment and suffering to the vast majority of people with ADHD. Experts in the European Union (EU) finally accepted Emotional Dysregulation (ED) as a Core Feature of ADHD back in 2018. It is still awaiting official acceptance in the United States (US). This is not because the often-devastating impairment from a lack of emotional control was in question in the US. It wasn't, and never has been, in question.

The problem has always been that Emotional Dysregulation is so hard to study using the traditional methods that brought ADHD validity in the scientific community over the previous 50 years. The diagnostic description of ED as it presents in ADHD had to be based on features that were visible to the researcher and, therefore, could be counted, and subjected to the statistical analysis that was needed to get an article published.

Once the invisible obstacles to even talking about the difficulties with the control of emotional expression were removed by the EU, the topic could be discussed and occasionally studied. But even though the experience of emotions is much more intense than it is for neurotypical people, the study of emotional life and regulation in ADHD doesn't happen for a number of reasons:

- These intense and disruptive emotions are not always present. They are constantly changing and come and go according to that individual's perception of the current situation.

- Even when particular emotions are present, they often cannot be named and measured. This is a particularly difficult problem for study because the emotions themselves are usually normal in every way except their intensity. An ADHD person's emotions are triggered by events and perceptions just the way they are in neurotypicals; the mood shift matches the person's perception of the situation just the way they do in neurotypical people; and the shifts are instantaneous just the way they are for neurotypicals. It is only the overwhelming intensity of emotional over-reactions that makes them a disruptive problem.

- Often, emotions are embarrassing and overwhelming to the degree that people with ADHD either hide their feelings or consciously avoid letting people see their emotional storms due to the reality-based fear that others will see them as weak, immature or 'a head case'.

Acceptance that problems with controlling the expression of emotions had to be understood by different research methods has changed a great deal how people defined and understood ADHD. The method that has worked the best so far is to define ADHD in terms of the *lived experience* of people with the condition.

Unfortunately, 'lived experience' was disregarded as being unscientific and somehow not even real. Even though traditional methods had not worked, or produced a single non-medication-based therapy that had lasting benefits, experts refused to see the reports of people with an ADHD-style nervous system as valid and reliable.

Everything was frozen until the EU accepted that any progress was going to require more flexible ways of thinking and understanding. Now, finally, the study of the unique experience of emotions in people with ADHD has become acceptable. It has become more mainstream to talk to people with ADHD and learn from their lifetimes of living with a type of nervous system that is fundamentally different from the more common, importance-based neurotypical nervous system.

Alex Partridge is the *right person* in the *right place* at the *right historical moment.* He has ADHD himself and has managed it by becoming fully immersed in the available information from both science/medicine and from lay literature. His podcast, *ADHD Chatter,* has gone well beyond the limited medical information and provided the opportunity for Alex to interview most of the top thinkers in the domain of ADHD. This has provided him with access to a level of lived experience that is beyond almost everyone else in the field.

Most of all, it has given him real-world insights into the subject of the book you hold in your hands, *Why Does Everybody Hate Me? Living and Loving With Rejection Sensitivity Dysphoria*, providing up-to-date information that is hard to find anywhere else.

The experts on ADHD have always been the adults with ADHD who still experience significant impairment (children do not have much in the way of self-appraisal abilities and are not able to tell their families and clinicians what they are experiencing, or the effects that they see from the medications they take). When people with ADHD grow up to be articulate and self-aware adults, they are our best source of information. The trouble is that our researchers study ADHD from a detached point of view but rarely talk to people with ADHD to get their information from the source firsthand the way that Mr Partridge does. It has come to the point that the people who study ADHD are almost completely separate from the people who have ADHD, and the people who treat it.

Alex Partridge, however, has spoken in depth to thousands of people and has synthesized what he has heard into a description and understanding of the most difficult aspect of ADHD to grasp: Rejection Sensitivity Dysphoria (RSD). This is a commonly used concept among those with ADHD for one reason. When people hear about RSD, they immediately identify with and accept it, because it accurately and completely describes the experiences of their entire life. They know that the RSD concept is valid because it matches their *lived experience* exactly. It is comforting to discover that this great source of shame and embarrassment is shared by many others and that they are not alone with this part of themselves that they have kept hidden from the rest of the world.

Consequently, only a person like Alex Partridge – who has spent years listening to so many people talk about their experience of RSD – is prepared to understand and communicate to others the multiple facets of Emotional Dysregulation in general and RSD in particular. Even though most of the publications of the last 5 years have agreed that ED is probably the most impairing aspect of ADHD at all ages, there are just a handful of formal articles in print about what ED and RSD are and how they can be treated.

Indeed, much is still to be learned and books like this one are where the progress is going to be made.

Bill Dodson, MD, LF-APA

Introduction

It was 3 a.m. I was lying in bed scrolling social media.

I glanced at the clock. I needed to be awake in four hours. I triple-checked my alarm and placed my phone on the bedside table. I rested my head on the pillow and closed my eyes.

That's when the thoughts came flooding back:

> 'Everybody hates me. Everyone thought I was too much. Why did I overshare… again?'

I had returned home from an event five hours ago. My brain was replaying last night's social interactions over and over and over again.

'That person didn't need to know about my trauma. Why did I talk so much? I can never show my face to those people ever again. I need to move to a different part of the country and start a new life, far away from these people who now definitely despise every fibre of my being.

'Why do I always over-explain things? I said hello. I went off on three tangents, I told four minutes of unnecessary backstory and I ended without finishing my original thought. I wanted to keep the conversation going but why did I share all my personal secrets?

'I literally just met that person and I dumped all my demons on them like they were my therapist. Everyone's going to be talking about how annoying I was. I told myself after the last social event that I would stop oversharing, but I've done it again and this time was worse! At this point in my life I'm so socially awkward I might as well legally change my name to Hagrid. I shouldn't have said that. I should not have said that!'

This was the worst feeling you can imagine, a haunting, gut-twisting agony that would last for weeks, if not months. My life was full of moments like this, an episode of catastrophic painful overthinking, convincing myself that everyone was mad at me, accompanied by a double dose of anxiety and shame.

My boss once asked me for a 'quick chat' on Monday morning and didn't give me any context. The pain was instant and visceral. It felt like a dagger to the chest with someone twisting it. My brain went into meltdown and immediately jumped to the worst-case scenario:

> They've finally realized I'm terrible at my job.
> I'm getting fired!

I didn't sleep all weekend. I also didn't get fired.

It is a never-ending list of disproportionate responses to reasonable situations.

Some more examples

Someone replies to a text message with a 'thumbs-up' emoji, or simply with 'Okay', so my heart sinks and I'm convinced that person now hates me.

I'm having a conversation with someone and I sense a minuscule change in their tone of voice or a shift in their facial expression, so I take it as a signal they they're fed up with me and now hate me.

I'm showing someone my favourite movie and they stand up to use the bathroom without asking me to pause the movie, so I take that to mean they hate the film and therefore hate me, too.

It creates a physical pain within me that's near impossible for anyone else to understand.

Friendships are hard for me to maintain because as soon as I perceive the smallest of slights, I cut cords with the person in an effort to protect myself from the pain. I know I should communicate with them, and accept that maybe I said something that justified a slight push-back, but it's simply easier for me to ghost that person and assume they're terrible for me because they made me feel truly awful – and the thought of initiating a mature dialogue is too risky because it will expose me to more discomfort. I can't bear to be called 'too sensitive' yet again.

I isolate myself from people because it's safer. I've lost count of the number of bridges I've irrationally burned.

I've always taken stuff very personally, and consequently been called 'too sensitive'. It's a horrible feeling, living in a constant state of hyper-alertness, scanning the room for threats and picking up on the slightest sense of rejection.

Introducing our new 'friend', RSD

I received my attention deficit hyperactivity disorder (ADHD) diagnosis three years ago, and in my journey towards understanding the complexities of a brain that's wired a bit differently, I stumbled across the term 'RSD'.

RSD stands for rejection sensitive dysphoria, and it explained my entire life.

Now it all makes sense, I told myself. I'm not crazy, I'm not imagining this feeling, it's real and it's got a name. And, more importantly, there's a reason why it happens.

I dug more into the research, led by William W. Dodson, who first coined the term 'rejection sensitive dysphoria' in 2017. Dysphoria comes from a Greek word meaning 'unbearable', a fitting one to describe the feeling of RSD when it's triggered. This was backed up when I asked the listeners of my podcast, *ADHD Chatter*, to describe how RSD feels and they said,

It's a symptom of the emotional dysregulation often seen in individuals with ADHD and it's massively under-researched and misunderstood. Individuals who experience RSD describe it as an emotional tsunami, a physically painful reaction to real or perceived rejection or criticism.

A pain so extreme that we go to extreme lengths to avoid it. We people-please, we say 'yes' to things we don't really want to do, and we put other people's wants and needs before our own, all in an effort to avoid offending anyone and therefore exposing ourselves to criticism.

Why is rejection so painful for individuals with ADHD?

Research shows that children with ADHD receive 20,000 more negative comments during their childhood than the average child – 20,000 more criticisms and micro-corrections:

> 'Stop being so sensitive! You're too much! You're not enough! Stop fidgeting! You're so lazy!'

Those 20,000 comments compound in that individual's brain to create an adult who is very fearful of rejection. One could easily argue that the effect of so many negative comments creates a cumulative trauma that, when triggered by a minor criticism as an adult, creates a physically painful response. It confirms our internal belief that we're useless.

This response can be very confusing for the person to understand. Why do I feel so enraged by that tiny comment? Why have I instantly become overwhelmed with internalized sadness or rage? Why have I verbally lashed out at my partner, or that person who cut me up in traffic?

The emotional response is seldom proportionate to the comment or event that triggers the reaction.

But what if I told you that it wasn't that little comment that triggered you? You are not crying or raging because someone cut you up in traffic, or because your boss made a tiny correction to your work. You're feeling such overwhelming emotion because that tiny comment reminded your nervous system of those 20,000 negative comments you received when you were a child.

The unpredictable nature of RSD, not knowing when it might strike and the disproportionality of your response can cause immeasurable shame. When triggered, your mood can alter significantly and instantly, resulting in people comparing your behaviour to that of such extreme examples as Dr Jekyll and Mr Hyde.

The impulsive nature of an individual with ADHD adds an extra layer of risk, often resulting in nasty comments being exchanged in the heat of the moment. The act of saying unpleasant, disproportionate things can later weigh on the individual's mind, creating a downwards spiral of guilt, shame and regret, and, in the worst cases, heightened feelings of low self-esteem and even suicidal ideation.

RSD can dictate the life of someone who experiences it without them even being aware of its influence. It can create someone who is fearful to leave their home, go outside or interact with people in any way. Why risk being rejected when you can stay inside the safety of your home?

Or, on the contrary, it can create someone who overcompensates and becomes extremely extroverted, the life and soul of the party, in an attempt to conceal any vulnerability.

However, whichever route you take, both come with an abundance of people-pleasing. Like an AI robot, you read people, scan their personality and adopt it. You mirror people and pretend to like what they like, talk about what they want to talk about and disagree with nothing.

The impact of RSD on people's lives is huge.

I wrote this book to put a name to a feeling that many people with ADHD experience. I truly believe it's the worst part of ADHD. A tiny comment can incapacitate me for days.

It's a huge injustice that RSD is not currently included in the diagnostic criteria for ADHD, or at least considered in the assessment process. We, as a community of neurodivergent people, can be kinder to ourselves when we understand that our executive functioning challenges can be explained by neurological differences, but we also must understand that those same neurological differences cause significant emotional dysregulation.

Further, I didn't write this book to put a new hopelessness in you, or some new knowledge of yet another part of you that is broken. Rather, I wanted to sprinkle you with awareness, that RSD is real, that it's not your fault, and that, with awareness, you can put strategies in place to manage it, take away its power and ultimately live a significantly happier and easier life.

Without this understanding, we can beat ourselves up, ruminate over our behaviour and believe we are crazy.

This book is loaded with tangible strategies to help minimize the effect of RSD on your life. I have personally felt a significant increase in my happiness since adopting the tools on offer within the pages of this book.

However, my primary motivation for writing this book is bigger than making us all happier; it's making us all realize that we're not crazy, that the doctor who diagnosed us with an anxiety disorder was wrong, and that the ex who told us we were 'too sensitive' was incorrect.

It's the most liberating feeling to finally realize that RSD is part of the ADHD experience and not just a defect of character or insecurity, that there is an explanation for the part of you that harbours the most shame, the part of you that makes you feel truly different.

I want every single ADHD person to recognize RSD for what it is: a devastatingly painful experience but one that is not their fault. Too many of us have felt deeply flawed for too long, going through life without an understanding of RSD and the power it has over us.

This book is calling RSD out. It's us holding it at arm's length, staring it directly in the eye and saying, 'I see you.'

And when we see something for what it is, we take away its power and remove its grip over our lives.

RSD makes us feel like we're crazy, but we're not; we're different and there's a neurological reason in ADHD brains to explain that difference. You were never 'too sensitive', you were never 'too much' and you were never 'not enough'.

You live with RSD, and it will try to convince you every day that you're not enough and not worthy of any approval, but through this book, I'm saying 'sorry' to the millions of people who, like myself, are only discovering RSD now, after a lifetime of tiptoeing around others, people-pleasing, avoiding putting yourself out there, being too afraid to take steps towards your goals and being convinced everyone hates you.

But RSD is wrong – everyone doesn't hate you. In fact, the only person who hates you is you. We say things to ourselves that we wouldn't say to our worst enemies.

We've been conditioned to believe that we're flawed; however, as you will see as you progress through this book, that conditioning was never our fault, we never asked to be criticized 20,000 more times than the average child. And when you understand that those comments were not your fault, that they were a consequence of our unique differences, then you can begin to control the power they have over you today.

Without an understanding of this, RSD has the power to destroy lives, but when we give meaning to those three letters, we take back control of our emotional responses because we're able to remind ourselves that we're not broken, just different, and we have always been enough.

1

20,000 tiny cuts
What is rejection sensitive dysphoria?

I've always felt extremely vulnerable to criticism and rejection, as if I could be totally crushed at any moment, like an orange whose flesh is exposed without its peel. A tiny negative touch from the outside world and I can implode into a sticky mess, dripping with sadness and rage. It could be the smallest of criticisms, but my brain turns it into the most heart-wrenching comment ever uttered or heard.

It doesn't even matter if the criticism isn't verbal. I can pick up on tones of voice or facial expressions that remind me of a past criticism and it feels like someone is pressing down on a deep, agonizing wound. A wound with no obvious cause, no rationale behind its existence. But I know it exists because I can feel it, and I know the feeling is real because people witness my response to it and call me 'too sensitive'.

It's always amazed me how some people can receive a criticism and not be totally crushed by it. A critical comment seemingly bounces straight off them and they carry on with their day, relatively unaffected. I once witnessed someone explain why they weren't concerned that their friend had not texted them back. They said, 'I'm sure they'll text back in a few days; they must just be busy.'

How can someone be that casual over such a devastating situation? What did they have that I didn't? How was their peel still intact?

By contrast, if someone doesn't text me back, I catastrophize and overthink every single reason why that person now hates me:

> Am I crazy, or simply different?

ADHD vs neurotypical

I was diagnosed with ADHD at the age of 34 – finally, an explanation for my racing mind, that internalized hyperactivity that felt like 17 highly caffeinated squirrels barrelling around in my brain. The never-ending internal chatter, a thousand simultaneous thoughts, each one pulling me in a different direction.

ADHD certainly makes you different – in many positive ways, and many more challenging ones. I've always felt different. My earliest memory of someone confirming this was when I was six years old. I was in the playground at school and someone said to me, 'Alex, you could be one of the cool kids if you weren't so weird.'

Since starting my podcast, *ADHD Chatter*, I've spoken to more than 100,000 individuals with ADHD. They all describe almost identical early memories of feeling different. It was no surprise to me, therefore, when I found out that it's estimated that children with ADHD receive 20,000 more negative comments than your average child:

> 'Why are you being weird? It's not that loud in here.'
>
> 'You're embarrassing yourself.'
>
> 'Why are you being lazy?'
>
> 'Stop biting your nails!'
>
> 'Why are you crying?'
>
> 'You've let me down!'
>
> 'You're really rude.'
>
> 'Stop being weird!'
>
> 'Stop fidgeting.'
>
> 'Calm down.'
>
> 'Be normal!'
>
> 'Stop it!'

Sometimes the comments come from school, sometimes they come from home, and sometimes they come from both.

This is due to the neurological differences in an ADHD brain – a child with ADHD has an executive function developmental lag of two to three years, meaning the way they process the world and respond to it will be different from that of a child without ADHD. From the way a child with ADHD perceives the passage of time, to the way in which they concentrate or remember things, their processes and responses will be unusual and therefore attract lots of critical feedback.

Comments like 'Think before you act!' and 'Why are you handing in your homework late, again?' become commonplace.

This heartbreaking exposure to critical feedback makes us feel like we don't belong, as though we are not accepted as we are, and ultimately that we are broken. And when a child feels like they don't belong in a tribe, their survival instinct subconsciously kicks in. They adapt and change who they are in order to be accepted by the pack. They copy behaviour and mirror personalities.

ADHD children are the world's best actors, we are experts at playing a character called 'normal', and this highly effective survival strategy works well in the short term, but the long-term consequences are troubling. It creates a massive disconnect between our inner self and the version of ourselves that everyone else sees.

Our constant desire to be likeable is exhausting. We mirror people's body language and tone of voice. Everything we do must be perfect because being perfect is the ideal camouflage for our core self that we've been told, 20,000 times, is deeply flawed.

The flaw in the plan

Rejection is very painful for people with ADHD because it feels like our efforts to hide our core self have gone to waste. Someone has seen right through our camouflage.

In our minds, masking is something we're very good at. We pride ourselves on our ability to do it, so when we are criticized it feels like we can't even mask properly. We can't even do THAT right!

In a moment of critique, our core self, the inner child that we deeply believe to be broken, our biggest shame, has been exposed to the world.

How and why are we triggered

When a human being experiences a rejection, certain brain regions help them to navigate the experience and regulate their response to it. The prefrontal cortex (the area of the brain directly behind the forehead) and the amygdala (located deep inside the brain) are both key features.

The prefrontal cortex is responsible for decision-making and the amygdala is responsible for processing emotion. Both parts of the brain work together to:

A analyse the rejection

B create a response to the rejection.

A neurotypical person's brain, using the neurotransmitters dopamine and norepinephrine, can efficiently transmit signals throughout the prefrontal cortex and the amygdala to accurately assess the severity of a rejection and also regulate a proportionate response to it.

An ADHD brain processes dopamine and norepinephrine differently, and this difference creates glitches in the prefrontal cortex and the amygdala. These glitches make it hard to control how much information enters the brain and how the brain regulates that information. They can make many areas of life challenging for someone with ADHD, but they mean that rejection is especially brutal because of one more ingredient that neurotypicals do not have: those 20,000 negative messages.

This difficulty in regulating emotions, coupled with an already existing belief of being broken, creates a person who is constantly 'on edge' and in a heightened state of anticipating rejection.

The impulsive nature of ADHD and our fast way of thinking can quickly work together to assume the worst-case scenario, often turning a benign comment from someone into a monumental personal attack.

Isn't everyone sensitive to rejection?

Of course, nobody likes to be rejected, but there's a difference between rejection sensitivity and rejection sensitive dysphoria.

Every human being is sensitive to rejection. After all, it's in our DNA to want to be included in a group of other humans. When we lived in bands or clans 2.5 million years ago, sensitivity to rejection served an evolutionary purpose. Being cast out of the band would be fatal as we would no longer have the pack to protect us, so our bodies would naturally alert us to any sense of rejection, an early warning sign of imminent abandonment and subsequent death. Today, rejection isn't fatal. Humans can survive independently, but our nervous system has not adapted to this new reality, so it still alerts us to the danger of abandonment.

It's not nice to feel like we have been cast out.

However, a neurotypical (someone who wasn't exposed to 20,000 extra negative comments when they were a child) is able to recognize the rejection, rationalize it, feel bad about it and then move on fairly quickly with their day. RSD is different. It's physically painful, all-consuming and disproportionate to the event that triggered it. It feels like a bull has charged at you and headbutted you in the chest.

Another key element of RSD that separates it from general rejection sensitivity is the tremendous amount of shame that comes with it.

Normal rejection sensitivity vs rejection sensitive dysphoria

Rejection sensitive dysphoria is similar to ADHD in the sense that most people will experience the traits of both.

For example, you'd be hard pressed to find a human on this planet who has never forgotten something! However, the difference between how a neurotypical person experiences forgetfulness and how someone with ADHD experiences forgetfulness is the frequency, intensity and duration of the experience.

The same applies to separating rejection sensitivity from rejection sensitive dysphoria. A neurotypical person will feel sad or angry when rejected, but this feeling will occur only when there is a real rejection. The feeling of sadness or anger will also be proportionate to the size of the rejection. For example, being picked last for the sports team, someone breaking up with you or firing you will all trigger emotional responses because these are real rejections and part of the natural human experience.

RSD is different because someone who lives with it will experience two things:

1 Intense pain, both emotionally and physically, that is disproportionate to a real rejection

2 Intense pain, both emotionally and physically, in the absence of a real rejection.

The second point makes life very hard because if you are triggered in the absence of a real rejection, if you perceive non-rejections as rejections, suddenly everything becomes a threat and the frequency of your pain increases significantly.

Additionally, the intensity of RSD separates it from normal rejection sensitivity. Instead of simply feeling sad or angry, RSD is all-consuming, creating feelings of absolute devastation or rage, followed by crippling embarrassment and shame.

The final difference between normal rejection sensitivity and RSD is the duration of the effect. When a neurotypical person experiences a rejection, the emotional response will be short-lived, and because their response to the rejection is proportionate, they will not ruminate over their embarrassing response to it. When someone with ADHD experiences a rejection, the emotional response can last for hours, days or even years, and can be triggered again and again if the person is reminded of the event. Even after the intense feelings of sadness or rage have passed, the mind can re-enact the encounter over and over again, each time creating more intense emotions and shame.

Normal rejection is like when someone slaps you on the back – it hurts but you are able to move on fairly quickly. RSD is like being slapped on the back when you have really painful sunburn, an immediate all-consuming, long-lasting sting; the sting then attracts a million wasps, each one landing on you, stinging you, injecting you with shame.

You shout, you scream and you get angry. You lash out at those you love. You cry. You're unable to think about anything else other than the pain, it's the most horrible version of hyper-focus, and it's only when the RSD passes that you're able to reflect and realize that nobody actually slapped you.

What does RSD feel like?

RSD is the worst pain I have ever experienced. Someone once told me they were too busy to attend my birthday party and it felt like they had swung a bag of bricks into my chest. The pain was instant and nothing could distract me from it, the worst type of hyper-focus, turning everything else in my proximity into my worst enemy. Every background noise suddenly sounded angry, each nearby footstep sounded like someone slapping me; the roar of every car engine turned into an angry voice screaming at me, each one further tightening the vice-like grip on my heart.

You want to die on the spot. Rational thinking leaves you entirely. It's like a dark cloud suddenly appearing above you, showering you with immense feelings of self-loathing and sadness, leaving you vulnerable to seemingly disproportionate responses, a flood of tears or a rage-filled outburst for which you will later be made to feel immeasurable shame.

RSD makes you feel as though there is truly something wrong with you, like you're unable to operate as a human being, too afraid of making moves in the world because everything exposes you to the possibility of being triggered. So, you play it safe and make yourself smaller, hide away from the world, retreat into your safe place where no rejection can reach you.

Those 20,000 negative criticisms we received as a child have given our nervous system an in-built bias towards negativity. We will be very sceptical of positive feedback and crushed by neutral feedback; genuinely negative feedback will obliterate us.

What are some common triggers for RSD?

1 Impulsive communication styles

ADHD people are deep, emotional communicators and this can lead us to say something personal to someone we don't know very well. This style of communication is wonderful, but also unusual for many people, and may be met with a negative response or one where the other person retreats. The ADHD person will immediately sense this withdrawal and feel immense feelings of embarrassment that can trigger an RSD episode.

2 Social media

Most people use social media to showcase a version of themselves to the world. When someone creates a social media post and it doesn't receive the number of 'likes' they expected, the ADHD person can immediately feel rejected.

3 Being misunderstood

When someone misunderstands us or misrepresents what we say, it can be very triggering.

4 Confrontations and arguments

Confrontations happen because of disagreements, not because someone dislikes you, but the ADHD brain is highly sensitive to rejection so will confuse the two and assume an argument means the person actively dislikes you and doesn't approve of you in any way. This means the ADHD person will likely be triggered in an argument, which can quickly escalate into a volatile situation.

5 Being excluded from a social event

Being left out of any form of social interaction is not nice, but it's particularly painful for an ADHD person. The ADHD person might not be aware of their exclusion perhaps until they see a picture on social media of a group of friends on a night out, or a message in a WhatsApp chat, and this will create an intense RSD episode.

6 Constructive feedback

Any feedback that isn't overly positive can be cripplingly painful for someone who suffers from RSD. The person – whether a boss or a friend – means well and wants to help us progress in our endeavours, but we can perceive the 'constructive feedback' as an attack on our character. It causes an instant physical pain because it feels like our best efforts were not enough.

7 Feeling unseen

Being ignored after you have shared something that means a lot to you, or when someone replies with a closed reply that's barely connected to what you said, can trigger RSD. It's painful because it's as though you've offered up a bit of yourself, been vulnerable, let the mask momentarily slip, and it's been cast away as unimportant, discarded and ignored.

8 Drifting apart from close friends

Have you ever drifted away from a group of friends and then felt too embarrassed to message them because you think it will be too awkward? I've experienced this in my life over and over again. I've been really close with some people, thinking we were lifelong friends, but then I don't reply to their text message one day, and it weighs on my mind, and suddenly six months have passed and I feel like the friendship is finished. I didn't mean to sabotage the friendship, but maintaining it was too difficult. The RSD trigger happens when I see a picture of that past friendship group on social media and I instantly feel as though they have cast me out and never liked me, even though it was me who didn't put in the effort to nourish the friendship.

Or when someone you have known for years starts speaking to you in a different style of voice, a formal voice, using the kind of speech you use when you meet someone for the first time, overly polite and without the nuance of any 'in jokes'. Remember that time when you met up with an ex for coffee and it was as though you hadn't been dating for years – no more playful exchanges, flirty back and forths, as if you don't know each other anymore?

RSD feels awful here because that person was your whole world and now they're a stranger.

You used to be their whole world, too, the person they cuddled on the sofa and lay in bed with, and now they're speaking to you across the coffee shop table like you're a passer-by, a nobody, with little acknowledgement of the past you shared together or the future you once fantasized about. It's the saddest feeling in the world.

9 Being snapped at

This is particularly painful because, when someone snaps, it feels visceral, like a rabid dog barking at me. Nothing makes me burst into tears faster than suddenly being snapped or shouted at.

10 When you 'come out of your shell'

When someone with ADHD meets a new person, it's common for them to make themselves smaller or bigger than they really are. They put on a toned-down, introverted version of themselves or they put on a blown-up, extroverted version of themselves. This is sad, and a result of years of being told they're 'too much' or 'not enough'. Over time, the ADHD person feels comfortable around the new person and they allow their true self to show. However, this can sometimes be met with comments like 'You've come out of your shell' or 'You were quiet last night'. This can be devastating for the ADHD person because it makes them feel like the masked version of themselves had social approval and their true self does not. This can trigger RSD and make maintaining friendships and relationships challenging.

Ned Hallowell's RSD experience

RSD is horrible because it forces you to focus on the negative rather than the positive, even when the positive is greater than the negative. I'm going to share here the story of Dr Ned Hallowell's Harvard speech. Ned is the global figurehead of ADHD advocacy, a hero of mine, and I was lucky enough to interview him on my podcast, *ADHD Chatter*, where he shared this story with me.

He was giving a speech at Harvard, his old university, to a packed-out auditorium. The audience listened with great interest and erupted with applause and cheers at the end, giving him a standing ovation.

But as Ned was listening to the cheers, he noticed a woman in the upper left-hand corner of the auditorium who had remained seated and was scowling.

Driving home after the talk, Ned felt crushed; all he could think about was that scowling face.

When he arrived home, his wife asked him how the talk went. He told her about the scowling woman, and his wife replied, 'Ned, are you crazy? For all you know, that woman could have had indigestion! Was anyone else scowling?'

Ned replied, 'No... Well, actually, I did get a standing ovation.'

The consequences

RSD feels awful in the moment; however, its impact is felt in all areas of my life. In fact, I would go as far as to say it has always controlled my life; I just didn't know it.

As I look back at my life through my new lens of RSD, I can contextualize it to explain the following behaviours:

- Not being dependent on others. (When you perceive everyone as a threat, it's easier to remain alone.)

- Delaying or avoiding communication with others, not wanting to take phone calls and avoiding contributing to social conversations. (If I don't contribute, I can't be criticized.)

- Being a 'yes man' – people-pleasing and putting other people's needs before my own.

- Unnecessarily saying sorry, especially when the other person is the one who should be apologizing.

- Forgetting to breathe because I'm too busy over-explaining myself. I can turn a 'yes' or 'no' question into a five-minute monologue. I can't relax until I have explained every detail and given a clear rationale for my decision, otherwise I fear the other person will misunderstand me and think I'm nasty.

- Holding myself to impossibly high standards. This creates a lot of stress because everything I do has to be perfect, and if I fall short of those high standards, I mentally torture myself.

- Hearing 'I need to talk to you' and instantly feeling as though I'm going to be told off.

- Feeling pressure to smile at strangers as I walk past them in the street – you know, the type of smile where your mouth doesn't open – and being convinced they hate me if they don't smile back.

- Ghosting people out of my life, even long-term friends, because I sense they now hate me and think I'm a terrible person.

- Removing people from my life before they get a chance to reject me.

- Becoming non-verbal for the rest of the day if someone calls me 'too loud'.

- Seeing 'Come if you want?' at the end of a party invitation and assuming the host doesn't want me to come at all. 'Come if you want?' feels cold and as though they would be annoyed with me if I turned up. I need to be explicitly invited to events or I feel like the person doesn't want me there.

- Seeing a group of friends laughing and immediately jumping to the worst-case scenario: they're all laughing at me!

On one occasion I was sitting in the middle of the carriage on a London Underground train on my way to a haircut appointment. I only had two stops left before I needed to get off. However, the train stopped at the next station and loads of people poured on. Suddenly, the passage from my position to the train door was blocked by people. I was trapped. As we approached my stop, I wasn't brave enough to ask people to move out of the way, so I stayed where I was and missed the stop. I sat there for an unnecessary 20 minutes simply because I was too afraid to say, 'Excuse me.' I missed my appointment and lost my deposit.

A few years ago, I was working in an office job and I said 'Good morning' to a colleague with a smile on my face. The colleague didn't match my energy; in fact, they simply said 'Morning' under their breath with a cold tone. I felt RSD instantly. I couldn't stop thinking about the interaction all day. I was lying in bed that night thinking about it.

I replayed the interaction over and over. Did I do something to offend them? I ruminated over it all week. I couldn't focus on my work, until my boss noticed, called me into a meeting and asked me what was wrong. I said I thought I had offended my colleague and asked if they knew anything about it. My boss said the colleague had recently lost a family member and that it was nothing to do with me and I should not be 'so sensitive'.

I felt terrible and instantly filled up with shame. I never went back to that job.

I was always told the two certainties in life are death and taxes, but with an ADHD person, the third certainty is shame, a feeling of being totally deficient and worthless to others around you. You feel like everyone else on the planet is something, but you're something different, an outcast who deserves nothing but ridicule, and with every episode of RSD, the constant feeling of being unwanted, useless and revolting deepens even further.

RSD is always ready to pounce, like a coiled spring, ready at any moment to be triggered and provide you with even more irrational responses to ruminate on and feel shameful about.

We receive an email that doesn't start with 'I hope you're well' and we feel like the sender hates us.

Our boss might ask, 'When are you free to talk?' and we hear, 'When are you free to be fired?'

A tiny change in the tone of our partner's voice and we ask, 'Are you mad at me?'

We sense the tiniest hint of rejection and we need to understand what we've done wrong. It's impossible to simply let it go. 'They definitely hate me and I need to immediately understand why.' So we self-sabotage and start an argument to interrogate the person in an attempt to find out why they're mad at us. But they're never actually mad at us; we created a problem that wasn't there and now we carry even more shame over yet another overreaction.

It feels like we upset people just by existing. If someone comes into work in a grumpy mood or someone in a café is angry, I immediately feel as though they're mad at me, like I have personally pissed them off.

RSD is the most challenging aspect of ADHD

I was diagnosed with ADHD some years ago and immediately dived into the research, attended lectures and watched every video I could find. I was desperate to understand what having ADHD actually meant.

I was told it was a neurodevelopmental disorder characterized by traits of impulsivity, disorganization and forgetfulness. I smiled and thought, 'Tick, tick and tick.' I immediately felt seen and validated after years and years of feeling different and misunderstood.

However, it felt that the narrative surrounding ADHD was incomplete, that there was a missing jigsaw piece that had been excluded from the diagnostic criteria, criteria that primarily focused on the executive functioning challenges of ADHD. Forgetting to empty my washing machine, running late for a train or forgetting why I had walked into a room were all extremely frustrating experiences, yet there were tools I could implement to manage these challenges.

I truly believe that in order to fully appreciate the complete ADHD experience we must also talk about the emotional dysregulation challenges that come with having an ADHD brain, and the epicentre of these challenges is RSD. In my experience, both personally and from interviewing thousands of people with ADHD, RSD is the hardest part of living with ADHD, and it is shocking that it has not been explored in depth before this point.

Left unchecked, RSD can cause more damage to someone's life than any of the executive functioning challenges that come with an ADHD brain. In fact, it's easy to assume that the most damaging aspect of RSD is when RSD is triggered and the physical and emotional pain that comes with that moment, but in actuality the most painful and the most damaging thing about RSD is what comes immediately after the trigger: a feeling of immeasurable shame.

The working memory challenges that come with ADHD mean we can forget many things in life, but the one thing that we never forget is the feeling of shame, that perpetual inner voice that continuously tells us we're worthless, disgusting and not deserving of happiness or love.

Shame is one of the most horrible feelings a human can experience, and when we recognize RSD as the cause, we will go to extreme lengths to avoid it. It's in these extreme lengths that RSD does the most damage, in the subconscious choices we make in order to avoid being triggered and therefore to avoid adding more memories to our back catalogue of disproportionate emotional responses.

Avoiding RSD

How many times have you said 'yes' to something that you didn't want to do simply to avoid upsetting someone?

How many times have you not texted a friend, not because you didn't want to speak to them but because the thought of them not replying was too painful, or you're still traumatized from the last time you arranged to meet a friend and they messaged you at the last minute, saying, 'I'm so sorry but something unavoidable has come up and I need to reschedule'? (They even gave you options for alternative dates, but your RSD was triggered and you told yourself they think you're a worthless piece of crap and never want to see you ever again.)

How many times have you suppressed a crush on someone or not made a romantic move, not because you didn't want to find love but because the thought of it not being reciprocated was too unbearable?

How many times have you stopped yourself playing a song in your car or putting on your favourite movie, not because you didn't want to play it but because the thought of someone not liking it was too much to bear?

How many times have you wanted to say 'no' at the end of a first date but you said 'yes' because you didn't want to offend the other person?

How many times have you decided to self-isolate, even when you're chronically lonely, simply because putting yourself out into the world feels too risky?

The effect of RSD is everywhere – it's huge. It's living in a constant state of walking on your own eggshells, waiting to be caught out for something but you don't know what, and it's not until we become aware of RSD that we see the impact it's having on our lives.

You're not crazy

I burst into tears when I first read this definition of RSD on the Cleveland Clinic medical website:

> Rejection sensitive dysphoria (RSD) is when you experience severe emotional pain because of a failure or feeling rejected. This condition is linked to ADHD, and experts suspect it happens due to

differences in brain structure. Those differences mean your brain can't regulate rejection-related emotions and behaviours, making them much more intense.

I always thought having frequent emotional outbursts was a 'me' problem and that there was something inherently broken about me. The validation that comes from knowing your deepest shame has a medical explanation, and is a part of ADHD and not a part of you, is massive. All those nights lying in bed, overthinking the previous day, worrying that I was 'too much' or that I'm hated by everyone close to me, happened for a very real reason.

The perfect RSD storm

We have a brain that struggles with regulating emotion – every emotion is either 'all' or 'nothing'. We don't have an 'in-between' setting. We were also criticized 20,000 times more than the average child – and I don't apologize for repeating this here and elsewhere: it's something you need to know and to fully absorb. This early exposure to continuous criticism means our nervous system is always anticipating another criticism, even when it's not there, and our impulsive nature automatically assumes it's personal, activating the only emotional response we're capable of: 'all'.

My 'all' story

I could tell hundreds of stories where my RSD was triggered, but for now I will share one involving pancakes. I shared this story in my first book, *Now It All Makes Sense*, but it is so relevant to RSD that I think it bears sharing again here.

Ever since I was a young child, I have always enjoyed making pancakes. I hold a huge amount of pride in my ability to make them. A few years ago, I thought it would be a nice idea to make pancakes for my partner. I drove to the shop, bought the ingredients, came home and started mixing the flour, eggs and milk. I was euphoric with excitement!

But then something happened. My partner, who was watching me as I prepared the pancake mix with a huge smile on my face, said, 'You've added the ingredients to the bowl in the wrong order.'

As fast as a heartbeat, my euphoria evaporated and was replaced with sadness and pain, like someone had tied my heart in a knot of barbed wire. There was no pause, no warning, nothing to prepare me for the immediate shift in my emotional state.

I retreated into the bedroom, filled with shame, and just like that, the night was over.

I come back to this story and what happened next during Chapter 3.

I've analysed this event – let's call it 'The Pancake Trigger' – over and over in my quest to assign meaning to emotions and have concluded the following:

> ADHD means I'm different and these differences have exposed me to a lifetime of judgement towards my behaviour and abilities, far above what a non-ADHD person would experience. It is therefore harder for me to find something that I'm truly proud of because I hold a deep belief that I'm useless at most things.
>
> So when I do find something that I'm truly proud of – making pancakes – I place so much of my value as a human being in that thing, and when someone comments on my ability (or otherwise) to do that thing, it feels like they are attacking everything about me. It's unbearably painful.

Those 20,000 extra criticisms I received as a child make it so much harder to hold positive feelings about myself. Where some people might be able to derive confidence from lots of different things, I'm contending with an internal monologue telling me I'm not good enough, resulting in me dumping all my positive feelings into one thing: my ability to make pancakes. When the pancakes are criticized, I don't have a reserve of other qualities or skills to fall back on, so I collapse.

It's not you, it's RSD

RSD isn't a deliberate choice we make in the moment; we don't consciously decide to fill up with rage or sadness, but we do, it's there in our minds before we even know it's there. It's automatic and a response to years of being told we're 'not enough'. It's not deliberate, but it is real, and when not understood, RSD will control our lives.

It's exhausting going through life expecting to be tripped up at any given moment, genuinely believing that there's something wrong with you and being fearful of your own reaction when you perceive someone to have pointed out your wrongs.

RSD can make you totally withdraw from relationships because it's easier to hold the intense emotions inside rather than risk saying something back and escalating the moment.

We don't only fear an RSD trigger, an out-of-proportion response, we also fear the shame that follows. The gut-twisting embarrassment and guilt because we've shouted at someone we love again, or burst into tears in a moment that was meant to be joyful. We fear the lying in bed at night alone, staring at the ceiling, convinced our entire family is speaking about our outburst behind our back, detailing how we ruined yet another family event.

Taking RSD seriously saves lives

For far too long, people with ADHD have not been taken seriously. Even today, RSD is brushed aside in many therapy rooms, leaving the person feeling even more shameful. This is a travesty.

We are not a community of 'overly sensitive' people, we are a community that has been told 20,000 more times than an average child that there's something wrong with us. The ADHD community deserves to know about RSD, to understand who they are and why they acted the way they did in past situations and, most importantly, to understand that RSD is not their fault. This is the only way to eradicate the compounding effects of shame.

RSD is more than simply not liking rejection, it's totally consuming and debilitating, but with knowledge comes an ability to manage it and an ability to take away its power over you. RSD doesn't need to control your life, there is a way out of its grip, and you holding this book in your hand, or listening to it through your earbuds, is the greatest threat to RSD's vice-like grip on your life.

2

Approval addicts
RSD and people-pleasing

People-pleasing nearly cost me my life

It was 2013 and I was asked to sign a business contract in relation to one of my social media companies. I remember the day vividly. I met two people in Brighton, a sunny seaside city on the south coast of England. But something felt off that day. My intuition was screaming at me, telling me not to trust these people.

I met them at the train station and we walked to a nearby restaurant to discuss the contract, but with every step of that walk, I sensed threat, a loud voice inside me saying these two people were not to be trusted. But that voice inside me was unknown to anyone else because I had a massive smile on my face. I was pressing the traffic light buttons on their behalf. I even opened the door to the restaurant for them and smiled as they walked in. I remember saying 'I'm sorry', as I felt I hadn't quite opened the door wide enough to allow them enough space to comfortably pass through.

We sat down, and the contract was placed on the table. 'Run!' said my internal voice as I noticed a shiftiness about them, the way they periodically made eye contact with each other, the way their smile seemed forced in an effort to hide their true intentions.

'I'm so sorry,' said my actual voice as I pulled my chair in, causing a small scratching noise against the floor.

'What do you think?' one of them asked as he slid the contract closer to me.

'It looks great!' I replied as I looked down at the piece of A4 paper in front of me.

My eyes moved from left to right, taking in the first few words of the contract. They had got to the end of the first paragraph when I realized I had not taken in anything that I'd just read. My eyes were moving through the words but my mind was elsewhere, unable to focus on any of it.

'Do not sign this bit of paper!' said my internal voice as I looked up and caught a certain type of look passing between the two men.

'Where do I sign?' said my actual voice out loud, followed by, 'I'm so sorry it took me so long to read it.'

I was removed from my company shortly after signing that contract. It triggered a five-year legal battle to win it back. The stress turned me into an alcoholic. During one of the many alcohol-related hospital visits, a nurse told my mum, 'If Alex had had one more drink last night, he would be dead.' Apparently, my blood alcohol limit was at near-fatal levels.

So much stress, simply because I didn't know how to say 'no'. I was terrified of upsetting those two men, so I signed a bit of paper simply because, in that moment, it was the easiest thing to do. It was the path of least resistance, and also one that protected me from feeling criticized and judged.

> ## A tip
>
> Rather than saying 'yes', say, 'Can I let you know tomorrow?'
>
> This initiates a gap between the ask and your response; it bypasses the knee-jerk reaction of saying 'yes'.
>
> You can then reply from a place of regulation and calm. And when you're in a place of regulation and calm, the answer is, more often than not, 'no'.

RSD feels horrible in the moment, but it's the actions we take in our lives in an effort to protect ourselves from that horrible feeling that do the most damage. The almost-incessant nature of putting other people's wants and needs ahead of your own, even when the consequences of those actions are terrible for your own wellbeing. It's simply easier to comply, say 'yes' and agree than it is to even consider doing something that could expose you to the horrible effects of RSD.

People-pleasing is a phrase we all know, but the term doesn't sufficiently describe the behaviour of someone living with RSD. It's very human to want to please others, but what separates the actions of someone who lives with a standard level of people-pleasing from someone with RSD is the control it has over their lives. The pain of RSD is significantly more than the pain of normal rejection, therefore the actions we take to avoid it are also greater.

Over the following pages you'll find some examples of highlighting the differences between 'normal' people-pleasing and 'RSD' people-pleasing.

Struggling to set personal boundaries

A 'normal' level of struggling to set personal boundaries may be when someone feels uncomfortable setting a boundary but will set one anyway. Setting a boundary is often not liked by, or is resisted by, the person on the receiving end, so setting one can lead to an uncomfortable feeling – but the person setting a boundary understands that just because the other person doesn't like the boundary doesn't mean the boundary is wrong. A person who is able to do this is someone whose self-esteem is secure enough for it not to matter that they or their behaviour are not, in that moment, liked by another person.

Someone who lives with RSD will struggle to set personal boundaries in a different way: they won't be able to do it at all. The potential pain of RSD will act as a constant repellent to the act of standing up for yourself. Someone with RSD craves a feeling of safety, and putting a boundary in place is, in their mind, threatening that safety because it's exposing them to the biggest threat of all: not being liked. RSD will make them betray their values in order to be accepted by someone else.

If someone texts you at 7 p.m. and asks if you want to go out at 7.30 p.m. and you're exhausted and would rather not go out, RSD will make you reply, 'Yes, let's go out!', even though 'Apologies, I'm too tired. Can we maybe go out next week?' is what someone without RSD might say.

If someone cancels dinner plans at the last minute, RSD will make you reply with 'No worries!', even though 'Sure, that's a shame – I do need to tell you that I have just spent money on a taxi getting to the restaurant, so cancelling last minute is disrespectful of my time' is what someone without RSD might say.

If someone tries to take a kiss further when you're not feeling into it, RSD will make you stay silent, when someone without RSD might say, 'We should stop.'

When someone with RSD learns how to set boundaries, they will notice it upsets people close to them. Pay close attention to who these people are because these are the same people who benefited from you not having any boundaries in the first place.

Always deferring to someone else's opinions or perspective

A 'normal' level of people-pleasing looks like deferring to someone's else's perspective in minor ways in order to avoid offending someone you care about. For example, saying you enjoyed your partner's favourite movie when in fact you thought it was mediocre, or saying you thought a meal was delicious when in fact you thought it was average. When someone you care about shares something of value to them with you, or has gone to considerable effort to make something for you, adopting an agreeable and grateful position even if that's not an accurate reflection of your opinion is a common response. It keeps any unnecessary friction at bay.

If humans always adopted a disagreeable tone with every encounter, the never-ending confrontations would make sustaining any level of connection with anyone impossible.

Someone who lives with RSD will experience deferring to someone else's perspective in a different way. Instead of occasionally adopting an agreeable position in order to protect someone they care about, they will always adopt an agreeable position, even with complete strangers, and even when the consequences of this position put them at risk.

For example, someone with RSD might accept a cigarette, a vape or drugs from someone else, even though they know these things are harmful. They might attend a dinner party and agree with someone's political views, even if they're in stark contrast to their own values. They might be a vegetarian but agree to eat a beef burger at a social event.

Adopting an agreeable position and taking on someone else's opinion or adopting someone else's perspective can be soothing in the moment because it protects you from any criticism, but it can also leave you exposed to embarrassment. I've lost count of the number of times someone has asked me if I like a particular band and I've instantly said 'Yes!', even though I have never heard of them. The next question is invariably, 'So, what's your favourite song?' and I always go red in the face and reply, 'Oh, how funny, I can't think of any right now.'

Constantly deferring to someone else's point of view is also very expensive, especially when you're standing in front of a salesperson. You could be buying a car, a new item of clothing or even a new paint for your home. It's the salesperson's job to nudge you towards spending more money than you need to. They could be talking about a flashy add-on feature of the car, a more expensive clothing fabric or a state-of-the-art, scratch-resistant paint, all of which you know you definitely don't need – and yet you're standing there nodding and smiling. So you agree to the expensive extras, pay for the goods and walk out of the store. A car salesman once made me say 'yes' to installing a £400 fridge in my car. I didn't want the fridge! I used it once to store a banana. (I only remembered the fridge was there two months later when I smelled the mouldy banana.) RSD trigger avoided, bank account wounded.

RSD amplifies people-pleasing. For example, when someone without RSD is walking home alone at night and they see someone ahead of them, they might cross over to the other side of the street. Their survival instinct kicks in and they decide to put distance between them and the unknown person. However, for someone with RSD, crossing the street might be perceived as an unlikeable thing to do, causing offence and making them seem unkind, so they won't cross the street or take any potentially evasive action. The RSD trigger-avoidance instinct is greater than the survival instinct. A person with RSD has a need for external validation and approval that is much stronger than any internal warning system that's trying to alert them that they're in danger.

Frequently serving as an emotional outlet for others

Offering someone a supportive ear during a moment of need is a nice thing to do, particularly when it's a friend or a family member who is experiencing hardship. Acting as a free therapist for those close to you is valid and can be extremely rewarding for both you and the friend in need.

However, someone who lives with RSD needs to be careful that their kindness isn't abused. What started with a 'Yes, come over and have a chat' can morph into a 'I don't know how to tell this person I don't want to listen anymore'. After all, for someone with RSD saying 'no' can feel impossible.

Most humans are very good at noticing vibe signals and picking up on other people's moods; someone who truly cares about you will recognize when you no longer want to offer a supporting ear and stop, or at least pause, in the act of asking you for help. They will notice you hesitate momentarily before you say, 'Yes, come over again and let's have another chat.' They will assert a boundary on your behalf and say something like, 'Actually, let's chat another time. I don't want to burden you.'

Some people, however, will notice your hesitation and still ask you for continued support. They will detect the people-pleaser within you and sense an opportunity to exploit you. Venting our problems feels soothing and, to some people, finding someone who will never *not* want to listen feels even better. This relationship is toxic because the person suffering is abusing your inability to say 'no' and they are aware that they're doing it. They can further manipulate you by turning up with gifts or a bottle of wine, knowing that this will make it harder for you to say 'no' in the future.

People-pleasing abusers are easy to spot because you will rarely hear anything from them unless they are asking you for something. If someone is frequently asking you for support but not reciprocating the offer, or giving any form of friendly back-and-forth without asking for something, this should be taken as a red flag. They have spotted your inability to assert boundaries and are walking all over you.

The perfectionism that many people with RSD experience further complicates the matter. We may not want to be a supportive ear, but we also feel the need to be the perfect supportive ear. If we're going to be roped into being an unpaid therapist, we're damn sure we're going to be the best unpaid therapist the world has ever known!

We can hyper-focus on the moment, listening intently to the person's woes and interjecting with the perfect responses. These responses will come from our naturally heightened empathy. The sufferer will sense that authenticity and feel you are happy to be an ongoing supportive ear, thus continuing the cycle of asking. In this case, the person suffering isn't being abusive because they are reading your signals and your signals are saying, 'I want to help!' when you're in the moment, your empathetic nature doesn't let you not want to help.

You're also extremely good at being an unpaid therapist because you've had experience of, or can relate to, the emotions that are being shared with you. You will be able to hyper-focus on the stories and spot the patterns in the character's behaviour because you probably would have encountered similar situations yourself.

Being a continuous comfort blanket for others is exhausting, both physically and mentally. You will put so much effort into the chats that it will leave you drained and unable to put effort into other areas of your life. The chats will be deep and complex, often leaving you overthinking what you said, sometimes not even remembering half of it, but always leaving you with an anxious feeling of 'they think I'm terrible as a therapist and a friend, and now they hate me'.

Quickly accepting blame, even when undeserved

Accepting blame when you haven't done anything wrong is one of the biggest forms of self-betrayal there is, but sadly it's also one of the most common behaviours of someone who has RSD. Saying 'It wasn't my fault' or 'I didn't do anything wrong' feels impossible because we are hardwired into thinking the opposite. Those 20,000 early criticisms, unfortunately, have made us believe that everything is our fault and we can't do anything right.

Even if the sense of injustice is too much and we feel compelled to say 'I didn't do that!', the fear of triggering the RSD and feeling immense sadness, rage and shame will overpower it and we will accept all responsibility or simply stay silent.

I was once accused of queue-barging in a supermarket. I joined the line to pay for my food but didn't notice that I was stepping into a gap in the queue and not at the end of the queue. The person behind me shouted,

'You can't jump in the line!' and I immediately said, 'I'm so sorry!' while I scurried to the back of the queue. I was raging inside because my mistake was innocent, but I had been accused of deliberately queue-jumping, a social crime. However, rather than explain that to the person and easing my feeling of injustice, I instantly apologized, went bright red in the face and moved to the back of the queue, filled with shame.

I once had a minor car crash that definitely wasn't my fault, but the other driver was shouting at me, so I immediately apologized and promised I would be more careful in future. The thought of standing up for myself and creating social unease was too unbearable. I knew it wasn't my fault – I'm a great driver – but someone pulled out of a junction without looking, which didn't give me enough time to brake, so I drove into the side of their car. Instead of blaming them, I jumped out, smiling apologetically and shouting, 'Oh my gosh, I'm so sorry!', to which they responded, 'You idiot!'

I took the blame, even though I knew it would penalize me financially because I would have to claim on my insurance. Even in the face of obvious injustice, my need to be friendly, polite and cheerful at all times cost me dearly.

I can call the other driver an 'idiot' now as I sit here writing this book, sitting safely in front of my computer screen, years after the incident. In the moment, however, the feeling was very different. I was in survival mode.

Avoiding uncomfortable conversations

People with RSD go to incredible lengths to avoid uncomfortable conversations. To them, an uncomfortable conversation is always a confrontation. Their 'all or nothing', dysregulated way of thinking doesn't allow for nuance – it's either a comfortable conversation where the other person doesn't hate them or it's an uncomfortable conversation where the other person hates them. It's one or the other, and no middle ground exists. The reason for this is simple: they can't tell the difference between a confrontation and a rejection, and because RSD is so painful for them, they must protect themselves and therefore must never experience confrontation.

If you always avoid confrontation, you also avoid being loyal to your values. Every time you choose to avoid an uncomfortable conversation, you're reinforcing your internal belief that your voice doesn't matter and that your values are not worth protecting. You are strengthening the effect of those 20,000 criticisms.

Your inner child will scream out from time to time, reminding you of your values, begging you to speak up on their behalf. Every time you avoid speaking up, you're denying that inner child the thing they desperately need, the thing they were robbed of by those 20,000 criticisms: approval.

Instead, you prioritize the approval of others. In your desperation to avoid uncomfortable conversations you end up damaging the relationships that matter to you. You put distance between you and the important people in your life. You detach emotionally and choose the path of least emotional vulnerability. This disconnect with those closest to you creates confusion, misunderstanding and resentment, which, ironically, will increase the chance of a real rejection.

How to speak up on behalf of your inner child

Your inner child has been battered to pieces by those 20,000 criticisms. It's a mute presence somewhere inside you, heavily shielded by a thick mask, a mask that's able to change its identity at any time in order to fit in and avoid being criticized. But to minimize the effect of RSD, we must take steps to nurture our inner child and enable it to see itself in a new light.

The first step is acknowledgement, and this comes with an understanding of what masking is and why we do it. Before my ADHD diagnosis, I had no idea how much I was burying my inner child under layers of masks. Once diagnosed, I finally realized how much effort I was putting into concealing my inner child, my core, a place full of shame.

The second step is to reconnect with your true values. You showed those true values when you were younger and were told off for them, made to feel that what you stood for was weird, not necessary or not consistent with what 'normal' kids should stand for.

I remember reading about the political system at school and declaring, 'This wealth divide doesn't seem fair.' My teacher shut me down: 'You don't need to have an opinion on that; you just need to know the dates of when the political party came into power.'

We suppress our true values and adopt the value system of others in order to win their approval. This works well – after all, we're experts at masking – but it also creates a massive disconnect between our exterior values and our core values, and when our exterior values get approval, it makes us feel that our core values are worthless and deeply shameful.

Reconnecting with our core values is vital because it enables us to create a value system that is designed to seek approval from ourselves rather than from others. There will be clues as to what they are and they will come in the form of knee-jerk emotional responses. These are automatic emotional responses that our body will give off before our conscious self, the masked version of ourselves, has had time to respond.

For example, you may become aware of someone who is cheating on their partner and instantly have an emotional response of empathy towards that partner. You might want to immediately text them and say, 'You're being cheated on.' But the thought of the cheater finding out and the possible confrontation that might ensue overrides the desire to get involved. That knee-jerk reaction of empathy is your inner child screaming, reminding you of some of your core values: integrity and loyalty.

Or you might see an elderly person struggling to carry their bags up a flight of stairs and you immediately want to help, but your conscious self is aware that stopping will make you five minutes late for work, so you keep walking. You feel terrible and can't stop thinking about the elderly person all day. This feeling is your inner child reminding you of another core value, kindness – the value that was suppressed by the thought of having to confront your boss, needing to explain why you were late, and the RSD that would follow.

These emotional responses will happen all the time. It's good to be aware of them, take note of them when you can and remind yourself of them as often as you can. They are the voice of your inner child trying to get your attention, trying to remind you of who you really are.

Every time you listen to that voice, you're validating its existence. You're telling your past self that you matter and that you are enough. You're travelling back in time, metaphorically putting your arms around yourself and whispering into your own ear, 'Don't listen to those 20,000 criticisms. You don't need to change who you are, because you're not broken, just different. You are enough.'

If you repeat this activity often enough, you will add so much colour to the image of who you really are.

A once abstract portrait of yourself will become defined and populated with values, opinions and beliefs. You can then add the final detail: your new understanding of the differences that come with being an ADHD child. This is when you strip away all the shame because you now understand that those differences are neurological and not your fault, and therefore neither are those 20,000 criticisms. You did nothing wrong. The version of you that you hold so much shame over is based on a lie. The parts of you that you've desperately tried to hide all these years never needed to be hidden. You were misled by external messaging and expectations, a world with no understanding of ADHD and one that required every child to act the same way.

The little girl or boy inside you deserved a better life than the one RSD gave it. They didn't ask to be hidden away. They wanted to be loved, not betrayed. But it's not too late. You probably have a photo of yourself as a child somewhere – if you need to, print it out, then stick it on the wall, somewhere you walk past every day, and speak to it as often as you can. 'I'm sorry for hiding you. I'm no longer ashamed of who you are – it wasn't your fault. You deserve happiness. You deserve to be seen by the world; the world deserves to see you.'

Every time you validate your inner child, you reduce the shame you associate with yourself, and when you are less shameful about yourself, you are less fearful of other people seeing it – you take control of your internal narrative and stick up a metaphorical middle finger to those 20,000 criticisms.

How to set boundaries when you have RSD

People with RSD are hardwired to feel as though they're always letting someone down. They've been made to feel that way their whole lives, and as a consequence of this, they have always felt as if their needs don't matter. In fact, they go out of their way to accommodate everyone else's needs. It's exhausting. If someone has a problem with them and starts a confrontation, they can have a total meltdown because they don't have the tools to deal with the situation in the moment. They can't argue back in any way because they truly feel that their needs aren't worth defending and don't matter.

Learning how to set boundaries when you have RSD is a three-step process.

1 Cultivate RSD awareness

My life became so much easier when I learned about RSD. I was able to put a name to my extreme emotions and see a reason for them. This enabled me to chip away at the mountains of shame I associated with them. RSD is caused by those many thousands of criticisms in the earliest years of life, and by having a brain that's wired a little bit differently. Neither of these things are my fault, and therefore my emotional reactions to them are not my fault. You have successfully already completed step one simply by reading or listening to this book.

2 Identify your boundaries and craft alternative responses

People who live with RSD have spent years masking and pretending to be someone other than themselves. They are living in a constant state of alertness, scanning their environment for threats while simultaneously expertly playing a character called 'normal'. They have one priority: avoiding pain at all costs.

Constantly being in survival mode makes it hard to think about what your boundaries are and what they are not, but there will be clues that we can take note of. We can reflect on these notes when we're in our safe place – mine is my bedroom – and use them to craft alternative responses to the event that triggered the note.

I carry a notebook with me (or I use my phone notes app) and I write down moments when I felt uncomfortable during the day, followed by how I could have acted in that situation if I had been able to assert my boundaries. This allows a version of me who isn't caught up in the middle of an RSD avoidance moment to give advice to a future version of me, for when they encounter a similarly uncomfortable situation.

Recently, I was walking my dog. I had a virtual meeting in ten minutes, so I was trying to hurry back to my car, but I have a cute dog, so people often want to pet him. I saw a man and a woman walking in my direction. As they got nearer, I could see they were wanting to say hello to the dog. My anxiety peaked: the thought of being late for the meeting was excruciating. I really wanted to say something like 'I'm really sorry but I've got to dash as I'm late for a meeting,' but I didn't. I stood there and smiled for two minutes and made awkward small talk as they petted my dog.

I was two minutes late for the meeting and felt terrible all day. So I wrote 'dog walking disaster' in my notebook, followed by 'I hope you don't mind if I don't stop but I've got a meeting in ten minutes' as a suggested alternative response.

You can't undo years of RSD avoidance in a few days, so don't worry if you don't immediately get clarity on what your boundaries are or if you're not immediately able to deliver alternative responses to similar situations. This exercise isn't designed to deliver huge changes overnight; it's an ongoing exercise that will enable you to rehearse, and eventually deliver, more assertive responses to uncomfortable situations, something those 20,000 extra criticisms made you believe wasn't possible.

3 Communicate your boundaries in a safe way

Lots of existing advice around setting boundaries says you should set your boundary in person – in other words, face to face with the receiving person. This feels risky for someone with RSD because the receiving person is likely to respond in a negative or angry way. The explosive nature of RSD can make these situations escalate very quickly.

I like to use email or WhatsApp to set my boundaries when the receiving person is not someone I spend a lot of time with. (I will be covering RSD boundary-setting at work and in relationships later on, where this advice may not always be applicable.) For example, a simple email or WhatsApp saying 'I felt uncomfortable when you spoke about my lifestyle choices. If I feel that way again, I won't be able to spend any more time with you' will give you, and the other person, time to respond calmly, therefore avoiding an escalation.

However, even sending an email like this can feel extremely uncomfortable for someone with RSD. Standing up for yourself can make you feel vulnerable, so it's important to protect yourself during the process. The type of brain that causes RSD is one that gets easily overstimulated and overwhelmed, so it's important to choose the right location to send a message that will not overstretch your brain. A bedroom is my preferred choice. Wearing super-comfy, loose and non-scratchy clothing can also help to reduce overstimulation and make you feel safer.

No one can criticize me if I'm perfect

People with ADHD are the world's best problems solvers – fact! – because we've had to solve problems our entire lives in order to survive and fit in. It is, therefore, no surprise that the ADHD community quickly solved the problem of RSD: we became perfectionists! Being perfect at everything we do is the perfect defence against criticism.

We either have to totally commit to something or we don't commit at all; we're either 'all in' or we're 'all out' because we can't risk the possibility of someone catching us being mediocre at something. If we start a new hobby, let's say candle-making, we can't simply buy a beginner's set and make a crappy candle; we will buy the most expensive candle-making set we can and spend weeks watching every candle-making video on YouTube.

If we decide to clean the bathroom, we don't simply wipe down the surfaces; we will buy brand-new, state-of-the-art cleaning equipment and spend four hours deep cleaning the bathroom to a professional standard.

We simply cannot understand people when we hear them say, 'Done is better than perfect.' No, 'done' means it's probably not perfect, and not perfect means the person who witnessed me doing the thing is going to think I'm a useless piece of crap and will never want to see me again. Oh, and they'll hate me for ever, too.

People with RSD will find it easier to perform tasks when they are around other people – for example, if your friend asks you for help cleaning their flat, or someone at work asks you to help them with a project, you'll be able to jump into action and do the task to a perfect standard. In contrast, your home will be a bombsite and your own work projects will be neglected. This is because, when we are outside, we are masking, and the masked version of us is compensating for the internal beliefs that tell us we are not good enough. The mask's job is to hide our true self, a version of us that we believe is useless, so we massively overcompensate.

RSD also leads to a type of perfectionism that is applied before you've even started the task. This is a troubling type of perfectionism because it often means the task never gets started.

You might want to get fit and start running, but you tell yourself, 'First, I need the perfect weather, the perfect running location, the perfect trainers, the perfect outfit, the perfect socks, the perfect fitness app and the perfect mood.' Even if you join a gym, you'll tell yourself you need the perfect gym vibe. You put impossibly high standards on yourself, which are also impossible to meet, so you never start the task. You then feel a huge amount of shame because you can't even do a basic thing like go for a run.

Or your desire to be perfect sends you off on an exhausting side quest. Perhaps you're at home and you get a rare burst of motivation and want to start a work assignment, but your home is messy and you can't work in a messy environment. So, you put on the rubber gloves and get cleaning. Three hours later, your home looks like a show home, but you no longer have any energy to start the work assignment. A week goes past and you get another burst of motivation for your work assignment, but your home is messy again, so the cycle continues.

Sadly, it's also common for people with RSD to require perfectionism in their physical appearance. Those 20,000 criticisms have made them believe that their true self is not worthy of approval so it must be covered up with designer clothing and make-up. Their hair must also be perfect. Even their mood must be perfect. They hold themselves to impossibly high standards which cannot always be met, so it's common for the RSD person to isolate for long periods of time, or to cancel plans at the last minute because they looked in a mirror and decided their hair or outfit wasn't perfect. This can cause a complete meltdown because cancelling plans at the last minute also comes with a risk of RSD because you need to let someone down.

The pursuit of avoiding RSD can be dangerous

Perfectionism means people will place impossibly high standards on themselves. If those standards are not met, the person will bully themselves, calling themselves names they wouldn't wish on their worst enemy.

> 'You stupid, horrible, worthless idiot!'

Another danger arises when the lens of perfectionism is focused on your own body. Sadly, even today, people are still exposed to imagery that depicts the 'perfect' body type for men and women. Billboards, movies and TV commercials are dominated by a particular body type. This creates a dangerous comparison for everyone, but especially so for someone with RSD.

The ADHD that caused the RSD also comes with an ability to hyper-focus, which means the person will zoom in on what they perceive to be imperfections in their body, hyper-focus on them and enter problem-solving mode. As mentioned, people with ADHD are the world's best problem solvers – they've had to solve problems all their life – but when the ability to solve problems is paired with a hyper-focus on what they perceive to be an imperfect body, they enter dangerous territory. The perfectionism within them craves control, so they become hyper-focused on their diet, they obsessively exercise, and they hate themselves if they eat too much or they skip the gym. If left unchecked, this pattern of behaviour and these looping thoughts can lead to devastating mental and physical health complications.

The burned-out perfectionist

The executive functioning challenges that come with ADHD mean that an ADHD person needs to work ten times harder than a neurotypical person to achieve the same task. Couple this with the perfectionism that comes with RSD and you have a person who is constantly burned out because they have never learned how to say 'Stop, I need a rest'. They overwork because they fear a lack of progress.

Sometimes the human body will force you to rest, but even if the person with RSD lies down for several hours, they are still exhausted because their racing mind hasn't enabled them to actually rest. Your body will tell your mind you need to rest, but your mind will tell your body you're in danger of rejection if you don't work.

With RSD, we become very aware of the symptoms of burnout, which themselves cause significant stress because we feel we aren't doing as much as we should or could be doing. Any tasks that don't take us a step closer to perfectionism, like making the bed, showering or brushing our teeth, feel impossible. RSD perfectionism is always an exercise in winning approval from others, so we neglect our self-care unless the consequences of self-care make us look better in the eyes of other people.

RSD turns a person into the type of perfectionist who goes through life in a constant state of worrying that they're going to be found out. This feeling of impending exposure keeps you from taking your foot off the gas, like you're always being chased by the next criticism. At school, this level of perfectionism looked like overworking to get the highest grades. In adulthood, this level of perfectionism looks like not being able to make decisions without asking other people for their advice, or spending weeks researching every single thing there is to know about every single option, and even then you're not sure!

The idea of booking a holiday stresses me out because I know I'm going to spend weeks researching the different hotel options. I need to pick the perfect one, with the perfect facilities and in the perfect location.

Instead of making a decision, you work and work and work until you burn out and abandon the idea altogether.

How to limit your perfectionism

I used to let RSD turn me into an extreme perfectionist. I had no ability to tell whether a task deserved 10 per cent of my effort or 200 per cent of my effort; my default was to give 200 per cent to everything. It felt like the safest way to protect myself from any criticism, any feeling of not being enough or looking stupid.

When I learned about RSD, I wanted to challenge my internal narrative, the one that told me a task isn't worth doing unless it can be done perfectly. All the self-help books told me I shouldn't compare myself to others, but I was convinced this level of paralysing perfectionism wasn't normal. So I went looking for comparison. I found examples of other people's finished tasks, ones similar to those I was trying to do. I watched people's social media videos showcasing their finished projects, and do you know what I realized? I did a better job than most people! I was better at the things I was interested in than most other people were. In fact, I was a lot better. Even when I only gave 50 per cent effort towards a task, it was better than most people's 100 per cent. It was the first time in my life I had listened to my beliefs and thought, 'You sound like an arrogant idiot!'

But it was true, and I really believe this is true for other people with RSD. When we find something we are interested in, we don't have a deficit of attention (despite the term 'ADHD' telling us we do); we have an abundance of attention towards that thing, we can hyper-focus on that thing and we can become the best in the world at that thing. In fact, hyper-focus enables us to be very good at everything we do. However, RSD is also a powerful force stopping us from believing in our ability, so we overwork to compensate. Sometimes, seeking out comparison to others will give you the evidence you need to remind yourself how brilliant you are.

I write this evidence in a journal, but I also write it on things that I use frequently. If the evidence isn't in my sight, it will leave my mind. To combat my poor working memory, I have a coffee mug that has 'Your 50 per cent is better' on the side. My coffee mug is always with me when I'm working on my laptop, so it acts as a constant reminder. Sometimes, I look at it and realize I've done enough: this email is finished and doesn't need any more read-throughs, and I click 'send'.

I once met someone with ADHD who had a tattoo on their arm that said, 'My 50 per cent is better than most people's 100 per cent.' I thought it was brilliant!

My biggest realization since discovering RSD is that shame needs isolation to grow stronger. When we actively seek out comparison, we discover that everyone else, even the people we see as authority figures, slips up and falls short of their own high expectations.

Since starting my podcast, *ADHD Chatter*, I have met hundreds of intimidating people, celebrities, politicians, world-renowned authors, Harvard-educated psychiatrists, people I would normally be scared of and had assumed were always perfect in everything they did. None of them were. They all emailed me after the podcast recording with concerns such as 'Was the interview good enough?' or 'I hope I gave you the interview you hoped for?' Everyone was worrying about not being perfect. They were all motivated by shame. When I discovered this, that the most apparently perfect people were not perfect at all, I realized that being perfect is impossible. It's an illusion created by shame in order for it to hide itself.

A mindset shift

My default mindset used to be that I had to complete an entire task perfectly or none of it. The shame, rage and sadness that would happen if I felt someone thought it was subpar were too painful to even consider, so big tasks felt too risky and never got started.

My mindset shift allowed me to break down big tasks into smaller ones. Rather than 'I need to clean my bathroom today', which felt overwhelming and put me at risk of RSD because I might lose interest in the task halfway through, I told myself, 'Today, I can pick up the towels off the bathroom floor.' Picking up the towels is a relatively easy task that comes with very little risk of attracting criticism. It's also not overwhelming and has often enabled me to find the momentum to carry on cleaning the entire bathroom.

Looking at tasks as icebergs and breaking them apart into ice cubes has enabled me to avoid becoming overwhelmed by big tasks. It also creates lots of mini-deadlines, and there's nothing my brain loves more than a deadline. Also, it's important to note here that some days even an ice cube task seems too daunting. That's when you need to get out your metaphorical hammer and make crushed ice.

For example, if clearing out your email inbox feels like an iceberg and looking at each individual email is an ice cube, then simply turning on your laptop would be a bit of crushed ice. There's nothing wrong with making the first step towards a larger goal as tiny as possible. RSD hates it when you achieve the tiny wins – it likes to keep you in a state of procrastination, bed-rotting and safe from threat of any criticism towards any imperfection – but if you can create perfection in the tiniest tasks, simply to get you over the start line, you immediately take away RSD's power.

How I beat my perfectionism

After COVID, I wanted to get fit, so I decided to join a gym. I bought the protein shakes and the latest gym gear and went every day for a week. I saw videos on social media of people's gym gains and knew I needed to achieve the same.

I also bought myself a food blender and a cookbook because I wanted to try home cooking. I saw videos on social media of people cooking delicious meals in their kitchen and I knew I needed to achieve the same.

I gave myself two options:

> Success = Achieve the same as the people in the social media videos
>
> Failure = Don't achieve the same as the people in the social media videos.

There was no middle option. My outcome was success or failure, all or nothing, approval or shame.

I held the definition of 'success' at another person's standard, when that other person probably cared about the gym or home cooking a lot more than I did. That person had already gone on their own journey of experimentation, trying and quitting different hobbies until they found one that truly connected with their core passions.

And then it hit me: the truth was that I had already achieved success by stepping into the gym and buying the blender, because each one taught me what I was not interested in. Success didn't happen when I completed a task but when I tried a task that I'd never tried before.

Having ADHD meant I had masked for so much of my life, motivated by shame, and often starting different projects because I was trying to seek approval from others.

I had spent very little time, if any, working on my self-awareness. I truly believe that ADHD is not a deficit of attention but a deficit of self-awareness, and when I realized this, my main objective suddenly changed – rather than needing to be the best at everything I tried, I needed to try as many new things as possible. I needed to conduct as many experiments as I could to collect the evidence that told me who I truly am and what I truly care about. And when you move the goalpost from 'being the best at something' to 'trying something' and you communicate this to those nearest to you, you remove the threat of RSD because the act of simply trying a new hobby or task is a massive win.

Trying something and stopping because you have lost interest in it is better, and will give you more approval from those closest to you, than doing something perfectly that doesn't interest you. You're now one step closer to finding your passion, and when someone with ADHD finds their true passion, that's when they become the best in the world at that thing.

ADHD is often thought of as having a deficit of attention, but I believe it is more a deficit of self-awareness. We typically have an abundance of attention towards things when we are actually interested in them. The many years of masking meant we became better protected from criticism. This, however, came with a consequence: we lost our true self in the process.

We learned that we were safer when we pretended to be something other than what we truly are. This solution, however, also came with a consequence: we became fearful of people seeing through our mask. The fear of letting people down turned us into expert over-worker.

With a little bit of knowledge, we can observe our own people-pleasing behaviour, our own struggles with setting boundaries and our own obsession with perfection. We can zoom out and spot which behaviours are serving us and which are not. Our journey towards putting our own wants and needs above those of others starts here, with an awareness of RSD and the control it has unknowingly had over our lives for far too long.

3

'Are you mad at me?'

Reading rejection in every pause

RSD can have an enormous impact on our relationships. It makes it hard for us to make the initial move because the thought of being rejected is too painful. We wait patiently for someone else to make the first move and when they do, we latch onto them because they have just given us something that we deeply lack: approval. This makes us vulnerable to abuse because the idea of approval is so appetizing, it overrules our inner BS radar that knows they might only be after one thing. So we say 'yes' to that first date and 'yes' to that home-cooked meal at their place because we crave what they are offering us, the opposite of rejection: acceptance.

The innate people-pleaser within us perpetuates this scenario. We then fall in love extremely fast. Our brain floods with dopamine. Someone thinks we're loveable and it feels incredible. There's nothing more addictive to an RSD person than when someone provides them with a narrative that contradicts their own internal one that tells them they're a terrible human being. It's euphoric. This person becomes their entire world, their new hyper-focus; they're an antidote to self-loathing. They want to see this person all the time, and when they aren't able to be with them, they think about them obsessively and search for everything they have ever posted on social media. Our obsession with this person is all-encapsulating: we need to know everything about them, every town they have ever lived in, every job they have ever had and every person they have ever dated.

This is where we stumble. We find a picture of their ex and our RSD instantly triggers because we always find a reason why they're better than us. Our inner critic starts laughing at us. 'Their ex is better looking than you! More successful than you! More adventurous than you!' Everything positive about their past relationships and everything negative about us are amplified. Every new detail we spot is another data point to validate our insecurities. A picture of their ex travelling around Thailand validates our insecurity of not being outgoing enough. A picture of them at a fancy event validates our insecurity of not being well dressed enough. We imagine detailed scenarios based on the images we find, each one showcasing our new partner having a better time with their ex than they are having with us.

This tendency to negatively spiral makes maintaining the relationship challenging. It's difficult to enjoy the relationship because we're constantly questioning whether our partner still likes us. We analyse every word and every facial expression, scanning for signs that suggest our partner is about to abandon us.

The RSD mind has a bias towards negativity. If there isn't clarity over the meaning of what someone said, we assume the meaning is negative and jump to the worst-case scenario. For example, if our partner sends us a text saying, 'Can you call me?', we might assume that means they want to have a chat with us about ending the relationship. The dread of this imagined scenario will paralyse us into waiting mode; we won't be able to focus on anything else until we've called them. We desperately want to reply with, 'What did I do wrong? Please just tell me now!', but we're acutely aware of how crazy that will make us look. So we suppress it and wait, but this comes at the expense of what we are supposed to be doing.

The RSD love languages

There are lots of little things you and your partner can do in your romantic relationships that help with RSD. I have called these 'the RSD love languages' and they will be scattered throughout this chapter. The first one is to always disarm RSD first and then add clarity instead of uncertainty. For example, instead of the scary uncertainty of 'Can you call me?', you can ask your partner to add calmness and certainty by saying, 'Nothing to worry about at all, can we speak later about [insert specific thing here]?' The 'Nothing to worry about at all' instantly disarms RSD. It's short, clear and enough to put to sleep the irrationality. Next, making reference to a specific topic adds the clarity that the RSD needs to stop it spiralling.

For example, instead of 'Can you call me?', your partner could say:

> 'Nothing to worry about but can we speak later about the plumbing issue in the flat? We might need to buy a new boiler.'

> 'No need to stress at all but the dog's flea treatment has run out. Shall we talk later about that and whether we want to switch to a different type of flea treatment?'

> 'Sorry to bother you at work. Just wanted to ask if you're free tonight so we can watch that new film at the cinema.'

All these examples could have been communicated as 'Can you call me?', yet none of them is anything do with a conversation about ending the relationship. However, without the calmness and the clarity that each example gives, the RSD mind will always catastrophize and the person with RSD will assume their world is about to end. This very simple love language, which is really just a simple adjustment in how you and your partner communicate, will save a lot of unnecessary stress.

Are you mad at me?

It's not easy maintaining the fun and outgoing persona that your partner fell in love with when you're always feeling as though you've done something wrong or are about to be told off. It leaves you second-guessing everything, questioning your partner's every move and taking every little interaction as confirmation that they are mad at you.

It helps to be honest with your partner about RSD and let them know how much you would appreciate them letting you know when something outside of your relationship has caused them to be in a bad mood. For example, if you know they have had a bad day at work, you will understand that their low mood or energy levels are because of that and not because of you. If you knew they had just had an argument with a family member, you would appreciate that their lack of intimacy might be because they are stressed about the fallout from their family argument and not because of anything you've done.

Making it commonplace in your relationship to let each other know about things that are stressing you out will add vital context to actions your partner takes that might otherwise be perceived to be about you.

The triggering effect of unmasking

Do they still like me?

Dating as a neurodivergent person feels extremely vulnerable. Our past experiences in the world of romance have taught us that prospective partners either think we're 'too much' or we're not enough. We're aware how important first impressions are and we're also aware of how challenging we find making a good one. For example, we're either hyper-focused on the person and come across as clingy, or we are distracted and come across as uninterested.

We're also aware that our hobbies and interests change like the seasons. One day we might be interested in the solar system and another day we might be interested in the mating habits of blue whales.

Our minds move incredibly fast, and this can mean we talk very quickly or interrupt others because we need to do that before we forget what we need to say. Either way, we've been told enough times in our lives that this behaviour is not acceptable, so sadly we put on a diluted version of ourselves when we are meeting people on a first date.

A successful first date is evidence that they liked us and that they approved of the version of us that we turned up as. This gives us the signal to carry on acting as this character on the second date and so on. As the relationship progresses, however, maintaining this version of us becomes hard work. We might be seeing them two or three times per week, and we might eventually move in and live together. Unmasking in a romantic relationship is a gradual process that happens organically over time. You will both relax into your natural selves around each other as time goes by.

This can be a scary time for someone with RSD because the stakes are so high. It can be a very overstimulating experience because our racing mind is trying to make sense of so many different data points and so many tiny indicators of approval.

We know they liked one version of us, but will they like this new version?

Did that heavy footstep as they came into the kitchen mean they prefer the older version of me?

They haven't texted me in two hours. Was that little dance I did last night in the flat too much? Did it put them off me?

The kiss they gave me as they got out of bed for work was slightly shorter than normal. Are they annoyed at me because I info dumped everything there is to know about the *Titanic* last night? I thought they knew I like to talk in depth about different topics every month depending on what my current hyper-focus is. I thought our relationship felt safe enough now to reveal that side of me. Oh, damn, they definitely think I'm weird!

First, it's important to talk openly about masking with your partner, to have a joint understanding of it, how it's an adaptive survival strategy adopted by many ADHD people due to early experiences of social rejection. Forcing eye contact, suppressing special interests and internalizing physical hyperactivity gives ADHD people a protective layer which allows them to fit in where otherwise they might be outcast. But it comes at a great cost: exhaustion. This exhaustion in relationships will lead to physical burnout, leaving ADHD people wide open to emotional vulnerability.

If the relationship does not provide a safe environment that allows for unmasking, the ADHD partner will continue to pretend to be this fake version of themselves until the effort of pretending becomes too much and the mask falls away. In these moments, the ADHD partner might withdraw socially or snap at their loving partner. The ADHD partner will be very aware of how different their unmasked version is compared with their masked version of them. They will also detect the shock in their partner as this unsettling new version of them suddenly appears. This can trigger feelings of rejection in both partners. The ADHD partner will fear rejection from their partner because they have shown a new side of themselves and they don't have any evidence to reassure them that this version is loveable. The non-ADHD partner will feel rejected as they will interpret this sudden character change as an attack on them, or feel as though they have done something to upset the ADHD partner.

Relationships should be a place where you both feel comfortable to slowly reveal your true selves to each other. This needs to be done in a safe and fair way, one in which you can both take steps to communicate your thoughts and, over time, feel secure enough around each other to let the mask slip without judgement.

Awareness, patience and compassion are required to explore this process together. Dedicate time as often as you can, weekly or monthly, to openly communicate with each other about recent events and how you showed up at each one. These events could be social events or personal events that happened just between the two of you. Ask each other questions like:

- What happened last week that made you anxious?
- What happened last week that made you feel happy?
- When did you feel safest to be your true self?
- When did you feel unsafe?
- What did you say 'yes' to that you wanted to say 'no' to?
- What did you say 'no' to that you wanted to say 'yes' to?

The thought of someone not liking us or being disappointed in us causes extreme emotional pain. In the early stages of a relationship, this fear turns us into perfectionists. It's simply easier to pretend to be someone else as we know this version is more likely to be approved by the world and therefore our new romantic interest.

However, this is not sustainable or healthy in a long-term relationship. The masks need to come off, but this comes with its own risks of rejection if not done in a controlled way.

Communicating with each other about unmasking will allow you both to openly transition between the masked version of you and the unmasked version of you. Doing it slowly will also provide an opportunity for each of you to give validation to every new trait, mannerism or interest that your partner reveals along the way. A slow, controlled and openly communicated journey to becoming your true selves around each other is key. Receiving positive feedback about each baby step towards authenticity will starve RSD and enable you to reclaim so much energy that otherwise would have been spent on playing a character called 'Normal'.

I spoke to a late-diagnosed ADHD woman who shared a story with me about how she masked an RSD trigger from her partner early in their relationship. She had just met him. He lived in another town about half an hour away from where she lived. She was getting ready in her flat and was about to make the ten-minute walk to the train station. As she was leaving, she received a text message from him that said, 'I've just looked on the weather app and can see that it's raining where you are. If it's too wet, don't worry about walking to the train station. We can meet tomorrow instead.'

She instantly felt RSD and burst into tears. 'He hates me,' she sobbed at herself. 'He doesn't want to see me because he's met someone else and thinks I'm ugly!'

She wrote a reply that said, 'Why don't you want to see me? Have I done something wrong?' But she didn't send it. She looked at it, thought, 'This sounds crazy', and deleted it. She forced herself to write a nice message that said, 'It's really kind of you to think about me like that. Okay, let's meet tomorrow, x.'

They're tolerating me

RSD makes you believe that everyone close to you is politely tolerating you. It's easy to slip into 'don't speak unless spoken to' mode after you've sensed your partner is ignoring you. Depending on how emotionally vulnerable you are – maybe you've had a volatile day at work, for example – the tiniest interaction with your other half could give you all the information you need to convince yourself that you're not worthy of their attention, nor do you deserve any kind of romantic relationship.

In this state of mind, it's common for someone with RSD to sabotage the relationship. We instinctively want to push the person away before they push us, even if the threat of them pushing us away isn't real.

The signal, compliment and distract technique

In these moments, it helps to have a code word or a code signal that the triggered partner can use to inform the other of their emotional state. A tap on the shoulder or a secret word that's only known to the two of you can be used to invite the other to compliment you.

The immediate antidote to RSD is to receive praise. As outlined earlier in this book, it is instantly soothing to hear or read good things about yourself in the immediate moments after being triggered. It especially helps when the compliment leads you into a conversation that will distract you from the event that triggered you.

For example, in my relationship, my partner knows that I like to talk about the progress of my podcast, *ADHD Chatter*. It's a topic that's full of positive points about me as a person; I am ambitious, creative and passionate about the podcast. When something triggers me in our relationship, I say, 'I'm feeling blue right now', and my partner understands that this is a time to compliment me on the podcast. She'll say something like 'The podcast is breaking records right now.' It acts as an immediate compliment that snaps me out of the horrible feeling of RSD and redirects us into a conversation about my podcast, a topic that's packed full of further compliments. Depending on how long you and your partner have been together, a code signal might not be necessary. Your partner might know you well enough to spot when you have been triggered and immediately implement the signal, complement and distract technique.

This technique can be discussed in your weekly or monthly couple communication sessions so you both understand what signal will be used.

Reading between the lines and assuming the worst

I'm 36 years old as I write this and still replay tiny interactions with my partner in my head for days. What did they mean by that? Why did they slam the plate cupboard when they spoke to me? They revved the car engine louder than normal when they pulled out of the drive for work this morning. They're going to break up with me when they get home.

RSD wants you to believe that you're crazy. It finds clues to disaster where there aren't any. It makes you incredibly lonely because you feel unable to ask for help, unable to seek reassurance because last time you did that you got shouted at and told that you're too sensitive. You've told yourself a million times that you can't reach out and ask for help because this is all your fault. However, when we have a true understanding of ADHD and RSD, we see that this is not our fault. We're not too sensitive or too dramatic; we're just really good at scanning for little signs of danger because our nervous system has learned to expect them. Having this realization means we can ask for help without shame and without fear of ridicule. We didn't deserve those 20,000 horrible messages.

A successful relationship in which one or both partners live with RSD is one that understands and accommodates the complexities of RSD. It understands that simple adjustments can really help remove the damaging effects of RSD from the relationship.

A powerful adjustment is when we make it commonplace to say, 'What did you mean by that?' Without this question, many people with RSD will second-guess the meaning of something their partner has said. With a bias towards negativity, this can be disastrous and lead to the start of a spiral that can turn into a row. However, when we use the phrase 'What did you mean by that?', we remove the uncertainty and the catastrophization that comes with it.

How many times have we started a fight with our partner because their body language was communicating something different from what they just said? Sometimes all we need is one extra sentence to add another layer of meaning to what they said to us.

For example, to someone with RSD, 'I'm tired. I'm going to bed' could mean 'I'm bored of your company. I'd rather spend this time on my own in bed on social media.' When we ask, 'What did you mean by that?' or, in this case, 'Why are you so tired?', the answer might be something like 'I've had a long day. I'd really love to spend another hour on the sofa with you, but I promised my boss I'd do the morning shift tomorrow.' This adds clarity to the reason for being tired and reassures you that it's not personal.

Your nervous system knows how to survive in love, not feel safe in it

The 20,000 negative criticisms you received when you were a child have taught your nervous system to expect danger at every corner. You can't relax because your normal state of being is 'fight or flight', always scanning for the next attack. Many people with RSD go through years of their life in this heightened state without even knowing it. They are experts at appearing normal, but under the surface they are primed to defend themselves when confronted with a rejection. This defence comes in the form of a sudden snap, an outward burst of emotion or an inward shutdown.

This constant state of being in 'fight or flight' makes sustaining a loving relationship challenging, but not impossible. Even with the knowledge that it's not our fault, we need to be aware of the signs that tell us that maybe the relationship simply isn't a good fit.

Five signs you're confusing love with RSD avoidance

1 YOUR INTUITION IS SCREAMING AT YOU, TELLING YOU SOMETHING DOESN'T FEEL RIGHT, BUT YOU'RE TOO AFRAID OF CONFRONTATION TO ADDRESS IT

This is very common in people with RSD. The underlying intuition that comes with ADHD is a powerful force. It has learned how to immediately detect danger. This ability turns you into a human lie detector. It picks up on tiny fluctuations in tone or mood before your conscious mind has a chance to see them. It knows how to read people and can spot BS a mile off. It's an expert judge of character. It sees the sliminess in that person who everyone loves and, when they turn out to be a wrong 'un, you're left saying, 'I told you so.' When your gut says something feels off, listen to it – it's been hit with 20,000 nasty comments in the past and it knows how to spot another one before anyone else sees it coming.

Your gut could be telling you something is off, but your fear of addressing it and therefore triggering your RSD stops you from saying anything. It's easy to bury your head in the sand and convince yourself that the relationship is strong and full of love; it's the easier path to take and one that avoids the pain of RSD. But it's important to recognize when you are avoiding RSD at the expense of your self-respect. It takes tremendous courage to confront your partner over your suspicions even if it means you're likely to be triggered by a confrontation.

2 YOUR PARTNER MAKES YOU FEEL MISUNDERSTOOD

It's common to feel misunderstood at the start of relationships because you're both new to each other and you're having fun getting to know one another. As time goes on, however, if you're still feeling misunderstood by your partner, it may be a sign that your RSD avoidance is blocking your true self from coming out. You may be subconsciously hiding your authentic self out of fear of it being criticized by your partner.

3 YOU MAKE YOURSELF SMALLER TO AVOID UPSETTING THEM

It's not real love if you're having to tone yourself down when you're in their presence. You're not too much, and neither should you feel like you are. It may not be a fault in your relationship. Many people with RSD have been told off for being too loud or too energetic in their past. It feels safer to suppress our true self from the world. Relationships, however, should be a safe space for your true self to slowly come to the surface. Having an honest dialogue with your partner about your fear of unmasking should enable you to make small steps towards being your true self.

4 YOU PRACTISE CONVERSATIONS WITH YOUR PARTNER IN YOUR HEAD BECAUSE YOU'RE ANXIOUS ABOUT HOW THEY WILL BE RECEIVED

RSD avoidance will make you rehearse every social interaction before it happens. In our RSD mind, it's easy to think that we won't be criticized if we've carefully planned every single sentence that comes out of our mouth. This level of overthinking is common before social gatherings with people we don't know very well, but it shouldn't be common in our romantic relationships. Unmasking is all about being able to speak organically and without rehearsal around each other.

5 YOU HAD MORE SELF-CONFIDENCE BEFORE THE RELATIONSHIP STARTED

A loving relationship is one that creates a safe space for both people to be their true self, free from the anxiety of being judged for their quirks. Our self-confidence should grow as our relationship progresses: our partner should praise us every time we show them a little bit more of our true self, and vice versa, over time creating a couple who are confident in being their true selves around one another. If the opposite happens and we become less confident in ourselves as the relationship progresses, this could be a sign that RSD avoidance is winning — our true self is being withheld for the sake of pleasing our partner.

The solution

Ultimately, RSD avoidance is masking. All humans mask to a certain extent in an effort to protect themselves from criticism and to progress through life. However, in our romantic relationships, we want to be able to be our true selves. If you're masking too much in your relationship, this can be a sign that you're not compatible with each other or that there is a lack of honest communication between you. It helps to discuss with your partner scenarios where you weren't showing up as your true self.

To do this successfully, you need to understand what your true self is. Ask yourself this question: How do you show up when no one else is watching, when there is no one around to judge you? Look for scenarios in your life that enable your true self to come to the surface.

Here are some examples in my life:

> When I'm driving on my own down country lanes. The music is blasting out of my speakers. I sing my heart out. I turn into the next Beyoncé. In my head, at least.
>
> When I'm in the shower. I'm still surprised I don't have a record deal.
>
> What I wear when I'm on my own. What clothes give me the least sensory issues?
>
> How I move when I'm on my own. What are my facial expressions, stims and movement patterns?
>
> What I say 'yes' to when no one is putting pressure on me. My true self knows what truly excites it.

What I say 'no' to when no one is putting pressure on me. My true self knows how to protect my energy.

What I do when no one is watching? Sometimes I dance, sometimes I stare blankly at the TV while thinking about my special interest. Ultimately, I do whatever I have the energy for. I rest when my body tells me it needs it, even if it looks lazy to other people.

I pay attention to what I say 'sorry' for. What apologies did I truly mean and what ones did I give simply to please someone else?

Showing up as your true self, honouring your own needs and creating a world that fits you rather than squeezing into one that doesn't shouldn't feel like an inconvenience. The same applies to your relationships, too. Pretending to be something you're not simply to avoid the pain of RSD is exhausting and will wear you down. When you truly understand who you really are, you can have meaningful conversations with your partner about masking. These conversations will allow you both to highlight moments in your relationship that feel like a betrayal to your true selves and moments that honour your true selves. As more of these conversations happen between you, and as you both build a better understanding of your true selves, you'll notice your relationship feels more aligned with the unmasked version of you and less like a constant performance designed to avoid the pain of RSD.

We are hypersensitive to our partner's feelings

Living with RSD makes you extremely empathetic to other people's emotions. We feel people's emotional pain as if it were our own.

It's easy to assume everyone else experiences life with the same sensitivity as us, so when we witness someone else being criticized or victimized, it's as if we get third-party RSD – we feel the effect of their trigger, too. This means we're extremely good at comforting people when they are in need. However, this ability to understand our partner's emotional state can lead to us obsessing over it, constantly reading their body language for clues to whether they are happy, or mad, with us. It feels stimulating when we find one. A sign that they are happy with us feels great, a sign that they're mad with us feels terrible, but both feelings can be stimulating enough for us that we seek them out regardless of whether they are positive or negative.

We find ourselves obsessing over our partner's needs because that lowers the chance of them rejecting us. Our wants and needs become less important to us as keeping them happy becomes our main priority. This focus on your partner's needs can be appealing to them at first, but over time it can erode the respect between you as the dynamic shifts from a romantic relationship to a parent – child relationship, with you being the child.

This shift in dynamic can lower the attraction levels in the relationship, which can lead to a real rejection.

It's easy to spot a change in your partner's mood when their mood is your primary focus. The smallest shift in their energy can leave you wondering if the relationship is about to end. It's easy to find a sign of something being wrong if you're determined to find one. If you find a perceived sign right before they leave for work in the morning, you can spend the day spiralling into a flurry of negative thoughts, each one bringing you closer to the belief that they are going to break up with you when they get home from work.

These spirals are horrible. You convince yourself that your partner is mad at you, is going to break up with you or they are cheating on you. You look through their social media accounts, searching for signs, a new follower, a 'like' from someone attractive, anything to validate your fears. You look through all the people they follow on social media and check if your partner has engaged with any of their content. You doubt your entire relationship.

You have brief moments where you're able to reassure yourself that there's nothing to worry about, no real evidence that points towards infidelity. You sit in peace for a few moments as you breathe a sigh of relief. But then it comes back: the way they left for work this morning, the way their kiss goodbye felt more hurried than normal, the way the front door slammed slightly harder as they closed it behind them.

They're in a rush to get to work so they can flirt with someone, you tell yourself. You can't focus on anything else all day. You sit down at your computer, desperately trying to start the work you need to do, but you find yourself back on the social media apps, scanning through your partner's profile, hunting for clues.

You play back every negative thing your partner has ever said to you. That time they said they play your voice notes on double speed, every argument, every raised tone of voice and every microagression – it all gets repeated in your brain as if being amplified with a megaphone. All the positive things they have said cease to exist. The rage builds inside you as the thought of them breaking up with you becomes clearer and clearer. You desperately need to seek reassurance or challenge them on their infidelity. It's the only thing that's going to scratch this excruciating itch. You resist the urge to call them. You're fully aware that only 7 per cent of communication is verbal – the actual words that come out of their mouth – the rest is tone, facial movement, body language, speed of speaking, eye contact (among other things), and a phone call won't allow you to absorb all this vital information. You simply have to confront them in person.

Hours go past. Your partner walks through the door. You don't want to appear crazy, so you internalize your rage. You continue to scan their every movement and every word for further information, signs that they're mad at you and no longer love you, but your partner has intuition, too, and immediately senses that you're acting differently. So they ask, 'Is something wrong?'

Two things can happen at this stage. You either explode and release all your suspicions or you retreat into the relative comfort of RSD avoidance and hide your suspicions. You confront them until the itch has been scratched and you feel reassured or you go quiet and become distant, desperately wanting to punish your partner for their disgusting behaviour but also not wanting to engage with them as this will expose you to a real rejection.

Any reassurance you get will be short lived, providing a few days' respite before the sinister thoughts of abandonment come creeping back, and with every mission to seek reassurance, every new conversation you have will only plant more seeds for future rumination. 'They looked fed up with me this time,' you'll tell yourself as you know this new piece of information will be the source of your next spiral.

How to overcome the RSD-fuelled spirals

RSD makes us expect rejection. Our nervous system, pumped with those 20,000 negative comments, expects it, even from those closest to us. It won't let us be happy in our relationships without a fight.

We need to be totally honest with our partner about our RSD and the intrusive thoughts that come with it. Not many people understand the complexities of RSD, even the ADHD community is only now beginning a journey of awareness into the topic, so there's a very high chance your partner won't have an understanding either. Many people assume that the thought patterns that come with RSD are considered and optional, that we have a choice and decide what thoughts enter our mind, but for a lot of us, there is an obsessive nature to the thoughts – we're unable to choose whether or not we spiral.

There will be a lot of people in your life who don't understand RSD. They will give you advice that makes you feel like you're crazy. Only take advice from people who understand it. Picking up this book is a brilliant first step.

The 'all or nothing' mentality of ADHD means we either feel love from our partner or we feel hate from them; they are either happy with us or they are mad at us; they either want to marry us or they want to break up with us. There is no middle ground when it comes to the RSD-riddled ADHD mind. Your partner needs to understand the extreme mood swings that come with this 'all or nothing' mentality.

Saying to your partner 'I have RSD' is the first step towards managing it in your relationship. This step can be as brief or as long as you feel comfortable. You can say something like this:

You know I have ADHD, right? A big part of the
ADHD experience is something called RSD. I
received lots of horrible bits of feedback from the
world when I was younger, so I react really badly
to criticism as an adult. I always feel like I've done
something wrong and that important people in my life
are going to walk out and abandon me. This makes
me hypervigilant to everything you do and say,
and sometimes that means I turn innocent actions
into sinister ones because I have a bias towards
negativity, because that's what my nervous system
expects. This means I might get angry or sad at you
sometimes because I took something you said the
wrong way, or perceived something you did as a sign
of you falling out of love with me. All these actions
come from a place of fear. I'm terrified of having my
past repeat itself. I want to love you and I need a little
bit of extra reassurance from time to time.

Having this conversation with your partner isn't as scary as it sounds. All humans experience a level of sensitivity to rejection, so hearing you open up about yours will be relatable to them and make them feel more comfortable, too. They will no doubt reveal that they also experience some level of rejection sensitivity and that they can empathize with what you're experiencing. They can now react to your behaviour with an empathic level of curiosity rather than judgement. Now, when they sense any coldness from you, rather than get frustrated and confused they can offer you a cuddle and reassure you that they still love you.

The details don't matter – RSD will find a way

Honest and open communication with your partner about RSD is the best way to make sure it doesn't completely derail your relationship. This communication should include details of how irrational RSD can be sometimes. These details will allow your partner to contextualize the event that triggered your RSD and provide you with reassuring evidence that contradicts what your brain was telling you at the time.

For example, your partner might be driving you somewhere when another driver cuts in front of the car. Your partner gets irritated at the other driver and sounds the car horn. A few minutes later, you ask your partner an unrelated question and they reply with a short answer and a tone that suggests they are in an irritable mood.

Many people with ADHD experience an unusual perception of time – they either live in the 'now' or the 'not now'. In other words, they hyper-focus on things that are happening right now but struggle to pay attention to things from their past or their future. This creates something called 'object permanence', a difficulty in remembering things that aren't happening right in front of them, and therefore a difficulty in bringing details from the past into their current moment. This struggle means they can't add important context from past events into how they react to moments in their now.

In the example of the other driver cutting in front of your partner's car and then your partner being short with you in your next conversation, many people are able to assume that the mood change within your partner is connected to the event that happened a few minutes ago – the annoying driver cutting in front of them. The RSD brain, however, struggles to make this connection. It only focuses on the information it has right in front of it, which in this case is your partner speaking to you in an abrupt tone. The brain makes an instant leap from your partner's abrupt tone to the thought of your partner being annoyed at them. The ability to rationally connect their abrupt tone with the annoying driver simply isn't there; instead, you believe they hate you.

Moments like this are an example of what should be discussed with your partner during your communication sessions. These can happen weekly or as often as you both are able. They are used to reflect on your RSD triggers during a particular time period. Making a note of your triggers throughout that period can be helpful to remind you of them.

For example, writing down 'driving trigger' will remind you to ask, 'Do you remember that time you were driving me somewhere and you snapped at me? That made me feel really sad.' Your partner can then add important context which will explain why they sounded abrupt in that moment.

While regular communication sessions are useful to help you both understand previous triggers, it's also good to have tools to manage RSD in the moment. RSD, if not managed in the moment, can escalate into visceral verbal attacks which can cause significant damage to your relationship. As a couple, you can agree on a code word that can be used to let each other know that you've been triggered and you need time to decompress. You can use this time to reassure yourself that your partner loves you. You could, for example, play a pre-recorded voice note from your partner. RSD grows with criticism and shrinks with praise. Asking your partner for a short voice note in which they express their love for you and explain the reasons why they love you will help bring you out of the RSD spiral. It's useful to bring this up in your communication sessions and ask if a new voice note could be recorded once a month, or more often if your partner has the time. The RSD brain will likely have a sneaky way of convincing you that voice notes recorded a long time ago no longer hold their meaning.

Work with your partner to understand what triggers your RSD

People with RSD live their life in fear. Your RSD is triggered when you are confronted with that fear. Coming face to face with what scares you makes you react in volatile ways as you try to fight it or run away from it. However, most people with RSD don't truly understand what their fears are; they only know when they have been triggered by one.

Part of your RSD healing journey is to understand what your fears are. You could be fearful of being cheated on, abandoned, made to feel like you're not enough, like you're too much, or as though your partner's previous relationships were more exciting for them than their relationship with you.

You can identify your fears when you reflect on your previous RSD triggers:

- What do they have in common?
- What topic were we talking about when I was triggered?
- Where were we when I was triggered?
- What objects were near us when I was triggered?
- What music was playing in the background?
- What were we watching?

Having an understanding of your triggers can enable you to limit your exposure to them in the future. It also enables both you and your partner to pre-empt triggers if you suspect a particular environment or conversation might include one.

It's overwhelming trying to make a relationship work when you're always worrying that your partner doesn't really love you and that they're simply tolerating you until something better comes along. This fear gives you an undesirable power: it enables you to assume the worst every time you detect an interaction with your partner that isn't overtly positive. RSD makes you assume the worst when there is ambiguity, and any lack of extreme positivity gets translated into extreme negativity.

Reminding yourself in these moments that you're not crazy, not alone and not broken is important.

'I'm upset with you right now'

As a couple, it's useful to have a rule whereby you say it will be made explicitly clear if one of you is annoyed with the other. If one of you is genuinely mad at the other one, there will be no ambiguity around communicating this; it will be made totally obvious.

You need to both be comfortable saying 'I'm mad at you right now' when you are mad at each other or 'I'm upset with you right now'. This clarity of communication will create some uncomfortable moments, but it removes far more uncomfortable moments because you will know when your partner is genuinely upset with you.

RSD triggers happen in the absence of clarity. We assume our partner is upset with us when most of the time they won't be. When we know moments of true upset will be clearly communicated, it gives us the power to disarm RSD in moments where that clarity is not given.

Know when it's time for a 'state change'

An RSD trigger can quickly escalate an otherwise benign conversation with your partner. Your partner may have said something minor that triggered you and that meant you retaliated in a disproportionate way.

It's important to recognize when you're triggered and understand that state change is often the best course of action. A state change is when you physically remove yourself from the current situation.

For example, if you're both in the kitchen and your partner says something that triggers you, your state change could be to take yourself out of the kitchen and into the bathroom. If you're both in the garden and something triggers you, your state change could be to take yourself for a walk.

A strategically adopted state change allows you to put physical space between you and your partner. It also creates a time break between the trigger and what is said next. Without this physical and time gap between the trigger and what happens next, it is likely that the RSD person will snap back at their partner, or withdraw emotionally, both resulting in the other partner shouting back or asking, 'What's wrong with you?' Both outcomes will cause an escalation – the RSD person will take their partner's escalated response as another criticism and the RSD will be further triggered.

The state change strategy works only if you both understand it. Otherwise, the RSD partner walking away from the situation will look like a passive act of retaliation. Their partner might think, 'They've just stormed out of the room. They must be mad at me', when in fact they've left the room in order to avoid reacting in the moment.

When you have your communication sessions with your partner, you can say something like this:

> When something is said or done that triggers me, I will say, 'I've been triggered. I'll be back in 30 minutes.' This will enable me to communicate to you that something you have said or done has been perceived as a criticism and, rather than me react in a heightened emotional state, I am going to state-change to give myself the opportunity to decompress before I come back to you. We can discuss the trigger and why I was upset at a later date.

Without this state change we run the risk of escalating the situation and acting in a way that exposes us to a real rejection. Extreme fear of rejection often leads to behaviour that makes a rejection almost certain. However, we can intercept this behaviour by removing ourselves from the situation for long enough that we're able to come down from our heightened emotional state.

Signs you're entering a high emotional state include the following.

THOUGHTS ABOUT YOUR PARTNER ARE HYPERCRITICAL

We project our insecurities on to others when we sense we are under attack. Our mind is powerful and it often gets ahead of an incoming rejection. Our subconscious mind stores patterns of behaviour. It remembers what actions your partner took right before they were last mad at you. This enables us to pick up on the actions when they happen again. This display of intuition is an automatic body function, but your mind will react to it by directing critical thoughts towards your partner. Your brain is essentially loading its ammunition, ready to retaliate when the predicted criticism from your partner lands.

YOU START BEING HYPERVIGILANT TO THINGS AROUND YOU

Things that don't normally annoy you start to do so. The sound of a bird chirping outside, the beeping of a lorry reversing, or your partner chewing sends you into a fit of silent rage. Entering into a heightened emotional state places us in a 'fight or flight' mode which makes us very aware of dangers around us. We observe every single detail of our surroundings, often leading to overstimulation and a very low tolerance for unwanted noises.

YOU WITHDRAW EMOTIONALLY

People with RSD hold a lot of shame within them around how they have acted previously during other heightened emotional states. Maybe they said something they didn't mean, or stormed out of the house and slammed the door behind them. Acting this way often enough can create a fear of those actions being repeated, so, when triggered, that person will deliberately internalize their emotions in an effort to protect themselves from future shame.

The brain of someone who has RSD is intense. It's what makes us fun and exciting to be with. The person without RSD will often sense when their partner is approaching a state of heightened emotion before they know it themselves. They, too, will remember patterns of behaviour from previous triggers within their partner. Sometimes it takes the non-RSD partner to be confident enough to suggest a state change before their partner does.

It needs to be understood by both partners that when either one of them signals it's time for a state change, it is not intended to shut anyone down or to be patronizing but it is for the benefit of the relationship. Nine times out of ten, when a state change happens, the couple forget about the heightened emotional state and move on quickly, having successfully avoided a big RSD-fuelled argument.

Being aware of these early warning indicators will enable you to notice when you're at risk of being triggered. There should be zero shame associated with recognizing these traits within yourself, and rather a sense of empowerment because you have a new understanding of yourself.

I don't deserve to be in a successful relationship

The inner critic that lives in the RSD mind is a stubborn beast. It tells us things that we wouldn't tell our worst enemy. This constant internal monologue makes it hard for us to believe that we are worthy of love, despite our best efforts to prove it wrong.

Even with the best intentions that come with setting aside time to openly communicate with our partner, our tendency to forget things that we feel are not important to us can hinder our progress. For example, we can have lengthy chats with our partner about RSD and our known triggers but we can also forget what we say in these chats. Our brain has a habit of deprioritizing self-care because it believes we don't deserve to improve ourselves. The echo of those 20,000 negative criticisms acts as a reminder to not store the contents of our chats in our subconscious. Our brain applies object permanence, when something is out of sight and therefore out of mind, to the work we do with our partner. It's one of the drawbacks of the ADHD tendency to always live in the current moment.

To combat this, we need to work a little bit harder to store the memories of our chats. For example, if we chat with our partner about an event that triggered us and we determine that a particular topic was the cause of the trigger, rather than trying to remember that that topic triggers us, we need to write it down in a visual way. For me, this involves colourful notes that I stick on the wall in my home. Over time, you can identify what topics are the cause of the most triggers. These will be your core topics to avoid talking about.

For example, your core trigger topics could be your partner's ex, a particular location, or a particular object that reminds you of a previous rejection. When you've put in the work with your partner to identify your core trigger topics, you can find a way to keep the memory of these topics in your mind. For me, this is by giving each trigger topic a nickname and putting each one on a bracelet charm. The nickname is known only to myself and my partner. You can also keep your nicknames printed on the side of mugs or within pictures hanging on the wall in your home. The point is to keep them on show in places that you see regularly. This will help minimize the 'out of sight, out of mind' effect and keep them front and centre so you remember not to bring them into a conversation.

Make a relationship that enables you to let go of the you that existed only to please others

Communication is the biggest tool in your arsenal to combat RSD. There needs to be absolute clarity within your relationship surrounding RSD and what it actually is. Without clear communication, your partner might believe RSD means you think they don't pay you enough attention when in fact it means you genuinely believe you're a crappy person and that your partner would be happier if you were not in their life. RSD is thinking your partner hates you when you haven't done anything wrong. It's an intense feeling of disliking yourself and believing everything you do is not good enough. The feelings of shame are overwhelming and can completely debilitate us.

This description seems bleak and difficult to fully comprehend for people who don't live with RSD but it is the truth for many that do. We shrink ourselves down in order to make other people feel big – our own wants and needs are put at the back of queue and we put everyone else's first.

When you open up about your RSD with the person closest to you, you invest in yourself and your relationship. You take the shame away from something that hides in the shadows and you give your partner the context they need to make sense of behaviours that might otherwise feel confusing.

If you are both kind to yourselves and share a commitment to growing your relationship, you will be able to deepen your connection and strengthen your bond.

PS: They're not mad at you.

4

Taking the sting out of RSD

Some practical strategies

I had a massive RSD episode this week. I was due to record a podcast with someone I really admire, and as I was sitting on the train on my way to the studio, they emailed me saying, 'I won't be able to make it today. A family emergency has happened and I need to be there. So sorry for the short notice.'

My chest tightened and my mood changed instantly, from feeling relatively content (if a bit anxious about the podcast recording) to seething rage and shame. I looked up and made eye content with a few of the other train passengers, fully aware that none of them would have any idea about the monumental mood shift that had just occurred inside me. I could feel my heart thumping and my back beginning to moisten with sweat. And then came the thoughts:

'They are lying. They hate me. Someone has told them I'm a nasty person and they no longer want to be associated with me.'

I reached down for my phone and looked at the email again. My eyes were moving through it, but nothing was being absorbed by my brain. I was suddenly in survival mode. I needed to find out why they had really decided to cancel.

I clicked 'Reply' and wrote, 'Hi, so sorry to hear that, thinking of you. Can you let me know if I've done anything wrong or if I've upset you in any way?'

I hovered my finger over the 'Send' button and moved it closer to the phone screen, but then I stopped myself as I remembered everything I had learned about RSD. Every single piece of advice I had been given came flooding back to me and just as this happened, I realized – I had successfully accomplished the first step of RSD management.

Step 1: Give your strong feelings a name

It's impossible to manage something without defining it first. I call RSD my 'ADHD nasties', and when I'm triggered, I tell myself they have come to the surface. I then remind myself that I am wired differently from many people and that there is a neurological reason for this feeling. With this reminder, it's easier to remind myself that I'm not broken, just different, and that I've always been enough.

This first step is often enough to bring me out of the initial intensity of an RSD flare-up.

But not always, and that's when Step 2 comes in handy.

Step 2: Remove yourself from the situation

The intense feelings that come with RSD are often too overwhelming for Step 1 to work, so it's important in those instances to remove yourself from the thing that has triggered you. In this case it was an email, so it was important for me to put my phone down and stop myself from responding when I was in a heightened state of sensitivity.

If a person triggers you, it's important for you to remove yourself from their proximity as soon as you can. This may not always be possible, of course, and I will outline what to do in that scenario later on.

When RSD is triggered, it's common to see red and lose sight of rational thought. Something has happened – a remark, or a boss making a small correction to our work – that has snapped our nervous system back to the memory of those 20,000 comments. The intense emotions that come with RSD are not directly caused by the thing that has happened to you just now but instead by the cumulative effect of being told you're 'not enough'. This can manifest as rage or bursting into tears.

It's important to remove yourself from the thing that has triggered you because your reaction will not be proportionate to the trigger. For example, if a friend texts you saying they are too busy to see you, replying with 'Do you hate me?' might seem like a disproportionate response. If your boss says they need to make a few alterations to the piece of work you've just submitted, it might seem disproportionate to rage-quit your job. Or in my case, if someone cancels on me at the last minute, asking if I have upset them in any way, would also not be a proportionate response.

Removing yourself from the situation allows you time to complete Step 3.

Step 3: Remind yourself how valuable you are

After you have removed yourself from the situation, it helps to flood yourself with evidence showing how amazing you are.

There are two ways to do this. First, you can look at a list on your phone or in your journal of your achievements and positive character traits. This is a crucial exercise in RSD management that I encourage everyone to do.

Find a time when you are feeling calm and relaxed. Write down as many positive things about yourself as you can. For example, you could write, 'I am good at my job because I am creative, intuitive, a great problem solver, empathetic, calm in a crisis and an out-of-the-box thinker.'

Or you could write, 'I am a good romantic partner because I am loyal, kind, spontaneous and adventurous.'

Having this list to hand when you are experiencing RSD can be really useful. However, the irrational nature of RSD can sometimes make us think that our own positive words and affirmations are not true. In this scenario, it can be helpful to implement the second way, which is to speak to a trusted person who can also remind you how valuable you are. This can be a partner, a friend, a family member or anyone with whom you have a strong bond, someone you can make contact with when you're experiencing the intense feelings that come with RSD.

Ultimately, these three steps in RSD management enable you to ground yourself. They provide an opportunity for you to self-regulate, to soothe your misfiring nervous system and to protect yourself from reacting on impulse.

Don't make decisions when you're raging

I've spoken to thousands of people who have RSD and I've heard so many stories of people reacting impulsively and making huge life decisions in moments when they were raging. Some of the most memorable examples were:

- quitting a job
- saying, 'I think we should break up'
- unfriending or blocking someone on social media
- leaving the group chat
- writing an essay response to the person who triggered them.

Holiday RSD and refund woes – two stories

Sasha told me about a holiday she had with her husband after surprising him with a long weekend away for his birthday. After they arrived at the hotel, they went for a walk around the town. Sasha said, 'Do you like the hotel I booked?' and her husband replied, 'Yes, it's lovely. The town's a bit quiet but it's a lovely place.' Sasha explained to me how she heard 'the town is a bit quiet' and her mind translated it to 'this is a terrible holiday'. She instantly felt rage and replied, 'You're never grateful! I stayed up all night researching this holiday. I desperately wanted you to have a good time and now you've ruined it. I think we should get a divorce!'

She explained how the next three hours were a bit of a blur in her memory; she vaguely remembered her husband looking shocked and asking, 'Have I done something to upset you?'

Sasha stormed back to the hotel room and packed her suitcase, with a confused husband in tow. Her next memory was of feeling the rage dissipate, but this was a little while later, as by then she was sitting on a plane, flying home on her own.

'I felt terrible!', she exclaimed to me. 'I wasn't thinking rationally. I called my husband and apologized. I got the next flight back to him. It was the most expensive RSD overreaction of my life! He met me at the airport with a smile on his face. We hugged and it was okay, but I can really lose control sometimes.'

Jay spoke to me about his most embarrassing RSD episode. He explained how he had to visit a shop to return a T-shirt because it didn't fit correctly. He approached the desk and said to the retail assistant, 'Can I return this T-shirt, please?' The shop assistant replied, 'Yes, of course,' but Jay told me he detected a slight tone of annoyance and condescension in the shop assistant's voice. Needless to say, Jay maintained his composure as he received his refund but then cried in the back of the taxi on the way home as he left a review on the shop's website saying 'RUDE STAFF. WILL NOT RETURN'.

We wish we could get a handle on our emotions when we're in the moment, but often we can't. What's important is that we're aware of this fact and put systems in place to stop us from taking actions when we're in this heightened state of irrationality. Because the truth is, when Jay 'came down' from his heightened state, he was able to rationalize that he probably wouldn't ever see that retail assistant again, the shop retail assistant wouldn't remember the tiny interaction with him, and a million other things could have been responsible for his apparently annoyed tone of voice. It almost certainly wasn't because the shop assistant hated a stranger he had just met.

The trouble with RSD is that it doesn't matter how irrational the situation is, it doesn't even matter how aware you are of how irrational the situation is, and it definitely doesn't matter how prepared you are to be triggered by something irrational. Once the trigger happens, the chain reaction has started, the tidal wave of disproportionate emotions begins and there's nothing predictable about what might happen next. The only predictable thing is your ability to remove yourself from the situation, not to act in the moment and not to make big, impulsive decisions when RSD is steering the ship.

How to remove yourself

If you want to avoid unnecessary expensive return flights to reunite with your confused husband, or avoid re-reading cringey reviews on shopping websites, it's a good idea to be prepared to act when RSD strikes. You need to plan your exits.

In other words, you need to have an escape strategy in place for when you're in a situation where RSD could happen. The plan needs to enable you to retreat from the primary source of the trigger and towards a safe place where you can regulate your emotions.

For example, if you're in a restaurant, the plan would involve saying, 'I need to use the loo/restroom.' This would allow you to stand up and disappear for several minutes. You can call your safe person or do breathing exercises while in the restroom.

If you're at a social event, the plan could be to pretend to receive a phone call. This would enable you to leave the situation. You can expand this plan by coming back to the social event and saying the phone call was an emergency and you need to leave.

Learning when and how to leave a situation is an important step in RSD management. It will always be possible to continue a conversation later on, but sometimes it will be impossible to repair the damage caused if you stay. Learning this simple fact was a huge step forward for me in understanding how to manage RSD.

After you remove yourself from the situation, it helps to know some basic regulation tools. For me, this looks like focusing on my breathing. I take a deep breath in through my nose, hold for two seconds and then exhale through my mouth. Doing this 5–10 times helps me to control my circulating thoughts and takes me out of the intense feelings of RSD. After all, when we are struck by very intense emotions, we lose control of many things – except our ability to breathe. It's in these moments that we need to seize back control in any way we can. By focusing on your breathing, you tell your nervous system that you're fighting back – only in a small way at first, but it will be enough to put you back in control and break the feedback loop that's feeding your anger.

Separate fact from fiction

It's also important to remember that anger is a normal emotion that has an evolutionary purpose. It is a defence mechanism used to protect us from danger and to warn us when our safety is being compromised. However, the RSD nervous system struggles to differentiate between a real threat and a perceived threat, so it's crucial to analyse what triggered the anger and recognize if the anger is justified. Nine times out of ten with RSD, the anger won't be justified and will be coming from a place of heightened sensitivity (because of those 20,000 early criticisms, of course).

With RSD, big emotional responses are rarely rooted in truth, so it's important to decipher between what's real and what's fiction, what's true and what your brain is making you believe is true. You can do this yourself after some time has passed and you have reached a place of regulation, or you can speak to a trusted person if you need to make that determination right now, when you are in the heat of the moment and RSD is blocking your ability to think rationally.

Another holiday RSD

Lee shared with me a story about the time he was on holiday with his wife. They had made friends with another couple in the hotel restaurant and shared a lovely meal together. Lee went down for breakfast the next morning and saw his new friends having breakfast with another couple. His RSD was instantly triggered. He said his mind instantly started racing with thoughts such as, 'Did I offend them last night?' He went back to his room, slammed the door and said to his wife, 'They hate me!' His wife, who was able to rationalize the situation, reminded him that their new friends had booked an early excursion and had to have an earlier breakfast so they could catch the bus. Lee had simply forgotten a tiny detail which, had he remembered, would have given him the facts he needed to make a rational judgement: they didn't hate him, they needed to catch a bus to go scuba diving. The two couples also ended up having another lovely meal together that evening.

Separating fact from fiction can be challenging for someone with RSD because you are trying to separate the cumulative memory of those 20,000 early criticisms from one tiny criticism today. In other words, you are trying to tell the difference between a suspicion based on memory recall and a suspicion based in evidence. To put it another way, is there any evidence to prove that the person who just triggered you actually hates you or has something they said reminded you of your early exposure to criticism?

To help you separate the two, it's useful to ask yourself the following questions:

- Has this person shown any signs of hating me before this moment?
- Have I done anything to cause this person to hate me?
- Have I said anything to cause this person to hate me?
- Does this person have a reason to feel threatened by me?
- Has this person lost out on an opportunity because of me?
- Have I done anything to make this person feel insecure?
- Have I ever done anything to hurt this person?

If you answered 'yes' to any of these questions, you will have evidence to prove that your suspicions about this person hating you are real. However, if you didn't answer 'yes' to any of these questions, you will have evidence to prove that your suspicions about this person hating you are imagined and not grounded in reality.

Testing RSD

Maya told me about a time when she used these questions to test whether her reaction was evidence based or if it was her RSD playing tricks on her.

Feeling hungry and finding herself wanting to go out for dinner, Maya picked up her phone and texted her best friend. The text said, 'Hey Olivia, I'm starving, do you want to go out to eat?'

Olivia replied immediately with 'Hey! Sounds good, we can go out for dinner if you like.'

Maya described how her eyes focused on the three words 'if you like' and she instantly felt the pain of RSD.

'She hates me,' her mind screamed. 'She doesn't want to go out for dinner with me. She's only saying she does because she doesn't know how to say "no". I've massively inconvenienced her life by asking.'

On this occasion, though, instead of reacting instantly, Maya asked herself the questions above and quickly realized that the feeling of rejection was perceived, not real. Her friend had never showed any signs of hating her and she had done nothing to offend Olivia.

Maya was triggered when she read the words 'if you like' because those three words did not specifically show that her friend was keen to go out for dinner and the RSD brain will interpret something that is not specifically positive as negative. The RSD brain wants to hear 'Yes, I would love to go out for dinner with you' because this leaves no room for perceived negativity.

Of course, it's not always possible to ask everyone to respond to you in explicitly positive ways — life doesn't work like that — but it does help to ask your trusted inner circle to have RSD in their thoughts when replying to you. A tiny change in how people communicate with you can make the difference between a fun evening and an anxious one.

For example, you could have a rule in your friendship group that nobody replies with 'Sure'. My RSD flares badly when I ask someone if they want to do something and they simply reply 'Sure'. But just turning 'Sure' into 'Sure, that sounds great!' can make a huge difference to the overall mental wellbeing of the group.

It's important to be aware of our RSD and take accountability for it, but it's also useful to adopt some basic tools in our trusted friendship groups. Having open communication within your tribe will enable the entire group to conquer the strong feelings that come with RSD. Simply turning 'if you want' into 'I want to' can save lots of unnecessary RSD-induced tears.

Don't reply to text messages with a thumbs-up emoji

For someone with RSD a reply consisting only of a thumbs-up emoji can be triggering, as it suggests to them that the sender really can't be bothered or has no time for them. Again, this can be avoided by having a no-thumbs-up-emoji rule in your group. Placing little communication rules like this within your friendship group sounds tiring, and it is, but it's also a sign that you're in the right friendship group. If you ask someone to adjust how they speak to you and they respond with something like 'Dealing with you is exhausting', then that's a sign you're not destined to be friends. A group of friends who understand the complexities of RSD will appreciate the rules and have no problem with putting in the tiny bit of extra effort to protect everyone within the group.

Time to recharge

It's hard navigating life with a constant inner belief that everyone is simply tolerating you, hiding their true feelings about you and secretly disliking you. When you experience a real or a perceived rejection that reaffirms this belief, your nervous system erupts like a thousand volcanoes erupting simultaneously. After this happens, let the dust settle and then remember that it's important to take a break before you do anything else. You need to lie down, talk, go for a walk, do breathing exercises, have a cold shower or speak to a trusted person. There's no shame in crying your eyes out, screaming at a wall or punching a pillow. You need to bring yourself back down to the baseline. It's common for us to want to simply carry on and move on to the next thing, but if we don't take a moment to recharge, we won't bring the best version of us into the new moments. We will bring a restless, agitated and unregulated version of ourselves, one that is more likely to be triggered yet again.

I use these recharging moments to remind myself of some fundamental facts: I am not broken, I am different, and I have always been enough. I tell myself that RSD is real, I'm not too sensitive, I am not weak or a bad person, I have ADHD and I am working tirelessly to manage it. RSD is the hardest part of ADHD, and I am doing a very good job at keeping my head above water. I am not alone in dealing with this. There are millions of us. We are the strongest collective tribe of emotional warriors the world has ever seen.

The exposed balloon analogy

Someone who lives with RSD experiences life as a balloon. For them, a criticism is like a pin. When the pin touches the balloon, it explodes in a sudden bang, often shocking everyone around them. It doesn't matter if the pin is tiny or if the pin is larger, almost like a nail, it will still explode in the same way.

A neurotypical person also experiences life as a balloon. However, their balloon is wrapped in thick masking tape. This tape acts as a shield between the pin and the balloon surface. When the neurotypical person experiences a criticism, the pin jabbing into them still hurts, but their protective layer of tape stops them from exploding.

People with RSD can adopt strategies to protect themselves from the explosion. We don't need to live constantly in fear of the pin. But simply acknowledging that we're sensitive to rejection sometimes isn't enough, so we need a reliable source of scaffolding to wrap around us instead.

Here are three things you can do.

1 SPEAK TO AN ADHD FRIEND

Many people with RSD speak to the wrong people for help. If you speak to someone who doesn't understand RSD, they will respond to you with comments such as 'You need to get a grip' or 'You need to stop being so sensitive'. Find people who understand RSD (there are a lot of us out there!) and have a chat with them about your experiences. This will also remind you that you're not alone or the only person feeling these big feelings. It really helps when you realize that it's a common ADHD experience and not something wrong with you.

> ## We're in this together: your stories
>
> I've spoken to thousands of people who live with RSD. Here are some stories that have helped me feel less alone:
>
> *Zara:* My parents like to give me weekly 'catch-up' meetings where we sit around the dinner table and they give me an appraisal. They normally focus on my career progression. I always end up shouting at them before running to my bedroom where I have to do breathing exercises to stave off a full-blown panic attack.
>
> *Kiran:* I always end my relationships whenever I sense any wrongdoing from my partner. The heartbreak is unbearable because, deep down, I know they haven't done anything wrong, but

> my RSD makes me genuinely believe they are cheating on me. I try to act normal around them but it's impossible because I hold so much resentment towards them, even though they haven't done anything wrong. I sabotage all my relationships this way.
>
> *Joe:* I was depressed for two months after rage-quitting a job. I didn't even like the job, I was going to hand in my notice anyway, but someone got a promotion instead of me and I got so angry with my boss for not seeing how hard I was trying.

2 CATCH YOUR STORIES

It's common for people with RSD to create elaborate stories in their mind to reinforce the feeling of perceived rejection. We construct full-blown narratives in our head to add credibility to the idea that someone has slighted us. For example, a friend might say they are too busy to see you tonight because they have to finish a work assignment. This is a valid reason to not see you, but your mind will create a narrative that tells you they are lying about their work assignment, and that it's an excuse to justify not seeing you because they have discovered something horrible about you and now despise every fibre of your being. Whoa, slow down! We need to be aware of when our mind runs away from us and creates these false stories because, if we don't, we can find ourselves reacting to the fake-story version of events and not the real-life version of events. If you don't catch your story, you could spend the evening ruminating over it, obsessing over the details, torturing yourself with anxiety and replying to your friend saying, 'What have I done to offend you?'

If you don't catch the story early, you become totally invested in it, wholly convinced by it; it puts you in a 'fight or flight' response mode. The more you ruminate over it, the more real it becomes and the more elaborate it grows, leaving you responding from a place of pure panic.

'I can't see you tonight because I need to finish my work assignment' placed a seed of fear in your mind and, left uncaught, it grew into a full-blown RSD attack and triggered a disproportionate response.

It's a good idea to challenge these thoughts at the earliest possible opportunity. You can use the list of questions included earlier in the chapter, but it also helps to look for any evidence that invalidates your intense emotions. For example, reading through recent chats with the friend who has just triggered you will most likely give you important context and reassure you that the friendship is, in fact, very strong. This simple exercise helps you determine whether your strong emotions are coming from a place of fact or fiction.

It also helps to zoom out of the situation and ask yourself, 'What advice would I give to someone if they came to me with this situation? Would I be able to look at it objectively and see that the friend who triggered them was simply busy with a work assignment and that it wasn't personal, or would I still think that the friend suddenly hated the person?' You can momentarily hack your nervous system by telling yourself that you're not emotionally invested in the situation, that you're a third party, an impartial fly on the wall, and although the clarity will be fleeting, you can use it to add context to the situation. This little bit of unbiased perspective will help you rationalise the event and reduce the likelihood of RSD being triggered. More often than not, this simple exercise helps me rationalize the situation and stops the negative thoughts from spiralling.

3 ASSUME GOOD INTENTIONS

Imagine you're driving and someone cuts you up, or you're walking up to a supermarket till and someone rushes past you and places their bag at the checkout before you. Both of these encounters would normally fill me with rage; however, it helps to assume that people have a valid reason for their actions. For example, the driver who cut you up could be rushing to visit a family member in hospital before visiting hours end, or the person in the supermarket could be having an allergic reaction to something and they're desperately hurrying to buy their medication.

Of course, we need to be vigilant about not letting people get away with being nasty to us, but starting at a position of 'Maybe they have good intentions' has helped me regulate my emotions during volatile moments. Over time, you begin to assume that other people think the same way about you, and therefore, when they make a comment that you perceive to be a criticism, you can say to yourself, 'Why would they say that to me if they still want me in their life?' This allows you to separate a criticism of a part of you, a trait or something very specific you've done from a criticism of all of you, and when you pair this with a real understanding of self, you can begin to see what criticisms have merit and which do not.

People with RSD need to work extra hard when trying to separate a criticism about a part of them from a criticism about the whole of them. For example, if their boss criticizes a piece of work they've done, someone with RSD will automatically assume their boss hates everything about them and not just the piece of work. Or if they show someone a song on YouTube and they lose interest before the song finishes, they're quick to assume the person no longer likes the whole of them and not simply that they didn't like that particular song.

I remind myself all the time that everyone is unique and that their interests, standards and preferences are also unique. It's incredibly rare, almost impossible, to meet someone who perfectly matches your level of enthusiasm towards everything you're interested in. Even if you could somehow meet an identical version of yourself, your enthusiasm towards the same things would rarely match up because you'd both be exposed to a unique sequence of events that led up to the moment of your interaction with each other and therefore your mood would never be in total alignment.

For example, if you went to the theatre together, when buying your tickets one of you might have a slightly more pleasant interaction with the ticket seller than the other and therefore your moods as the curtain rises will be different, making it likely that one of you will have a slightly better time at the theatre than the other. You will have slightly differing opinions about the show, opening the door for a potential RSD trigger as you sense the identical version of you disagrees with your review of the show. Of course, this scenario will never happen, but it demonstrates how impossible it is to meet someone who is not different from you in their mood, likes and interests.

It is inevitable that people will sometimes let you know about these differences. Maybe they will make a comment about your taste in music or fashion or your preference for dogs over cats. This is a normal part of human interaction, it doesn't indicate that the person has any negative feelings towards you other than the specific thing they have made a comment about.

In fact, rarely does anyone encounter a criticism that's about the whole of them and even when it does happen, it's normally coming from a place of insecurity or from feeling threatened.

I remind myself of this fact in the morning before I leave the house. I say in my head, 'Everyone is different and that means I am not going to agree with everyone about everything.' This lays a strong emotional foundation that enables me to be less fearful of encountering a criticism. It sets the tone for the day, one that plants a seed in my head that tells me that people are allowed to disagree with me about lots of things and that doesn't mean that they disagree with the whole of me.

People are able to reject a part of you without it being a rejection of the whole of you.

This is not you, it's RSD

I'm going to share with you one of my most effective methods of getting out of an RSD spiral.

My main trigger is when I'm feeling left out of a group setting or when I'm being excluded from a social event. Sometimes it's based on evidence – for example, I'll see a social media post – and sometimes it's based on pure fantasy – I've imagined a scenario and it's triggered that sensation of despair. In these moments, I grab a pen and paper and write down all my strong feelings, as though I'm screaming all my rage out onto the sheet.

It can look like 'How dare you forget about me!? Do you not like me anymore? Have I done something to offend you?'

Sometimes I even write an entire draft email or text message to the person who has triggered me, venting all my thoughts at them and making them feel guilty for causing my agony. I never send the messages – simply the act of writing them is enough to transfer the energy out of my mind.

The writing process also acts as a useful distraction from the irrational thoughts. I use these moments of clarity to tell myself, 'These intense emotions are not really you, they are your RSD. They are not caused by the person who has triggered you and they are not caused by you; they are caused by the 20,000 horrible messages you got when you were a child. Those messages were not my fault and neither is this feeling.'

This mantra helps to bring me out of the initial fog and see the intense emotions for what they are: traumatic memories from childhood. It enables me to catch myself from spiralling and pull myself out of the illusion that I'm a terrible human being and that I should curl up into a ball and cry because someone signed off an email with 'Regards' rather than 'Kind regards' – or didn't invite me to eat pizza with them.

When you want to show a funny video to a friend but they tell you they're too busy to watch it right now, rather than spiralling into an RSD 'they despise me' moment, simply remind yourself that your RSD is back – it's not you, neither is it them, and it will pass.

Ride the RSD roller coaster

RSD feels like a roller coaster that you can't control – you can't decide when it starts or stops. You could be having a conversation with a friend and you notice they didn't reply to your previous comment with the same level of emotional intensity and it triggers you: the RSD roller coaster shoots off the start line; you're off at a million miles per hour, zooming through the loops, heart thumping, unable to think about anything else other than the tsunami of shame.

Once we understand what RSD is, we can allow ourselves to ride out the RSD roller coaster. We don't need to panic because, although we don't have any control over it, we know what this feeling is: we know it's RSD and we know it will end at some point. By giving yourself permission to simply ride out the roller coaster, you're able to seize back control over it and remove what made it scary, its mysteriousness. RSD is ugly, it's painful and it's all-consuming, but when you're able to shine a spotlight directly at it, you're able to humanize it, and when you do that, you no longer live your life in constant fear of it.

Reduce or cut down on alcohol

RSD loves it when you're hungover. It senses your vulnerability and makes you much more sensitive to being triggered by minor things. You could be lying in bed with a sore head and simply the fact that no one has written anything in the group chat or texted you separately to thank you for a good night will send you into a spiral of assuming you did something terrible last night and everyone now hates you. Being hungover means it's harder to regulate your emotions; the coping strategies you normally use to manage RSD will be less effective.

When someone with RSD drinks alcohol, it's like pouring petrol onto a fire – you are more likely to lose your inhibitions and say something impulsively that you'll overthink the next day. You could have a vague memory of meeting someone new, oversharing your whole life, even telling them confidential details or extremely personal information, and this causes a gut-twisting anxiety the next day when you're sober, as you convince yourself that this new person thought you were too much. This is made much worse when you can't even remember the social encounter properly.

Alcohol is a mood enhancer. If you're in a good mood, alcohol will turn it into a great mood, and if you're in a bad mood, alcohol will turn it into a terrible mood. This also applies if your RSD is triggered. If your normal response to an RSD trigger is rage, then alcohol will turn you into an extremely angry person. If your normal response to an RSD trigger is to cry, then alcohol will turn you into a sobbing mess. This can compound the situation to create a version of you lying in bed with a vague memory of being triggered but without a clear image of what you did. An overwhelming level of anxiety will force you to pick up your phone and text everyone who was with you last night, saying something like 'I'm so sorry if I embarrassed myself last night.' And when people don't reply immediately, your RSD is triggered yet again, only amplifying the anxiety.

Your impulsivity will be heightened when you're drinking alcohol, so you may react a lot faster to a perceived criticism than the sober version of you would have. For example, if you're in a bar and someone bumps into you and says 'Sorry, mate' in an abrupt tone, you may perceive that comment as being disrespectful and you may respond with an insult. Or if someone accidentally spills their drink on your shoes and they don't apologize, the sober version of your brain that would be able to rationalize it as 'Maybe they didn't notice what they just did' would not kick in; instead, the impulsive RSD brain would jump to the conclusion that 'they know what they just did and I'm going to shout at them for not apologizing'. RSD can cause both situations to escalate into a verbal confrontation or even a fight.

Likewise, alcohol can magnify an otherwise benign comment from your partner, such as 'I'm too tired to go out for dinner tonight', into something that your not-sober mind translates into 'I don't want to go out for dinner with YOU tonight'.

I met Claire at an ADHD conference and she explained how RSD was destroying her marriage. She told me the following story. She came home from work and had a glass of wine. She turned to her partner and said, 'Shall we walk down to the marina and find something to eat?' Her partner replied, 'Do you mind if we stay here tonight?' Claire instantly felt a knot in her chest. She explained how her mind snapped to the thought, 'He's fallen out of love with me. He's having an affair. He's thinking about another woman. He doesn't love me anymore.'

Claire replied to her husband with 'Have I done something to upset you?' Her husband, looking confused, said, 'No, I've just had a long day, and I'm feeling tired. I'd rather stay here tonight.'

Claire explained how she sensed an abrupt tone in his last word, 'tonight', and that made her fill with sadness. Without hesitation, she said, 'You're having an affair, aren't you?' Luckily, she caught herself at that moment, immediately retracted the comment and apologized.

The topic of RSD was a common point of discussion in Claire's marriage, so they were able to laugh off the moment. She highlighted, however, how much harder just one glass of wine made it to control the impulses to react to an emotional trigger. Little moments like the accusation 'You're having an affair, aren't you?' served as a reminder of how badly alcohol had flared her RSD in the past. During the early stages of their relationship, pleasant evenings out together often ended up with explosive one-sided rows that appeared out of nowhere. They became so bad, in fact, that her husband voiced concerns over his ability to stay in such a volatile relationship, where he felt like he was constantly having to walk on eggshells for fear of awakening the monster that lived inside her.

The ability to catch our emotional responses is important when trying to manage RSD. To do this successfully requires a pause between the event that triggered us and our response to it. A degree of impulse control is necessary to create this pause. When we add alcohol to the RSD mix, we lower our ability to create this pause because our impulsive control is reduced, so rather than taking a breath and deciding not to react to a perceived negative tone from our partner, we react in the moment with accusations of infidelity.

Reducing or cutting down our alcohol intake will keep our impulse control stable, putting us in a much stronger position to bat off RSD-fuelled responses.

Never worry alone

I remember when I first heard about RSD. It was shortly after my ADHD was diagnosed. I was researching the ADHD life experience and quickly stumbled across a forum that contained stories from people with ADHD. They were all detailing their experiences with rejection and how their responses to it were all-consuming. I spent what felt like ten minutes but was in fact over two hours reading this forum, relating to every single story, every single description of RSD. When I got to the end of the forum, I closed my laptop lid, sat back in my chair and burst into tears.

I had never cried like this before; I had never read someone else describing my whole life experience with such accuracy. I cried for ten minutes, but it felt as though I was releasing 34 years' worth of pent-up shame, anxiety and feelings of being broken. I had found my people in this forum, other human beings who also experienced this crippling response to rejection, something I always thought affected only me because I was somehow morally defective. I wasn't defective or broken, I was different, and I had only just realized. I was always enough, and it was in this moment of realization that I saw the power in community, in finding your people – people whom you could relate to, share your experiences with, without feeling any shame or embarrassment.

Never worrying alone can take many forms:

1. Speaking to your friends, family or partner
2. Finding solace in a comments section on social media
3. Reading books about RSD (like this one!)
4. Listening to podcasts on ADHD and RSD
5. Attending seminars and talks on ADHD and RSD
6. Attending local ADHD meet-ups.

One of my preferred tools to help me never worry alone is to separate my non-triggered self from my triggered self and to use one to talk to the other. When my RSD is dormant, I can see clearly and think rationally, so I can use this version of myself to write down advice for my triggered self to read. This technique is useful because it removes the shame that often comes with talking about RSD with others. It's also always available because you can have your own advice on standby at any time of the day.

I wrote myself a letter to read in moments of rage or sadness:

> Hi Alex, I know you're experiencing big emotions right now, they're valid, but please know that you're not reacting to whatever that person has just said to you, you're reacting to those criticisms from your childhood. I know you're feeling sad, angry and shameful, and that's okay, it's good to feel these feelings, it's a reminder of the trauma you went through. You're not 'too sensitive', neither are you overreacting. Your RSD is consuming you right now and that's okay, it's part of who you are and it doesn't make you a bad person. It will pass. You are not broken, just different. You have always been enough. Alex.

Replying to your non-triggered self is a useful tool that will distract you during those moments of heightened emotions when you're triggered. Here's a reply I wrote to myself during my last RSD episode:

> Hi Alex, I'm feeling it right now, my hands are shaking, my appetite just disappeared in an instant, I feel sick and full of rage. I'm writing this letter to pull myself out of it. I hope it helps. Thank you for your letter, I just read it and felt it soothed by it.

When we're able to communicate our RSD experiences with other people, even if it's to ourselves, we're able to immediately remember that we're not the only person who experiences these things. This realization is extremely soothing because it reassures us that we're not alone and that these big feelings are not a reflection of a flawed part of us but a reflection of living in a world that's not experienced in dealing with difference.

Your brain vs you

It is useful to remember that, although RSD feels physically painful, it starts in your brain. As someone with ADHD, your brain is wired differently and has been exposed to different things. So when your boss corrects your work and you disappear into the loo to punch a wall, it helps in that moment to tell yourself, 'My brain is telling me that my boss hates me and that I should rage-quit my job, but I know that my brain is wired differently and is wrong about this.'

This tool is also helpful when you are trying to reconnect with someone whom you have not spoken to in a long time. To hack the shame and embarrassment of saying 'Sorry we haven't spoken in a while', you can be honest and say something like 'My brain was trying to convince me that you hate me. I don't hate you, but it was important for me to take some time out because it takes a little bit of time for the rational version of me to catch up and override the irrational RSD side of my brain.'

I used to be paralysed with fear, self-doubt and dread if someone left me on 'Read' on messaging apps, but now I'm able to remind myself that they are probably busy and will reply to me when they get a spare moment. Without this reminder, my RSD brain will convince me that they think I'm awful and no longer like me.

This reminder also helps me stop myself from asking people if they found their social interaction with me too awkward. After meeting someone new, I would often apologize for being 'off' or 'weird', sometimes even asking them how I was coming across and apologizing for being too much. In these moments, it really helps to remind myself about the difference between my brain and me.

The STOP strategy

When researching RSD coping strategies, I interviewed world-renowned ADHD mentor Matt Gupwell on my podcast. He shared with me his method to manage RSD: the STOP strategy.

The STOP strategy can be used immediately after you have been triggered. STOP is an acronym and it stands for:

S – Stop. Do nothing, say nothing and take a breath.

T – Take 30 seconds. Pay attention to your surroundings.

O – Observe everything that's happening, what happened before the trigger and what's happening now.

P – Process. Taking everything into account: whether your reaction is appropriate to the event that triggered you, whether you have misinterpreted something that was said or done, or whether your brain has added a detail that simply wasn't there.

Matt shared a real-life personal example of the STOP strategy. He was cooking dinner for his wife and kids. He was already in a grumpy mood because no one was offering to help him, but he continued to make dinner nonetheless. After they had all finished eating, Matt and his wife went to have a coffee in the other room. His wife stood up to get something from the kitchen and the first thing she said was 'It's never-ending – the kitchen is a mess.'

Matt continued to explain that, objectively, the kitchen *was* a mess. He told me how he had tried to tidy up after cooking but he hadn't done it perfectly. His wife's comment declaring how messy the kitchen was triggered Matt and he immediately shouted, 'I'm sorry! I've spent all afternoon cooking!'

His wife interjected with, 'Wait, I wasn't attacking you. I was simply saying that objectively the kitchen was spotless this morning and now it's messy. I'm very grateful for you cooking us all dinner.'

In that moment, Matt instinctively wanted to lash out even more, but he remembered the STOP strategy. He stopped himself, observed his surroundings and said, 'I'm sorry. I know the kitchen was spotless this morning.

I just haven't got around to tidying it up yet.' His wife immediately replied with 'I'm sorry. I didn't mean to upset you. I was just commenting on the facts.'

Matt was able to reflect on this RSD moment and rationalize that it was triggered by a response by his wife that didn't meet his internal expectations. He desperately wanted his wife to say, 'Thank you so much for dinner. It was delicious!', but instead the response was a comment about the kitchen being a mess.

The STOP strategy has enabled Matt to stop, take time out, observe and process an event before he allows his RSD to take over. (I've also used this method many times since my interview with him.)

Very few people care about you as much as you think

You will have people in your life, friends and family, that care deeply about you. However, the truth is that 99.9 per cent of people whom you meet in your life won't care about you as much as you think. Each and every one of them will be too busy worrying about their own reality and their own external perceptions.

I used to replay social encounters in my head over and over, convincing myself that I had said or done something that meant that person now thought less of me, or that I had upset them in some way. I would lie awake in bed, replaying a conversation with someone I had just met, a colleague at work or a stranger in the park, and focus on a single sentence. Worrying that I sounded silly, I would instantly fill with anxiety as I convinced myself that they thought I wasn't in any way credible. I would feel frustration bubbling up, telling myself, 'How could you be that stupid?', and then shift my frustration onto the person in the conversation, indignant about them daring to think I was stupid. I would have to play out an internal argument about a completely imaginary scenario between myself and an imagined response from a stranger who had probably completely forgotten about the interaction.

> They definitely hate me. How dare they hate me. I hate them!

It was exhausting. An internal venting towards a stranger for misunderstanding me. At least that's what it felt like. But the truth was often much less extreme. The stranger in the park or in the office didn't care about me as much as I thought. They were too busy dealing with the complexities of their own mind. They may even have walked away from our social encounter with similar anxieties over how they had come across; they were perhaps obsessing over a single sentence they had said that they now worried made them sound stupid.

But there wasn't anything they said that made them sound stupid; neither was there anything they said that meant I now hated them. The same was true for me. The flow of social encounters allows for ups and downs in tone, ebbs and tempo, and none of these things signifies that the person delivering them means anything by them.

I've lost count of the number of times obsessing over a social encounter got too much for me to handle and I've had to reach out to the person I thought my social awkwardness had offended. Time and time again, the person I thought may have believed me rude or awkward had no memory of the detail I was obsessing over. They didn't notice my eye contact wasn't consistent, neither did they spot how my voice tone dipped a decibel at the end of my last sentence.

It's easy to assume that, if someone doesn't explicitly state that they like you, it means they only tolerate you. It's easy to assume that, if someone doesn't explicitly invite you to a social event, or includes 'come if you want' when asking you to attend, they don't actually want you there, that your presence at that social event is a nuisance to them, and that they hate you, too.

My RSD brain translates 'come to the party if you want' into 'I'm inviting you because I feel like I have to, but I don't really want you to show up'.

The tiniest amount of dialogue from someone who isn't overtly positive, is unclear or emotionally neutral, can send us into turmoil. In these moments, it helps to remind ourselves that the person who has just delivered an RSD-inducing comment is probably contending with their own demons. The comment that just rocked our world was most likely simply a throwaway comment from them and they meant nothing sinister by it.

Stay ahead of the game

Removing yourself from the situation when a trigger occurs is a useful way to calm yourself down, but this technique can also be used to stay ahead of the game. RSD can make our emotions go from zero to one million in a heartbeat, and it's very hard to manage ourselves when this happens.

Instead, it pays to recognize when you're approaching a state of heightened risk of being triggered and take preventative steps at this stage.

For example, if you've had a terrible day at work and you're on your way home to spend the night with your partner (maybe it's date night) and you feel you are on the edge of having an emotional explosion, you could remove yourself from the situation before it occurs. This could look like sitting in your car for an extra few minutes before going inside, or taking a few extra deep breaths before turning the key to open your front door.

It helps to proactively remind yourself of your positives throughout the days when you have lots of things on and lots of social interactions to navigate. For me, this looks like reading my notes app on my phone, the same list that has my valuable points written on it. This simple exercise allows me to give my self-esteem a health check three or four times a day when I'm busy or even every hour when I'm really busy. There is no shame in doing this more frequently, even every ten minutes if you're feeling particularly vulnerable. Your notepad is your secret weapon, your ego armour that can protect you as you see fit.

It's also important to remember that, although I have referred to our emotional responses as 'disproportionate' to the event that triggered them, they definitely don't feel disproportionate to you when you are experiencing them. Your nervous system can't tell the difference between a minor comment today and a reminder of a lifetime of being made to feel weird – the same threat alert system is activated. The intense emotions are very valid. However, it is important to understand the difference. Without this knowledge, we won't have the scaffolding around us to pull us away from the thing that triggered us, and if we don't pull away, we are in danger of an emotional outburst that is disproportionate to the trigger.

When someone with RSD responds disproportionately, that very same response is often the source of considerable regret and shame. The person will lash out, scream and shout, or burst into tears, and this response will linger in their mind for years to come – they will ruminate over it and re-trigger themselves over and over. When we remove ourselves from the event that triggered us, we are lowering our chance of reacting in the heat of the moment, and we are therefore protecting our future self from another memory to ruminate over.

Personal growth is achieved when we have the self-awareness to do this, to realize that, while the intense feelings don't feel disproportionate to us when we are experiencing them, they are disproportionate to the event that triggered them, and by putting a gap between the trigger and our response, we are shielding ourselves from future shame.

RSD is by far the most challenging aspect of living with ADHD

It makes you always feel out of place and as though you don't quite fit in. You would think this feeling would make you retreat from the world and hide away, but there's a level of impulsivity that also comes with ADHD that enables you to socialize with people, even share jokes with them. It's a weird contradiction. However, these two parts of you clash to create an internal narrative that tells you people are simply tolerating you and that no one actually likes you.

You can appear confident and bubbly at a social event, maybe even the life and soul of the party, but with each different social interaction comes a new seed of doubt, one that will grow in your mind to create a rumination of thoughts that convinces you every single person at that social event found you incredibly annoying.

RSD can make you feel very lonely. The contradiction between how you appear socially and how your brain makes you think you appeared socially can make it hard for you to communicate about RSD because you worry no one will understand. You worry someone might respond with: 'I don't understand why you are overthinking your social encounters so much. You appeared to be having a great time!'

Your ability to mask and pretend to be normal is so good that it creates this confusion from others when you open up about your worries. People with RSD are experts at convincing everyone else that they're happy. We're very aware of this, so the idea of contradicting that illusion seems scary because we don't want to reveal our secret shame to anyone. So we internalize it, and it keeps us awake, staring at the ceiling, with a nagging churn in our chest, a feeling of not being good enough that won't dissipate.

RSD doesn't let you believe your worth. It creates an amplified version of the imposter syndrome that stops you from seeing yourself in a positive light. Someone could shower you with positive feedback after you achieve something amazing but the words bounce off you, unable to penetrate through the thick layer of RSD. You clap and applaud when others achieve things but you find it impossible to give yourself the same praise. In fact, you tell yourself 'That was a fluke' or 'I didn't work hard enough to receive the praise that people are giving me'.

You can't help but tell yourself, 'I'm in trouble,' even when you've done nothing wrong. Every single interaction with another human being turns you into an automatic body-language scanner, looking for signals that tell you their interaction with you isn't genuine, picking up on tiny micro signs of aggression that give away their true feelings about you. It's exhausting.

How to prevent an RSD spiral

As a late-diagnosed ADHDer myself, one of the most important things I've learned about managing my mental health is that there are some things that shouldn't be fixed. Trying to fix something takes up so much time and energy. RSD is a part of me. In the same way, every other human being on this planet has unhelpful parts.

Instead, I've learned how to manage it and, most importantly, how to prevent it from dictating my life. I've achieved that by following six principles.

Principle 1: RSD is a part of me, but it is not me

RSD is an irrational part of me that was born out of lots of early experiences of being different. It is not logical, nor does it react based on facts.

This principle, particularly the 'RSD is a part of me, but it is not me' part, helps me accept RSD, allowing it into my life rather than working tirelessly to push it out. Welcoming RSD into our lives gives us power over it. It enables us to see it for what it is: an irrational side of us that rears its head from time to time.

Principle 2: Reality does not trigger RSD, reminders of my past do

This principle grounds me in moments when I can feel a spiral about to happen. In moments when I am convinced that something I did caused someone else to dislike me, it reminds me that my past has made me hyper-aware of negative signals. These signals are an echo of my negative past and not a reflection of how this person feels about me today. Having an awareness of this differentiation helps interrupt the spiral before it escalates into a full-blown meltdown.

Principle 3: Stop apologizing for something that isn't your fault

RSD is reinforced when we apologize for our actions when we perceive they have offended someone. For example, when we turn up a few minutes late to a dinner party or have to pause a film to use the bathroom, RSD tries to convince us that these actions are massively inconveniencing the other person's life and causes us to over-apologize in an effort to avoid them withdrawing their approval. However, saying 'sorry' when it's not necessary to say it is only feeding the RSD compulsion to scratch the itch caused by our desperately wanting to avoid being criticized. Instead, say, 'Thank you for your patience' or 'Thank you for waiting.'

Principle 4: Distract the nervous system when RSD strikes

RSD is such an immediate, visceral reaction that it's sometimes impossible to stop the spiral happening. However, when you feel its intensity, it helps to distract your nervous system with something equally intense. For me, this looks like jumping under a freezing cold shower or biting down on crushed ice. Running my fingers under cold water also helps when I'm not able to jump into the shower.

Principle 5: Someone cancelling on you doesn't mean they hate you

If someone texts you and says something has come up and they're too busy to see you, RSD will convince you that your friend has discovered some secret about your past and they now despise you. However, it helps to stop yourself at this point and redirect your thought process towards trying to think of a couple of neutral explanations. Maybe they are having their own mental health issues, maybe they are experiencing a drama within their family, or maybe they are simply too tired because their workday was more chaotic than they expected.

Principle 6: The assumption of RSD is never accurate

Keeping a journal of your RSD spirals is helpful because you can compare the thought processes to the reality of what actually happened. For example, your partner may have a particular tone with you one evening and the RSD spiral will tell you they are mad at you. You might then discover they had an argument with their boss that day and they are simply in a bad mood because of that. You can journal it as evidence that your RSD spiral was wrong. Keeping a journal full of evidence of previous inaccurate RSD spirals can help to invalidate your current one.

There's one thing that RSD fears the most and that's when you become aware of its existence. For too many years, it's been able to fester inside you and control multiple aspects of your life. RSD hates the fact that you're holding this book in your hand. When we become aware of something's existence and truly become aware of where it comes from, we're able to hold it out in front of us, smile and say, 'You are only one small part, you are not all of me and you definitely do not control me.'

5

When every email feels personal
Navigating RSD at work

I was halfway through reading an email when the rage hit me like a freight train.

'Dear all, thank you for your attendance at the meeting this morning. I appreciate all of your contributions. After considering all your wonderful solutions to the problem, I have decided to implement Simon's idea…'

My eyes stopped absorbing any more of the email at that point. My chest tightened, my throat dried up and my palms went sweaty.

'Simon's idea?' I screamed internally.

My boss had called a company-wide meeting earlier that morning where she presented us with a problem to be solved. The company was having a PR crisis and we needed to put a statement out on social media to calm the waters. I had eagerly contributed a suggestion to wait two days before issuing our statement. Many other people had suggested alternative strategies, including Simon, who said we should release a statement immediately.

RSD is the reason so many people with ADHD have a history of storming out of jobs. Someone once told me how she couldn't stay in her previous job simply because her colleague had ended an email with 'Regards'. The absence of the word 'kind' before 'regards' made her believe that she was hated and that everyone in the office was speaking about her behind her back.

Mark told me how he once quit a job because his boss sent him an email saying, 'I don't think I've received your project. Can you send it to me by end of day?' Mark was so convinced he had sent it that he replied saying, 'I sent it to you last night.' The anger at the injustice made him stand up and leave early. As he was on the train home, he checked his 'Sent' messages folder and realized he had never in fact sent the work project. He was too embarrassed to ever return.

Fleur told me how she had to leave a job because of an energy mismatch in an email exchange. Her original email had contained an 'I hope you're really well', finished with a smiling emoji. Her colleague's reply didn't reciprocate the 'I hope you're well', neither did it have any emojis. Fleur immediately assumed her colleague now thought she was 'too much'. The awkwardness was too much to bear and she silently walked out of the office and never returned.

There are only so many times you can resend an email saying 'I'm so sorry, I forgot to attach it!' before your embarrassment becomes too much and you're forced to remove yourself from the situation.

RSD-induced rage can create very uncomfortable situations, too. I once spoke to a man called Thomas who received an email from a colleague, Sarah, that started with 'Hi Tomas'. His mind focused on the fact that she had missed the 'h' in his name and it triggered his RSD. 'That's rude, how dare she misspell my name! She must think I'm that irrelevant, she hasn't even bothered to spell my name correctly.' The only thing that would soothe Thomas's rage in the moment was to retaliate with an email that started, 'Hi Sara.' However, what felt like a good idea at the time turned into a source of overthinking that meant he could no longer spend another week in that office.

What RSD at work has done for me

I reflected on my employment history and wrote down six ways RSD had showed up:

1 I've lost count of the number of times I've spoken to a colleague about something and they've told me that 'I take things too personally'. I once felt left out because

three colleagues went for a client lunch and ended up going out in the evening, too. I was quite friendly with one of them and felt as though I had been excluded, even though the client meeting was not relevant to my job. I later bumped into one of those colleagues at the coffee machine. Impulsively I said, 'It would have been nice if you had invited me to the evening drinks.' She looked confused and replied, 'I'm really sorry, it wasn't anything personal. It was just an extension of the client meeting and I didn't think it was appropriate to include any more team members.' The response was rational, but my RSD didn't care.

2. Being late for work is excruciatingly painful. I either over-explain or simply shut down entirely. I then spend the whole day staring at my computer screen, unable to be productive in any way because all I can think about is how useless my boss and everyone else in the office must think I am. I don't mean to sound rude or dismissive – it's like I know I've done something that justifies a credible criticism and my nervous system enters 'fight or flight' mode.

3. I often get labelled as having 'extreme mood swings'. If someone changes a meeting time or if someone replies to my email and it doesn't match the energy of my original email, I snap into RSD mode.

4. I'm often unable to focus on my work because my brain is replaying a social encounter. Even if there is no evidence to suggest I came across as socially awkward,

my brain will convince me that I did and I won't be able to shake off the feeling of being weird, or that people are gossiping about me behind my back.

5 I often assume someone doesn't like me because they haven't replied to my email. In fact, generally any time someone in the office wasn't overly chatty with me, I assumed they didn't like me. I'd convince myself someone actively hated me if they walked past me in the corridor without uttering a few words of friendly small talk.

Fascinatingly, I've since learned that the assumption that someone dislikes me when they are silent comes from our early experience of having our emotional needs dismissed with silence. Human beings are very good at storing patterns in our brain, and if our early experiences of rejection are in the form of silences, we will always scan our environment for silence and we will assume we are under attack when we spot it.

6 Every second word that came out of my mouth was 'sorry'. I used to over-apologize a lot in the office because I always assumed that disapproval was imminent. Over-apologizing seemed like the sensible thing to do because it prevented anyone being too verbally aggressive in response to something I had said. If there was ever a pause in between my turn to speak and the other person's turn to speak, I would awkwardly fill it with another 'sorry'. If I was in a meeting and it was my turn to speak, I would start my sentence with 'This is probably wrong' or 'This

might sound stupid, but…'. My nervous system would constantly discredit my intelligence because it was hardwired to believe that everything that came out of my mouth was incorrect nonsense.

I once tried to count how many times I unnecessarily said 'sorry' during a normal working day. I had lost count by my lunch break, but I was already in triple digits:

> Walking into a busy lift: 'Sorry.'
>
> Pressing the 'floor 3' lift button: 'Sorry.'
>
> Getting out of the lift as it reached my floor: 'Sorry.'
>
> …And sorry, everyone, for wasting your time by stopping the lift so I could get out and carry on existing.

My day-to-day life in the office was ruled by RSD, but it wasn't simply the pain of being triggered that caused me problems; it was the barriers I put up around myself to protect me from being triggered. My perfectionism peaked at work. I never took on anything unless I was convinced I could complete it to an exceptionally high standard. This meant I rarely took on any projects that tested me or put me out of my comfort zone. I watched people around me as they were promoted simply because they were more 'ambitious' or because they were a 'self-starter'. Being able to self-start an emotional doom spiral of self-hatred wasn't a good enough criterion to be promoted, apparently.

So, what's different about the office when you have RSD?

It's hard to strive for career progression when you can't read an email without reading between the lines and translating every ambiguous comment into a signal that the sender doesn't value your existence, especially if the last time you met the sender in person the social encounter wasn't obviously positive. Even if the email doesn't trigger you, the euphoria from feeling like you've mastered your RSD and that you're 'normal' is so massive that you reply overenthusiastically, but then you're stuck in overthinking mode as you question whether your response was too friendly and therefore unprofessional. (Three laughing emojis was definitely too much to send to that client!)

RSD is particularly challenging in the workplace because you often feel you can't ask for help. The shame that comes with such volatile emotional responses is so big that it makes the person suffering it feel they might be mocked if they open up to someone. It's often the case, however, that many people in your office will be experiencing similar emotional responses, and all of them will be struggling in silence.

As a business owner myself for the past 20 years, I have always strived to create a safe environment for my team to open up about their mental health challenges, and I feel the most effective way for me to do that is to be open about my own mental health. If leaders openly share their personal experiences with RSD, or general sensitivities to rejection, it trickles down into the company culture and sends a clear message that says, 'This is a safe environment to talk freely about this topic.' When companies successfully create this culture, people start opening up because they no longer feel the threat of discrimination or bullying.

The companies that manage RSD successfully dedicate resources to supporting it. They create an internal support group and deliver company-wide training to educate everyone about the complexities of ADHD and RSD.

Companies will need to deliver performance feedback to employees, so it's important to give people plenty of notice before these meetings happen so that they can prepare themselves for the appraisal and strengthen their personal support systems in the lead-up to the meeting. For example, if you know you have a meeting on Monday where you are going to receive feedback on your performance, you can have your trusted person on standby and call them immediately after it happens. It's also important for companies to include lots of positive feedback along with any feedback that might be neutral or negative to help mitigate the effect of RSD.

It's difficult to maintain a job when you burst into tears every time your boss asks you for a 'quick chat', or you can't overhear two colleagues talking about how much they enjoyed lunch without ruminating over all the reasons why you weren't invited. It can be extremely triggering for you to witness someone who has been at the company for only a short time get a promotion over you, or for your boss to suggest you undergo a self-improvement plan. Waves of paranoia may hit you like a punch to the gut, a deep feeling of injustice followed by a suspicion about how you're being stealthily 'managed out' of the company.

'It's only a matter of time before I get called into a final meeting where I'll be told it's time for me to leave,' we tell ourselves. Overthinking the idea that we're about to get fired becomes common until we begin to make decisions that make the possibility of getting fired even more likely. 'I'm going to get fired anyway, so I might as well be ten minutes late for this meeting.'

RSD makes you feel as though nobody appreciates your hard work. Your boss makes a suggestion for how you could improve your work, your cheeks flush and you embarrassingly burst into tears. You also feel the utter injustice because you're working harder than most people, under more pressure than most people and delivering better results than most people, yet your boss only comments on one negative thing.

At this stage, it's useful to be honest with your boss and to openly start the conversation:

> I'm really struggling with the feedback loop you're giving me. I think a lot of my hard work is under-appreciated, and you only comment on my flaws. It would really help if you gave more general feedback that also highlighted my strengths.

Lots of managers will be empathetic to this approach and adapt to your request. Start with small asks, and make sure you acknowledge when your boss does things that soothe your RSD. This topic is new for everyone, including your manager; lots of people will want to accommodate people with RSD but simply don't understand how, or are worried they will come across as patronizing.

Things your manager could do to soothe your RSD include:

- coupling critical feedback with positive feedback
- giving clear written instructions that define expectations for a project
- bringing in neurodivergent speakers to deliver presentations on ADHD.
- saying, 'It's okay if you need the rest of the day off after this appraisal meeting.'

You shouldn't have to be constantly fighting off the urge to write your resignation letter because you're continuously living in 'fight or flight' mode; crying every night; being exhausted all the time; always in a heightened state of anxiety over your next performance review.

There's no shame in having a plan up your sleeve for when something triggers you. Take 20 minutes out to cry in the toilet or scream in your car. When you've come out of the rage, analyse the event that triggered you and separate the facts from your feelings. Re-evaluate whether the person meant any harm by what they said or did and reapproach the event with a calmer mind. Try not to react in the moment as this often leads to irreversible shame from an RSD-fuelled office outburst.

As soon as your gut says it's time to quit, it's time to quit

It's also perfectly valid to quit on the spot. If you've pushed back and still seen no changes, you can hand in your notice immediately or walk out of the door right now. No job is worth your mental health. No one deserves to enjoy the fruits of your neurodivergent brain if they can't be bothered to accommodate your challenges. It helps to fall back on your strong ADHD intuition here – it will tell you whether your manager is a horrible person or just unaware of how to help.

I appreciate that leaving your job is more complicated than I've made it sound, but it's also important to remember that mental health is more important than any job and there will be better employers out there for you. If your employer keeps berating you over every tiny mistake and never praises you, it might be time to start looking elsewhere.

You can't fix a terrible boss issue and you should never think it's a 'you' issue

From my experience of speaking to thousands of people with RSD, for many the motivation to escape the injustice of being incorrectly judged as not enough is greater than the financial need to keep a job. People with RSD would prefer to leave a job out of principle and be temporarily unable to pay rent than they would to stay in a job where they feel judged by their manager and inadequate in their role. It's the reason people with ADHD often have a history of consecutive short-term jobs. The moment they sense any form of criticism, they leave and never return, even if that means they will be anxious about how their volatile exit was perceived by others.

On my podcast I spoke to Chloe, who shared one of her work-related RSD stories. There was a broken-down car blocking one of the lanes on her normal route to work. Temporary traffic lights had been put in place to funnel cars through, but the delay meant she was five minutes late.

As she hurried into the office, her boss made a comment: 'Late again, Chloe? You should consider leaving your house earlier.'

Chloe felt totally belittled in front of the entire company. She retreated to her desk and sat in silence while simultaneously internally raging. She opened her emails, not to read them but to make it appear that she was doing something other than spiralling.

'How dare he speak to me like that in front of the entire company? He might as well have called me a worthless piece of crap in front of everyone.'

Chloe couldn't think of anything else other than writing her resignation letter. 'I was five minutes late and I had a valid reason,' she kept saying to herself. She wasn't able to complete any work all day. As the clock struck 5 p.m., she stood up, avoided making eye contact with anyone and walked out. She hurried to her car, unlocked it, got in and burst into tears.

Chloe shared with me how she understood that this was RSD, but also that it was out of line for her boss to talk to her like that in front of the entire company. She was still furious at bedtime and found herself unable to sleep, so made the decision to call in sick the following day. She desperately needed a break from that place, at least while she processed what had just happened and decided how best to proceed.

Needless to say, Chloe was still ruminating over it during her 'sick' day. She wrote an email to her manager saying, 'I think it was unfair the way you spoke to me in the office yesterday. You have really upset me. I would appreciate any future comments like that being made in private. If I have done anything that needs correcting, can you please call a meeting to speak to me? Thank you.'

She sent the email with a proud smile on her face. Hours passed, however, and the smile disappeared. The feeling of pride slowly morphed into one of anxiety and regret over sending the email. 'My manager is now going to treat me differently. He's definitely going to show my email to everyone and they will think I'm moody and oversensitive. I need to quit. I'm too embarrassed to ever show my face in that office again.' She never went back to that job.

This wasn't an isolated incident for Chloe. It really was a case of history repeating itself. Like Groundhog Day, it was a pattern she was very familiar with. 'I've had eight jobs in the past two years,' she proclaimed to me. 'RSD was the reason I quit every one.'

She detailed some of the reasons to me:

- being given unclear instructions and then being corrected when her work wasn't what was wanted
- being asked to 'jump on a call' without any indication of what the call was about
- being given a work assignment and told it was needed 'ASAP'
- finding out there was a group chat that she wasn't included in
- a colleague asking, 'Can we talk later?'
- someone saying 'Come if you want' when talking about a group lunch
- having one awkward conversation with a colleague and assuming they hated her.

The office RSD people-pleaser

Having both RSD and a compulsion to not let anyone down is hard because the people-pleaser within you makes it impossible to turn down work, yet the RSD within you makes it impossible to hand in work until you are certain it is perfect. Burnout is inevitable when you're constantly adding more to your work pile but also struggling to offload anything.

Having an understanding of RSD is a critical first step in overcoming this, but then we need to get better at saying 'no' when a colleague or a manager asks us to take on another work commitment. Shifting from a 'yes' mindset into an 'I'm too busy right now to take on anything' mindset feels extremely uncomfortable for many, but it is possible when baby steps are made towards it. For example, starting with 'I'll need to check my calendar and get back to you' or 'Thanks for thinking of me but I don't have any capacity at the moment' helps to put a pause between the ask of your time and your final response, which can come later in an email that simply says, 'I've now had time to check my calendar and unfortunately I'm unable to take on any more work at present.' You're only human, and there's only so much you can do.

Being a perfectionist is deeply ingrained in people with RSD, and it makes handing in work projects challenging because you're never satisfied with it, always worried that it will expose you to criticism. Your mind tells you there's always something that could be added or improved. It's useful to remember that nobody is going to be as critical towards your work as you are being towards it. Your worry about it not being good enough is based on the perception that your work is an extension of you, and that your feelings towards yourself about not being good enough are what's causing you to feel your work is not good enough.

It helps to remind yourself in these moments that your thoughts about not being good enough are based on a series of untruths: you were told so many times that you're not good enough, that you're lazy or 'too much', by other people who were simply judging you by their own standards of 'normality'. They didn't know you had ADHD, they didn't know you were simply different and that you work in ways that challenge the conventional way of working. When you remind yourself that your internal beliefs of not being enough are wrong, you can extend this belief to the work you produce. Your work isn't bad, incorrect or not enough; it's done, it's completed to a very high standard, and that's more than sufficient.

Turning up to work as someone with RSD is difficult to understand for someone who hasn't lived their life being told things like:

- Why can't you just remember?
- Why can't you just be on time?

- You just need to try harder!
- Just plan your time better!
- Why are you being weird?
- You're just being lazy!
- Stop being so clumsy!
- Stop making excuses!
- Stop being odd!
- Stop fidgeting!
- Pay attention!

Only someone who has experienced thousands of comments like these can understand what it feels like to live with never-ending worry that constantly makes you think you're about to be told off.

I feel sad when I think about the number of people who are making decisions in the workplace not because they want to take on, or have the capacity to take on, the task they've just agreed to do but because they don't want to upset or offend the person who has asked them.

In the workplace, where the next perceived criticism is never far away – whether it's a bit of feedback from your manager or an awkward social encounter at the coffee machine – it's really helped me to construct my own list, similar to the list above but full of positive declarations about myself. My list looks like this:

I am …

- someone who can hyper-focus
- calm in a crisis
- a good problem solver
- resilient
- creative
- empathetic to my clients' needs

I like to make an anagram of my positive declarations. The first letters of each line in my list spell 'SCARCE'. I went to the trouble of making a bracelet with six charms, each one with a different letter that spelled out the word 'SCARCE'. It worked because nobody else knew the meaning of my bracelet. Everyone's list of positive declarations will be unique and will create a different acronym. You don't necessarily need to create a bracelet – it can be anything that gives you an accessible visual reminder of your positive traits in moments of self-doubt.

It helps to also have a mantra you can tell yourself before you step into a situation where RSD is likely, such as a work meeting. Here is an example:

> I am enough. Seriously, I am. My ex who told me otherwise; my old boss who told me otherwise; my old friend who told me otherwise – they were all wrong. I am not broken, I am different. I have always been enough.

This simple mantra enables me to ground myself before stepping into a high-risk scenario. It elevates my self-esteem in the moment, increases my tolerance to being triggered and makes me less vulnerable to RSD.

It blew my mind when I realized how much RSD was controlling my behaviour at work. It made me scared to ask for help. I was terrified of people reacting to my request in a tone that would leave me worrying that they thought I was an idiot. I always felt like I was shooting myself in the foot because, on the one hand, I was scared to ask for help, but, on the other, it made me feel that I didn't know enough to complete the work to a standard that silenced my perfectionism.

I watched other less competent people get promoted above me simply because I was too scared to ask for a promotion for myself. It felt safer to just get on with the work I was given, fly below the radar as much as possible – putting my head above the parapet to ask for more seemed too risky. What if they said 'no'? I'd have to leave. There was no way my RSD could cope if my request for a promotion was declined.

I dreaded giving presentations. I worked in the creative industry so I had to do them occasionally. I thought nightmares only happened when you were asleep. I was wrong. I never slept the night before a presentation. Every time, I stood up on stage and rushed through the presentation as quickly as possible. At least then, if I got feedback that wasn't totally positive, I could blame the fact that I had prepared the presentation at the last minute and the RSD wouldn't hurt as much. I rapidly learned that it was better to give presentations without my glasses on. Keeping the audience blurry was a good idea. There was no risk of spotting a confused-looking face or someone using their phone instead of listening to me if I couldn't see anyone. (I still haven't figured out a solution for ignoring people who stand up and leave for the toilet mid-presentation.)

I went into self-preservation mode whenever I felt I was going to be told off or corrected. I'd leave jobs if I sensed I was about to be fired; I'd get ahead of the curve and seize whatever control I could by walking out the door before they could say they didn't want me there anymore. It felt safer to reject them before they had the opportunity to reject me. It felt easier to jump head first into unemployment than to put myself in the crosshairs of RSD. The pain of not being able to pay my rent for a few months was tolerable, the pain of RSD was not, so I always chose the first option.

Can't pee, must work

The RSD compulsion to overwork often overrides the human need for breaks. I've been known to hold in a pee for hours simply because I wasn't satisfied that my work project was finished to a good enough standard.

When we fear the pain of rejection so much, we create maladaptive strategies to avoid it, and one of those can look like ignoring our basic human needs to go to the toilet, drink water or eat. The idea of leaving our desk to go to the bathroom is scary because we can't trust ourselves to get back to our desk and carry on working at the same level of focus that we have currently. It's this level of focus that's needed to complete the project to the standard that's going to stop me getting criticized and I won't dare risk jeopardizing that, so I'll hold in this pee for another hour.

Three ways RSD gets people fired

1 We're unable to ignore our strong sense of social justice

Living with RSD feels like you're always tiptoeing through a minefield, not knowing where any of the mines are and not knowing when the next explosion will happen. When someone experiences life this way, it makes them extremely empathetic to others who live life in the same way. We know how painful a criticism feels and we wouldn't wish that pain on anyone else. We can't, therefore, stand by and watch as others are criticized unfairly – our internal sense of justice won't allow it. We have to say something and our impulsiveness makes it impossible to resist that urge.

This can create problems in the workplace because what's right and what's wrong is subjective and difficult to prove. Our intuition for spotting injustice, which is built on years of storing patterns of injustice, is difficult to use as justification for our callouts because most people won't understand how accurate our intuition is. When people with RSD see unfairness, they call it out for what it is and it gets them into trouble because people don't like those who have an internal BS radar.

2 Our colleagues think we're rude

RSD can turn people into blunt communicators, especially when we are speaking to someone we perceive to be in a position of authority over us – and for someone with RSD, this is pretty much everyone. We want to get in and out of social interactions as quickly as possible because, in our mind, the less we say, the less chance there is for us to be criticized, rejected or talked about behind our back. However, being a snappy communicator is often perceived as rudeness, which actually increases the chances of us sensing negativity from others. Our instinct to avoid criticism in the short term by saying less actually increases the chance of us being criticized in the future, and when we pick up on any of this perceived sense of people thinking we're rude, we walk out of the door.

3 We struggle to recognize social hierarchies, especially if we think they're unfair

People with RSD are extremely empathetic due to their heightened awareness of emotions. As a result, they tend to communicate with all human beings in the same way.

They don't speak to someone considered by society to be of 'high status' any differently from how they would speak to someone whom society considers to be of 'low status'. They would interact with a homeless person in the exact same way as they would interact with the president of a country. They view every human being as equal, regardless of their perceived position in life.

This worldview can cause friction in the workplace where there is usually a clear hierarchy. Speaking to your boss in the same manner as you speak to your colleagues seems natural to the RSD person but can cause tension in the workplace, which, when the RSD person senses it, will create a trigger.

Managing after-work burnouts

How someone with RSD comes across at work can be drastically different from how they come across at home. At work, they could be bubbly, happy and conversational. People love them. They might be the most friendly person in the office. They can exceed expectations and outperform many, if not all, of their colleagues. They can have the outward appearance of having their life in order, excelling in their work and getting promoted over other people.

At home, however, the exhaustion catches up with them and they burst into tears every time they walk through the front door. They've successfully completed another day of avoiding criticism, yet it's come at the cost of their emotional energy. They wish they could stop acting, stop playing a character called 'Normal', that they could come into work as their true self, but they know that's too risky, so they put on their mask of perfection. It's exhausting, and a total betrayal of their true self, but it does the job. It protects them from the pain of being reminded of those 20,000 comments. It gives them the camouflage they need to survive another day.

I used to be that person, crying on the floor in my apartment at the end of every day. A daily burnout, both physical and emotional, that nobody witnessed except my dog.

Closing the gap between your work self and your real self takes time. What really helped me in those moments of evening burnout was to repeat my earlier mantra to myself, but to also add an extra bit:

> It's fine to protect yourself from RSD. It's okay to wear a mask.

In those moments of crying on the floor, it's important to give yourself grace and to remind yourself that you haven't done anything wrong. Your natural instinct is to protect yourself from pain, and your extreme masking is your body doing exactly that.

Having said that, it's not healthy to mask to such an extent that it totally wipes you out every day.

Overcompensating and people-pleasing

Before my ADHD diagnosis, I would subconsciously overcompensate at work. I used to be overly friendly to everyone, offer my assistance to anyone who appeared to be struggling, volunteered to take on extra work, offered to give people lifts into work... pretty much anything that anyone needed help with, I was there offering it to them. People loved me, or so I thought. In hindsight, I'm sure people just loved what I did for them, how much easier I made everyone else's life.

After my ADHD diagnosis, I examined my behaviour at work and took three steps to help me make decisions that benefited my wellbeing rather than everybody else's.

I stopped asking, 'What do they want?' and started asking, 'What do I want?'

At the end of the day, when I had the energy, I asked myself some really simple questions:

- What did I do today that benefitted me?
- What did I do today that benefitted someone else?

I replayed my workday in my head and questioned who my actions benefitted:

- Staying an extra half hour to finish my work because earlier I had agreed to check someone else's email before they sent it: Someone else (they left on time).

- Impulsively saying 'I'll do it!' after my boss asked everyone in the meeting if they had time to take on this extra bit of work: Someone else.

- Bringing croissants to the meeting so everyone had a snack: Someone else.

I felt sad for my inner child when I saw how much people-pleasing was dominating my behaviour at work. I had totally abandoned him. It was never a choice, or a decision I was consciously making. It was hardwired within me to seek approval rather than respect my authentic self. Understanding RSD gave me an entirely new lens through which to see my past behaviours – I wasn't acting with intention or forethought, I was acting to stay alive.

It's hard to be authentic when you don't know who you are anymore

As I've mentioned previously, I don't believe ADHD to be a deficit of attention as the name suggests, more a deficit of self-awareness. This is caused by excessive masking and constantly abandoning your true self in order to be more likeable and to fit in.

To be more authentic at work and to understand what decisions serve your best interests, it's useful to practise some self-awareness exercises. The following exercise is really simple and can be done at any point of the day. I prefer to do it in the evening.

Stand or sit still and ask yourself the following questions:

- What happened today that made me happy?
- What happened today that made me sad?
- What did I do today that I struggled with?
- What did I do today that I found easy?
- What social encounters did I walk away from feeling happy?
- What social encounters did I walk away from feeling anxious?

Doing this simple exercise will, over time, give you a clear picture of your genuine wants and needs, and when you understand what these are, you'll be able to make better decisions that align with your true self.

Now ask yourself two more questions:

- What did I do today that made me lose value in the eyes of others?
- What did I do today that made me gain value in the eyes of others?

I ask myself these questions because, before my ADHD diagnosis, I thought being a people-pleaser was keeping me safe, that other people valued people who made them feel good about themselves and that asserting my own boundaries would make me less likeable to others. I now know the opposite is true. People may like people-pleasers but they don't respect them. Nice people are not necessarily valued or admired, and they are frequently being used by others (consciously or unconsciously).

Being a people-pleaser felt soothing because it put me in control over other people's reactions' it made the world predictable for me and that felt safe, but it was costing me the respect of everyone around me. It was actually costing me the approval that I had thought it was winning. I thought being the person who remembered everyone's birthday and brought cake into the office would make me likeable and keep me protected from anyone's disapproval, but what it was really doing was creating an expectation that I would always be the fun person in the office, that 'Alex is the bubbly employee', he's 'such a nice person' and 'we can always count on him'. However, the truth was that these labels rarely came with much respect, and would in fact hold me back from getting the promotions I really wanted.

Accept that it's a 'them' problem

Coming home every evening with a stubborn suspicion that everyone in the office hates you is exhausting. 'My boss was standoffish with me today,' we repeat to ourselves as we try to fall asleep. 'They must hate me!' In these moments, it's good to ask yourself if there is any evidence to prove that your boss is actually mad at you. Did you do anything that could have caused offence? Did you say anything that might have been perceived as unlikeable? The answer to these questions is rarely 'yes'.

Ask yourself, 'Is there a reason why I might have perceived a standoffish tone from my boss? Did my boss say that they were mad at me? Did they give me any real indication that they were annoyed with me?'

Again, the answer to these questions is rarely 'yes'.

When we run through these questions in our mind and come to the conclusion that there is no evidence to justify our fears, it makes it easier to assume our boss's tone was a 'them' problem.

Everyone will go through life events that make them appear grumpy on a particular day. Maybe they just opened a bill, maybe they are having relationship issues or maybe they are having their own RSD moment. Either way, the way they are acting rarely has anything to do with how they feel about you.

It's also important to remember that, when people are mad at someone, they normally let that person know that they're mad at them, so the lack of any direct hostile communication from your boss is more evidence that their being cross with you is only your perception, not reality. Ultimately, there may be valid reasons why you sense someone's tone is standoffish or that they appear moody or short with you, but most of the time it will be a 'them' problem, something going on in their life that is influencing their behaviour towards you. Things that happen in other people's lives are not your responsibility; neither should you allow these things to affect how you feel about yourself.

Unless someone specifically tells you that they have a problem with you, you should assume that they don't; assume instead that they are projecting a problem they are experiencing on to you. Trying to constantly second-guess whether someone's abrupt tone is anything to do with how they feel about you is exhausting. It's not your responsibility to understand how someone is feeling towards you because you're not a mind reader. If someone has a problem with you, especially in the workplace, it's their responsibility to let you know directly and until that happens, you should continue to assume it's a 'them' problem.

Managing RSD and our emotions at work

It's common for an RSD-fearing people-pleaser to burn out, lose their social filter and suddenly lose their temper with a co-worker. Our low impulse control coupled with our inability to appropriately regulate our emotions makes us vulnerable to regrettable outbursts.

We might be hyper-focused on writing an email when a colleague taps us on the shoulder and asks, 'Have you got a spare five minutes to help me with something?' It's already been a rough day. We got caught in traffic and were 15 minutes late to the office, we didn't have time to decompress in the car before entering the building, we have unresolved conflicts with our partner (see Chapter 3), and we didn't sleep well last night because we were arguing with the demons in our head.

'Can you please just leave me alone?' we snap at our colleague. It's always a core memory for an ADHD person when we witness someone experiencing the unmasked version of us for the first time. That colleague who just tapped us on the shoulder and asked for help has only ever seen one version of us: the happy, friendly, people-pleasing version of us. The answer they were expecting from us was, 'Yes, of course I can help you.' What they got was 'Can you please just leave me alone?' The look of shock on their face never leaves us.

These emotional outbursts happen for two reasons. The first is because we are overwhelmed at work. We have taken on too many projects. We have too many thoughts happening simultaneously. Our brain is working overtime trying to organize them into some kind of order. We snap when another thought tries to enter our mind, such as a colleague asking if we can help them.

Every action someone with RSD takes is subconsciously directing them towards one goal: avoiding a rejection. However, we know that every single action or work project we commit to comes with a risk of rejection. Maybe we won't complete it to a high enough standard, or maybe it will expose our lack of expertise to someone. Either way, we're acutely aware of this risk at all times.

The emotional outburst comes when our inability to say 'no' to another work project (because we're a people-pleaser) clashes with our fear of letting someone down (because we have taken on too many work projects). We can't bear the pain of saying 'no', but we also can't bear the pain of taking on another project because we know we don't have the capacity to complete it to a standard that doesn't involve letting someone down. It's the ADHD equivalent of the metaphor that describes what happens when an unstoppable force hits an immovable object. The brain doesn't know how to compute the situation, so it erupts into a verbal outburst, one final attempt at self-preservation. However, in our brain's final attempt at protecting us, it leaves us wide open to regret. How can we possibly stay in a job after this? The whole office will now be talking about me behind my back. Everyone will think I'm 'fake'!

The second reason for an emotional outburst is because our hyper-focus has been interrupted. Traditionally, having any form of hyper-focus interrupted is rage-inducing; however, it's even more volatile when we are hyper-focusing on something that needs to be done perfectly. Not only are we being interrupted from a hyper-focus that's giving us dopamine, but we are also being interrupted from our efforts to avoid being criticized. When that person tapped us on the shoulder and asked if we could help them, not only did it pull us out of our focus, it also increased the risk of us being criticized because it stopped us being able to work towards completing a project to a perfect standard.

When this happens, we subconsciously go into 'fight or flight' mode. We need to vigorously defend ourselves from the pain of RSD. This causes us to fight anyone who is threatening our safety from it, and in this moment that is the person who has interrupted us from our work. We require our work to be perfect, and anyone who interferes with our process to achieve that needs to be pushed away.

Emotional outbursts are how our dysregulated ADHD brains try to protect us from the pain of RSD. However, actions have consequences and these consequences are impossible to think about when we're in the heat of our impulsive moment. I wish I could stop myself from saying, 'Can you please leave me alone?' and instead replace it with 'I'll overthink the situation for weeks if I tell them to leave me alone. It's probably better if I don't say that.' But I can't. If we could put a pause before our emotional outbursts, they wouldn't be outbursts, they would be well-thought-out rational, emotional responses and we probably wouldn't have ADHD!

What we can do is own our emotional outbursts and communicate with the person who was on the receiving end of them. Previously, after an emotional outburst in the office, I would be so desperately anxious about it that I would have no choice but to leave that job. I've since learned that it's easier to take a pause, put some time between the outburst and taking the next step, and then take the next step – which is to communicate with the person who was on the other end of the outburst:

> I wanted to apologize for how I spoke to you yesterday. I'm currently trying to spin too many plates and you caught me at a bad time. You didn't do anything wrong.

More often than not, a simple exchange like this, in person or via email, is enough to settle any lingering anxiety that comes after an office outburst. I've received similar emails from people after being the person on the receiving end of the explosion, and I'm able to respond with empathy because I understand where they are coming from and why they acted like they did. More often than not, my emails are met with a similar level of empathy from my colleagues.

'I think you're making it harder than it needs to be'

I've lost count of the number of times I've heard this sentence in the workplace, or, worse, this one:

> Why have you done it that way? What a stupid way to do it!

It really hurts when someone looks at the way you're doing something and criticizes your method. It makes you constantly question your ability, too scared to phone in sick even if you are sick, feeling like you're always disappointing someone and left feeling as though you're not enough. It's a common experience for people with RSD.

We don't get support for the way we do things because our ways of doing things are different from those of most people. This is because people with RSD are the world's best problem solvers – we have had to solve problems our entire lives simply to get by, to fit in and to survive. As a child, we felt different and we solved this problem by masking. We realized we hated rejection, so we solved this problem by people-pleasing. This in-built ability to problem solve enables us to find unique solutions to difficult (and sometimes common) problems.

The world is scared of difference, however, so our unique way of doing things is often met with negative feedback. We mask to solve the problem of being different, but then get called 'too much'. We mask to hide our sensitivities to rejection and then get called 'not enough'. And when we find unique solutions to problems in the workplace, we get criticized for being 'difficult'.

We're very aware of tiny cues of social disapproval, the eyeroll, the crossed arms. Our nervous system is primed to read all of them as hostile. However, it's really helped me to reframe how I read these cues. Rather than assuming they are expressions of disapproval, I assume they are signs of fear. People are scared of change and will pull away from someone who is acting in a different way or doing something using a different method. Difference makes people uncomfortable because it risks making them look foolish if the new way of doing something proves to be a more efficient way of doing it.

In an environment such as the workplace, people's reactions to difference will be more extreme because the stakes are higher. If your new way of doing a task proves to be a better way, it will make your colleague look less efficient in the eyes of your employer and this triggers fear in their mind. They don't want to risk looking bad in front of anyone so they will reject any innovative methods of completing a task. It's in-built in the ADHD nervous system to assume this reaction is a rejection of the idea, but in fact it's a fear response from the person who is reacting.

All great ideas were originally considered to be crazy, and many of them never succeeded because the person who thought of the idea was put off from pursuing it by the immense amount of criticism they received over it. Humans are resistant to change – it's an act of self-preservation to dislike new ways of thinking, even if the new way is a better way. You can see, therefore, how the RSD community, a community of problem solvers but also a community of people who are terrified of rejection, needs extra support to pursue ideas.

My support looks like creating a journal of evidence that proves my ideas are good. My ability to tell myself, 'Your colleague doesn't think your idea is silly; they are worried it will make them look bad', was strengthened when I started journalling the results of the innovative ideas I had. Whether I was at work or home, when I had an idea that would solve a problem, I would write down the outcome of that idea. More often than not, it was a significantly more efficient way of completing the task than the method I had previously been taught.

Reading my journal of evidence reminded me that I need to reframe how I respond to people who shun my new ideas. My ability to have conviction in my own ideas changed when I realized they weren't rejecting my idea, but were defending themselves from a rejection.

It's hard when you have to ask for something you need and expect the answer to always be 'no'

I've spoken a lot about how RSD manifests itself as people-pleasing, overworking and being a perfectionist, but what's not spoken about enough is the things we don't do. We don't ask for help. We would rather sit at our desk and struggle than risk having a colleague reject our ask.

Asking for help is an extremely vulnerable thing to do. Many people struggle with asking for help over the big things like going to therapy or cutting down on alcohol, but someone with RSD struggles to ask for help with the tiny things. Asking someone if you could borrow their stapler, for example, comes with myriad risks for the RSD person.

> What if they don't have one either? I'll be in the same situation I'm in now, but now they'll also think I'm stupid. What if they don't turn their entire body to face me when they speak to me? What if they are short with me or snap at me?

All these scenarios are scary for someone with RSD. It's simply easier to avoid the risks and not ask for help, even if it means your day is going to be harder. Someone with RSD would rather push through and work harder instead of reaching out for support. This extra workload is exhausting and perpetuates the burnout cycle that many people with RSD experience.

Each and every one of us deserves to work in a professional capacity without constantly feeling we are treading on eggshells, forever feeling we're about to be told off or made to feel like we've let someone down. It's very empowering to understand this feeling has a name, RSD, and it was not something that we intentionally created. In other words, it is not our fault. And this is ultimately the most important thing to remember in moments when we think our manager is scowling at us, when we are working late into the night for fear of not delivering the perfect project, or when we are anxious because we have taken on too much work.

Take as many breaks as you need and hold no shame if you need to remove yourself from the office from time to time because RSD has taken hold. Breathe, repeat your positive self-talk, and then come back in your own time with all the amazing qualities that also come with ADHD, such as creativity, pattern recognition, resilience, intuition, being calm in a crisis and hyper-focus.

Always remember your worth and advocate for yourself in a way that minimizes the effect of RSD on your wellbeing. And remember, if your current workplace doesn't offer the support you need, you can leave and find another job that will value you, RSD and all.

ADHD makes us amazing at so many things, but of course it comes with a harsh side helping of sensitivity to rejection. Nevertheless, with the right strategies, we can take away the power that RSD has over us and thrive in the workplace.

6

Holding a heart
Supporting a person with RSD

The effects of RSD aren't just felt by the person living with it; the impact ripples through the lives of everyone around them. Any parent of an ADHD child knows the heartbreaking feeling when you witness your baby stepping away from their true passions, not because they are lazy but because they don't think they are good enough to start. Or starting and stopping a project, not because they've lost interest in it but because they don't believe they're good enough to carry on. Or watching their eyes fill with fear as you drop them off at school, feeling powerless to help them overcome their worry of not being good enough at making friends.

Navigating RSD as a parent (with compassion)

It's a devastating realization when you learn about RSD as a parent. You look back at all the times you tucked your child into bed, said, 'Good night!', and walked out of their bedroom, completely unaware that, the moment your footsteps were out of earshot, your baby got out of bed and continued to work on that piece of homework, chasing perfection, fuelled by a fear of letting you down.

RSD in kids can be confusing for both the parent and the child. For a parent, watching your child snap between being completely fine and completely devastated can be traumatic. The tiniest correction to their behaviour can cause them to lash out verbally or to withdraw emotionally. The volatility is amplified further when the parent also lives with RSD. The reaction from the child can trigger the parent, leading to a quick escalation.

The first step in managing RSD in children is to be absolutely transparent with them about how their brain works. Have the conversation with them about ADHD and RSD. Let them know that they have many brilliant differences, but that they also have a challenging aspect of their brain called RSD. Make sure to have this conversation with them during a period of peace, when their brain is regulated and clear of any stress, and definitely not during an RSD-triggered event.

Communicate clearly with your child about the way their brain perceives criticism from others and how their emotional responses to these criticisms will be intense. It's important to highlight that there is nothing wrong with this. An ADHD brain experiences the world in a more intense way than many people, and with that come many advantages such as heightened creativity, intuition and problem-solving skills. But this heightened intensity also applies to how we experience rejection, and these strong feelings are simply part of the complete ADHD experience.

The intense feelings towards rejection come from the same part of the brain that enables your child to get excited more intensely than others, to bond with others more quickly and more often, and to think of innovative ideas more often than others. RSD is simply part of the same parcel that makes ADHD brilliant.

The most important thing you can do when your child experiences an RSD episode is *not* to match their energy. As said before, this can be tricky if your child's reaction triggers your RSD – and we will come back to this – but it really helps to remain calm and grounded in the moment. We need to shower our children with praise and reassurance instead. It helps to have a list of qualities to hand that you can use to remind your child of their strengths in moments when they are convincing themselves that they are a terrible person and a huge disappointment to everyone.

An RSD event will fill your child with shame. They will feel awful about themselves in these moments, and despite the high emotion, they will still be embarrassed about how they are acting. Remaining calm in these situations provides your child with a clear message that their big emotions are not a problem for you. You are totally fine with their behaviour because you understand where they are coming from. They are not to be corrected or told off; they are to be listened to and replied to with reassurance. Your child needs to be reminded that, although their feelings are valid and very real, they are not a true representation of how they should feel about themselves.

Ultimately, as parents (and with ourselves) we need to accept and be open to the sensitivities that come with ADHD, rather than feel shameful over it.

RSD can feel very overwhelming and all-consuming. To help settle this feeling, we can remind our child that the strong feelings are temporary and that they will not last for very long. RSD triggers come on very quickly, but they also pass very quickly, and it's a good idea to remind your child of this fact.

You, your RSD and your child

If your child's RSD triggers your own, you can draw on the strategies we discussed earlier in the book for managing your responses, so that you are better placed to help them. It will not be a quick fix – your own RSD runs deep and it will take time to get to a place where you can cope with your own reactions to your child's moments of rejection, frustration and overwhelm. Things won't get better overnight, but with persistence and compassion (for yourself, more than anything) you will find that, often, you can calm your reaction for long enough to deal with theirs.

On reading this chapter, you'll also probably remember the many times you've snapped at them because they can't seem to 'get it together', or they weren't listening, or they were late for school, or they'd lost yet another water bottle – and you'll feel so much pain when you realize the pain that your frustration and criticism must have caused them. Remember, you're learning, too – it's natural to lose patience or not keep your temper, and you didn't know about RSD until now, so don't be too hard on yourself: you can take the next steps today.

Never too much

Louise told me about her child, Rose, who has RSD. Louise knew that she could not be with Rose all the time, so she bought her a teddy bear with the words 'Never Too Much' embroidered on the teddy's chest. Rose slept with the bear and took it everywhere she went. Louise explained to me that the purpose of the teddy was to remind her daughter that she was never 'too much', but also to make it explicitly clear that no topic or issue that the child wanted to speak about was ever too much.

Louise extended this reassurance by sitting down with Rose and writing down all her qualities. 'Kind', 'patient', 'caring', 'funny', 'witty', 'intelligent' and 'creative' were all noted, and each one was turned into its own framed picture. The pictures were then hung around their home. They looked great, but more importantly they acted as a visual reminder for Rose of how brilliant she is. Louise also got each quality engraved on a charm and made a bracelet with them that Rose could wear when they were apart or could carry in her bag when she went to school.

Praise effort, not progress or perfection

It's easy to assume that, as parents, we should celebrate the achievements of our children. While this sentiment is true, it's important that we carefully define the word 'achievement'. An achievement is not simply a completed piece of work; it's every little piece of effort that's been exerted to get to that point. If we only celebrate achievements in their entirety, your child will come to believe that their worth is defined by the quality of their finished projects and not by the effort that's got them there. They will learn that, in order to receive praise, they need to complete something to a perfect standard. We need to congratulate effort and not only outcomes.

Children with ADHD often work ten times harder than other children, but this extra work doesn't always lead to better results. If your child puts in all that extra work, gets a good grade and hears 'You're so clever!' afterwards, they may learn that the only way to receive good praise is to continue to put in all that extra work. To combat this, we need to make sure that we are praising effort just as much as the final output.

Rather than saying, 'Wow, your homework is perfect!' when they show you a finished piece of work, accompany the praise with lots of little pieces of praise throughout the time they are working on the project.

Here are some examples of things you can say to your child while they work:

- 'It's really great how you're trying so hard.'
- 'Well done for finding a solution to that tricky part.'
- 'I'm so impressed with how much effort you're putting into this.'
- 'I'm so proud of how hard you're working.'
- 'Your persistence is incredible!'

If we continually praise the final output and not the effort, over time this can create a fear of failure that may compound and escalate the effects of RSD. This can lead to the child being too fearful to start future projects or turn them into perfectionists as they assign their value to their output and not their effort.

Praising effort in your child rather than focusing on the output will build a growth mindset rather than one that pulls the rug from under them every time they embark on something new or challenging.

Focus on the patterns

Whether it's our child, our partner or a friend in our life who's living with RSD, it helps to remind them to focus on the patterns rather than on one single comment. If they hear only one singular negative comment, they can assume that it is just another person's opinion and therefore not a fact. It should not be taken to mean that there is objectively anything wrong with them.

If, however, they observe a pattern that involves lots of people saying negative things about them, then there may be objective evidence that something they are doing is causing an issue for lots of people. A singular comment can be viewed as an opinion, but lots of comments can be viewed as a signal, pointing at something negative that should probably be addressed.

A person with RSD can use this mindset during moments of heightened emotion. If the comment that has triggered them is a one-off event, if it's something that has not been said to them by anyone else, they can use this signal to remind themselves that the comment is objectively not likely to be true and is more a reflection of the mood of the person who made it. However, if they notice a pattern in how people are speaking to them, if they hear the same criticism from different people, it may be a sign that they need to pay attention to that criticism as it potentially holds some weight. In the absence of a pattern, they can assume the criticism doesn't stand up.

Acknowledging patterns in how people speak to us can really help when we are challenging our negative thought spirals. When we receive a criticism, real or perceived, we should first acknowledge our reaction to it but then compare it to the evidence behind it. Ask yourself the following question:

> Have I heard this criticism from anyone else?

If the answer is 'no' and the behaviour was similar to what you have displayed in front of other people, you can assume the comment reflected more accurately an issue with the person who gave it to you and is not a character flaw of yours.

When faced with a criticism, the RSD brain is quick to tell you that you're not good enough. In these moments, it's useful to fall back on evidence to contradict these thoughts. I use these moments to reread my journal, where I have noted down my positive character traits, personal strengths and qualities.

My journal is full of useful reminders that my self-worth is not tied to the opinions of others. Not everyone we meet has to like us back. If someone sees the world differently than you, has a different taste in food or films than you, or prefers a different style of art than you, it doesn't mean that your preference is less than theirs. Your worldview and personal opinions are in no way inferior to anyone else's. Someone's favourite ice-cream flavour might be chocolate when yours is vanilla. Someone might prefer horror films when you prefer romantic films. Neither choice is right and neither is wrong; nor is one better than the other.

Separate the real from the opinion

Separating objective from subjective feedback is critical for dealing with RSD. We should not take subjective feedback personally because other people are allowed to have a different opinion from us. If I paint my house white and someone says they would have painted it grey, I should not take this personally because subjectively both options are correct. It's my house, so I painted it white. They would have painted it grey if it were theirs. There is no wrong or right answer in the subjective realm.

If you're doing a maths exam, for example, and you write the answer to 10 plus 10 as, objectively that is the wrong answer and someone would be right to correct you. Even if we take this correction as a personal attack, at least we know it was an appropriate correction because we were objectively wrong.

It's easy for the ADHD brain, notably low in self-esteem due to those horrible 20,000 negative criticisms during the early years, to have difficulty separating the objective feedback from the subjective. It can sometimes feel impossible to do this in the moment because RSD is instant and a very visceral reaction. Our nervous system often doesn't have a 'pause' button to press while we decide if a piece of feedback is objective or subjective. However, we can reflect on our triggers and choose retrospectively whether a comment that triggered us was objective or subjective. Doing this regularly will train our nervous system to be able to instantly distinguish between the objective and the subjective in the crucial moments when we are exposed to feedback.

You can't be compatible with everyone and that's okay

The RSD brain will convince you that every time someone gives you a signal that they don't like you it means you are less than, not good enough, or lacking in some sort of moral quality. Being rejected as a person, whether romantically or as a friend, will hurt like hell, but it really helps to remind yourself in these moments that most people are simply not compatible with most other people. It almost never has anything to do with your worth, but means only that you are not a match.

Although we must always pay attention to RSD and acknowledge when it strikes, we also need to be kind to ourselves and remember that not every person will be the right fit for us and neither will we be the right fit for everyone else. People are different, and these differences expose every human being on the planet to a level of criticism that has nothing to do with their worth as a person but is merely a clash of different likes, tastes and values.

Everything someone says, every opinion or piece of criticism, is a reflection of what's going on in that person's head. You can't control people's thoughts and opinions, and neither should you assume they are anything to do with you or a signal that you are in any way 'less than'. We need to allow people to have differing opinions. If someone makes a comment about you, tells you you're not funny, you're lazy or you don't work hard enough, these comments are based on assumptions they have made about you, an estimation of you based on the pieces of information that they have chosen to process, not based in reality.

For example, an onlooker sees you, a person with ADHD and RSD, lying on the sofa, scrolling through social media. They might look at you and, using the information available to them (your physical act of lying on the sofa), call you lazy. This may be very triggering for you because it confirms the judgement of your horrible inner critic who truly believes you are lazy. However, what the onlooker doesn't see is the emotional overwhelm that is causing you such physical paralysis in that moment. You are lying on the sofa because of your mental load, because of ten highly caffeinated squirrels running around in your head, each one pulling you in a different direction, each one with a different thought, each one with a new idea. If the person who called you lazy could see these squirrels, the mental load that is so exhausting, they would have more information and would come to the correct conclusion that you are in fact not lazy but instead overwhelmed.

This shows how a simple RSD-triggering comment such as 'you're lazy' can be delivered without a true picture of what the person in question is truly experiencing. So don't take it as a valid criticism. You cannot pass comment on someone without all the facts, and often a lot of the facts — in this case, the mental overwhelm — are invisible.

In fact, so much of what anyone puts out into the world represents only a tiny percentage of who they really are or what they are thinking in that moment. So much more of who that person really is is hiding away under a mask. This is particularly true for the RSD community who have masked all their lives simply to fit in and avoid judgement. However, it is what people put out into the world that gets criticized, and because what we put out into the world is only a small part of us, we should not take one negative comment to mean it's directed at our whole personality.

Perhaps you go to a party and start chatting to a new person. You may feel you came across too strongly, or maybe you felt a bit socially awkward. This leaves you overthinking the social interaction and worrying that your new friend didn't like you. Let's really think about the version of you that the other person saw – and ask yourself some questions:

- Was I feeling anxious when I was speaking to them?
- Did I go to that party with pre-existing worries in my mind (maybe work or relationship issues)?
- How much energy did I have when I went to the party?
- Did I sleep well the night before?
- Did I say anything that misrepresented who I am as a person?

The answers to these questions will help you to see if you were truly being yourself during the social encounter that triggered your RSD. Maybe you were masking heavily, or maybe you were exhausted because, subconsciously, your mind was preoccupied with another area of your life.

You cannot give any validity to a criticism when the criticism, real or perceived, was aimed at a version of you that was not a true reflection of your true self. Until I learned this, I was always dragging myself down with thoughts such as 'What's wrong with me?' and 'I feel so misunderstood'.

In any example where you meet someone for the first time, rejection or criticism is almost never personal, simply due to the fact that it's very hard to show your personal side when you first meet someone. Being able to separate criticisms that are directed at the masked or 'coping' version of you from the criticisms that are directed at the authentic version of you was a huge step for me in managing RSD. It made me realize that most of the comments I was taking personally were not in fact directed at the person I am underneath.

You don't actually hate me, do you?

RSD is stubborn and it leaves us believing every small criticism that comes our way. A friend could casually say our apartment is 'a bit messy' or our boss might suggest we need to 'work on our presentation style' and we instantly translate these comments into a personal attack. *They think I'm a terrible human being*, we tell ourselves.

However, in these moments we need to pause and ask for clarification. For example, when your friend makes a comment about your apartment being messy (you could try a joke that, in fact, you have an unusual way of managing your wardrobe), rather than assuming it's a personal attack on you, why not ask questions in return, in an attempt to add context to the perceived criticism?

For example, you could ask, 'Have you got any cleaning tips?' or 'I find cleaning really boring, do you feel the same?' When we add questions to comments that we perceive as critical, we are inviting the other person to expand on their comment and, hopefully, to add clarity to their meaning.

Typically, when someone says something that your RSD would take as a rejection, like 'Your apartment looks messy', it means one of two things: either it is a hostile remark, or it is an indirect offer to help. In most cases, and certainly in this example, your friend is making the remark as a light-hearted way of saying, 'Do you need any help cleaning your flat?' However, your RSD mind will jump to the negative option and assume your friend disapproves of your flat, and also of you as a person – and that hurts like hell.

In these moments, it's tempting to let your need for validation take over and ask something like, 'Do you hate me?' In fact, the impulsive aspect of the RSD brain sometimes makes a question like this unavoidable, almost a knee-jerk reaction as you sense the ultimate abandonment that comes with social disapproval. But rather than asking, 'Do you hate me?', you can ask a clarifying question, the answer to which will almost always reassure you that the person who has made a comment does, in fact, not hate you; rather, they are trying to offer you some kind of help.

When human beings want to help someone else, it's common for the person offering help to mention first of all the thing that they think you need help with. So, in the example of the messy apartment, a conversation would usually follow the initial comment and lead to an offer to help.

A typical exchange for someone without RSD (let's call them Sally) might look like this:

> Sally's friend enters her flat and says, 'Your flat looks messy.'
>
> Sally: 'You're right, it does.'
>
> Friend: 'Do you need any help?'
>
> Sally: 'That's kind of you – could you help me grab these things off the floor?'
>
> Friend: 'Don't worry, I know how quickly it all builds up!'

The conversation rapidly progresses towards an offer of help – there is no escalation, there are no emotional outbursts and the friendship is not strained.

Someone with RSD, however, will hear the first comment, the 'Your flat looks messy', and be triggered by it. The rational part of their brain will instantly stop working: they will enter 'fight or flight' mode and be desperate for some form of reassurance. This is where we need to intentionally ask questions that give us reassurance in a controlled way rather than in one that escalates the conversation into rage or sadness.

Here are some questions you can ask to help you determine whether the person is being hostile or just trying to offer you support:

- Can you clarify what you meant by that?
- Do you have any advice that might help me?
- Do you have any experience in this area?
- Can you help me understand why you came to this conclusion?
- Was there anything specific that caused you to say that?

The RSD brain is fast to assume that a comment is a personal attack, but often that is simply how your negativity-biased brain has perceived it. Asking clarifying questions can get to the bottom of it and find out whether there were any malicious intentions behind the comment.

Ask for feedback

The RSD brain receives vague, blunt feedback like a punch to the gut and instantly uses it to validate every negative narrative a person holds about themselves. For example, your boss could write 'Needs improvement, come and see me' in response to a piece of work you've submitted. Your brain translates that into 'Come and see me because I know you're useless.'

In almost all cases, situations like this will end up with the boss giving you more specific feedback that contradicts your initial fear. The meeting with your boss could conclude with them saying how brilliant you are and, with a little bit of a nudge in the right direction, your amazing potential will come to fruition. They can see how much effort you put into your work and how, with just a few minor tweaks, it would have been an extraordinary piece of work. In other words, your boss thinks you're great and simply wanted to offer you some minor feedback as they can see a few areas where you could improve.

This cycle is common for someone with RSD. We get an invitation from someone to meet them so they can give us feedback. For example, your boss asks for a quick chat, or your partner says they would like to speak to you when you get home from work. It is the gap between the invitation to meet and the actual meeting that is torturous for the RSD brain. The anticipation of an inevitable rejection leaves us in a state of heightened anxiety, unable to sleep or think about anything other than the fact that our world is about to implode. Of course, this rarely happens, and the meeting that we so fear turns out to be nothing more than a casual catch-up or such a benign affair that it makes our earlier fears seem almost funny.

However, as often as this happens, it doesn't mean we should simply tolerate the emotional roller coaster that comes with the cycle, especially if there are a number of days between the invitation to meet and the meeting itself. We can make accommodations for ourselves by asking for feedback immediately, or at least by asking for more context as to what the meeting will involve.

7

Dark places
RSD and the worst of times

> This chapter contains discussion of suicidal thoughts, which some readers may find distressing. If you are struggling or feel overwhelmed, please remember that you are not alone – and there is support available.
>
> If you need help, consider reaching out to a mental health professional, someone you trust, or one of the following crisis support services:
>
> **United States**
>
> 988 Suicide & Crisis Lifeline – Call or text 988 or visit 988lifeline.org
>
> **United Kingdom**
>
> Samaritans – Call 116 123 or visit samaritans.org (available 24/7)

> **Australia**
>
> Lifeline Australia – Call 13 11 14 or visit lifeline.org.au
>
> Beyond Blue – Call 1300 22 4636 or visit beyondblue.org.au
>
> **New Zealand**
>
> Lifeline New Zealand – Call 0800 543 354 or text HELP (4357)
>
> Need to Talk? – Call or text 1737 (free 24/7 support)
>
> If you need to take some time out or skip this section, that's fine. Your wellbeing is all that matters.

The compounding effect of RSD and jealousy

I wasn't sure whether to write this section of the book because I didn't want to make bold claims without evidence, but the truth is that RSD is so woefully misunderstood and under-researched, there aren't any official studies to source. However, I have spoken to hundreds of individuals with RSD and built a community of over 5 million people with RSD, and I'm passionate about raising awareness on this topic.

RSD has the potential to intensify the emotion of jealousy to the point where the individual could even experience suicidal ideation. The hyper-focus that can make ADHD so powerful is also what makes this scenario likely. The RSD brain will see threats everywhere, the impulsivity will jump to the worst-case scenario, and the hyper-focus will zoom in on the detail and create a torturous thought loop.

I get very frustrated when people dismiss 'jealousy' as a normal emotion that 'everyone experiences'. Not everyone experiences jealousy in the same way. For someone with RSD, the intensity, frequency and duration of the emotion are significantly heightened.

The RSD mind can't tell the difference between a real threat to your relationship and a perceived threat to your relationship, so the intense feelings will be triggered all the time. The ruminating is also strengthened by the person's ability to reflect on their past behaviour, which tells them that the thing that will soothe the suffering is also the thing that will make it worse. For example, you may be hyper-focusing on the jealous emotions, consumed by them, and you want to confront your partner about their infidelity, but your past behaviour has taught you that, in doing just that, you will push your partner away, further increasing the chances of your suspicions becoming a reality. This makes the person with RSD feel trapped.

In that situation, you try your hardest to mask your suffering, to act like you're not bothered by the fact that your partner was talking to someone at the party last week, but you can't do it. The mask you are wearing turns you into someone who is able to be with their partner but not be present because your mind is racing with suppressed sadness and rage. You are convinced your partner is being stolen by an outside threat and you want to confront them about it, but you can't because last time you did that it only heightened the paranoia and now you think they think you're crazy.

So you bury the intense feelings. This takes a huge amount of willpower and energy, which means you don't have energy left over to act like the 'normal' version of you. This will be met with comments from your partner like 'Are you okay?' or 'You seem off today', and you will lie and say, 'Yes, I'm fine.' You want to shout at them or punish them for cheating on you, you want to cut them out of your life in revenge, but you know deep down that you're being irrational. The RSD will be fighting any rationality, insisting that your partner is thinking about someone else.

RSD intensifies the feeling of jealousy because every single chest-tightening suspicion is amplified by an internal monologue that tells you you're not 'enough'. You tell yourself, 'Of course they are cheating on me because I'm a useless person and they can do so much better' and 'Who can blame them? It was only a matter of time before they figured out how useless I am.' What started out as a normal level of suspicion and jealousy turns into an all-consuming spiral of rumination from which it feels that there's no escape.

My own spiral of jealous despair

One of my closest friends, Leo, has an ADHD diagnosis. Recently, his partner, Asha, came home from a party and told him that a man had been flirting with her. She told him she had immediately pushed back with 'I'm married!' and he stopped. Leo was filled with sadness and rage. He went inside himself and didn't speak for three hours other than to answer 'Yes, fine' when she asked him if he was okay. His mind had constructed a narrative and was hyper-focused on the imagery of the encounter.

Leo says the narrative went like this: Asha was flirting back, they were laughing together, dancing together, he was charming and persuasive and temporarily made her forget all about her useless husband.

Leo told me, 'Even though I wanted to act as if it wasn't eating me up all the time, the demon inside me took over and didn't let me think about anything else other than the picture I had created inside my head. I was so hyper-focused on the imagery, I felt I was actually there, watching it all happening in front of me, like I could reach out and punch the other man and stop it all, but I couldn't: I was forced to endure the imagery, like being chained to a cinema seat with my eyes sewn open as it repeated itself over and over.'

It's hard to remain stoic in your relationship when you have thousands of criticisms reminding you that you're not worthy of being loved. Even as an expert 'masker', the ability to hide the feelings of jealousy just isn't there and the effort to prove otherwise is more transparent than cellophane. It's also hard to be vulnerable, especially as a man, because you worry you will be perceived as weak if you open up about your insecurity.

The compounding effect of RSD and jealousy created a burden inside Leo's head that he couldn't escape – each route out led to negative outcomes. Half of him wanted to sabotage the relationship and jump ship, while the other half wanted to keep it all inside and remain quiet, both of which might be soothing in the short term but only leads to heightened paranoia in the long term. After all, there are only so many times you can impulsively leave your partner and get back together before they get fed up and leave you. And there are only so many times you can pretend to be okay before they get fed up and leave you anyway.

Over the next few days, said Leo, there were moments of respite when hanging out with Asha – being in her company and reading her body language was enough to reassure him that she did indeed love him. It was when he was alone that the thoughts escalated into full-blown illusions of infidelity. He was unable to focus on anything other than the image of Asha and the other man together.

He told me that he tried to separate the irrational suspicion from the rational thought. There was zero evidence to prove Asha was flirting back, so why did he feel so convinced that she was? It's hard to rely solely on the external tangible evidence, or lack thereof, when RSD creates a feeling inside you that feels so real. The imagery inside Leo's head was no less vivid than the experience of seeing the flirting in front of him. In fact, seeing it in real life would have been easier for Leo, because the associated pain would have lasted only as long as the flirting did. But this was inside his head and it followed him wherever he went: when he was driving, when he was working, when he was trying to sleep. It was relentless and, with each replay, Leo's malicious mind added another gut-twisting detail.

Leo's face darkened as he continued to tell me how events unwound. 'I never understood how someone could be in such a dark place that the best possible option for them was to take their own life. I mean, surely there's no pain great enough to justify such a permanent decision? But right at that point, I could empathize – this pain felt like a hundred swords stabbing me in the heart.

'It was the first time in my life that I had walked past one of those train station posters, the suicide prevention ones that say "It's OK to not be OK", and properly understood the importance of them.'

Adults with ADHD are five times more likely to take their own life, and this is one of the reasons why: the compounding effect of RSD and jealousy.

Never worry alone – men need to talk

I could tell how difficult it was for Leo to talk about this. I told him to carry on, if he thought he could. He continued: 'I always felt I couldn't talk about my feelings, as if it would make me vulnerable, a weakling, mocked by everyone around me. And if I'm really honest, it never before really felt like I needed to. But here I was, going about my daily life as a shell, physically present but emotionally vacant. It felt like I had tripped up and fallen in a deep hole and none of the things that brought me joy had fallen in with me.

'There were fleeting moments of solitude when something happened that distracted me from the thoughts, like when I nearly crashed my car because I wasn't able to focus on the road and someone shouted abuse at me; or when a pushy charity fundraiser was trying to stop me walking so they could convince me to donate. Normally these things are really unpleasant for me because I hate confrontation, but I actually welcomed them as a distraction, because they pulled my hyper-focus away from my incredibly dark thoughts for a bit.

'I picked up my phone and looked through my recent WhatsApp messages, searching for the name of someone I felt I could share my experience with, someone who understood the complexities of RSD and how hyperfocus can turn a benign thought into a high-definition nightmare.'

And, thank goodness, Leo found someone to talk to – me – and it was the catalyst for our brave, honest conversation.

He told me that he had typed out a message, deleted it, typed it again, deleted it again, over and over. 'It was like

I was holding a fizzy-drink bottle and every word I typed gave the bottle another shake, increasing the pressure, and tapping the "Send" button would be the final shake to cause the bottle cap to pop off, showering me with shame. It felt impossible. Men don't talk about their feelings, they bottle them up and get on with life, don't they?'

He said he was scared I'd think he was stupid.

He said he was scared I'd tell him to 'man up'.

But when he tapped 'Send', the bottle cap didn't pop off. In fact, it rotated, just enough to allow the pressure to fizz out and return to a stable level. I replied straight away, of course, and over the next few days we sent numerous messages back and forth, opening up about our experiences. We even went on a two-hour walk around London during which we spoke freely about men's mental health.

Leo's courage in speaking openly about a topic about which he held so much shame allowed me to realize that so many of us experience these intense feelings. These conversations made both of us feel seen, and validated, and that was enough to turn down the pressure cooker inside not only Leo's head, but in my own.

Please talk to someone

I was recently stopped in a train station by a man. He was well dressed and looked very confident. He said, 'Are you the guy who talks about ADHD?'

I said, 'Yes.'

We quickly arrived at the topic of RSD and he told me it was controlling his life and making work very challenging, and that he was struggling with his mental health because of it.

We only spoke for a few minutes, but it was clear that this man was not as confident or as happy as he originally appeared.

We shook hands a few minutes later, smiled and said, 'Take care', but it was clear we had had a profound impact on each other's day, simply by having a two-minute chat.

It's good to chat about mental health, even if it's for two minutes in a train station. We reminded each other that we're not alone, and we're not crazy.

Thank you for stopping me.

I am in a privileged position where I'm able to meet lots of ADHD men. I often speak at events and meet people afterwards, or sometimes someone will recognize me when I'm out and about and say hello, like the man above. Social encounters with other men always follow the same pattern:

A The man looks confident.

B The man lets his guard down.

C The man reveals that he is suffering with his mental health.

When we dig a little bit deeper on the third point, it is almost always related to RSD. I spoke to one man who revealed he was drinking every night because RSD was isolating him from his family. It was causing raging arguments. Family gatherings didn't happen anymore because he was being too stubborn to apologize after a fallout. He was battling with the guilt of being 'the problem' but also grappling with an internal desire not to back down, even though he knew it was irrational. He said, 'Just because the rage is disproportionate to the thing my cousin said, it doesn't make the feeling any less visceral. It's brutal and makes me hate him, even though I know he hasn't done anything wrong, even if my subconscious RSD brain feels like he has hurt me deliberately and wants to punish him for making me feel this way. The only thing that quietens my racing, shame-filled mind is alcohol.'

How to open up about RSD

Masking, the very survival instinct that makes people with ADHD so capable to adapt is also the chink in their armour because it creates such a disconnect between how they feel inside and how they present externally. As said previously, people with ADHD are chameleons, able to bury their true feelings under a flawless version of themselves, one where no one suspects they might be struggling.

However, with an awareness of RSD comes an understanding of the importance of opening up about it. Whether it be in our romantic relationships or in other areas of our lives, we need to intentionally make time to communicate with our loved ones about RSD.

It can be difficult for someone who doesn't understand RSD to comprehend the actions of someone who does live with it. For this reason, we need to communicate openly with our partners, families and friends about RSD in a way that allows them to understand us through a lens of curiosity rather than judgement. Sit down with your loved one at a convenient time and walk them through what RSD means, what causes it and the likely issues it may bring up in your relationship.

Opening up about RSD is not a one-off conversation. The topic can be revisited at times when either you or your loved one, friend or family member experience behaviour that is confusing to one or both of you. It's a good idea, often after an event that has triggered you, to revise the conversation of RSD so that clarity can be added to the interaction you just had together as a couple.

Ultimately, sharing your RSD with those closest to you is very important because it adds vital context to otherwise confusing behaviour. You're not too sensitive, neither are you crazy – you're living with RSD and both your relationships and your mental health will benefit when not only you but also the people around you have an understanding of it.

8

The best and only way to truly beat RSD

A five-step plan

You deserve respect, but you have none for yourself

As I travelled through my ADHD awareness journey, as I reflected on my own 36 years on this planet, as I spoke to over 10,000 late-diagnosed ADHD men and women in person and over 300 world-renowned experts on my podcast, *ADHD Chatter*, I came to realize my biggest learning yet: there is a clear link between how much someone respects themselves and how vulnerable they are to RSD.

People with ADHD spend much of their life masking, pretending to be someone they are not in order to fit in. They are experts at doing this, and it works well in the short term. However, over time, it creates an adult who doesn't truly understand who they really are.

With a higher propensity towards masking comes a double helping of self-disrespect. Deep down, we know how much we are betraying our true self (we are reminded of it every day by our constant state of exhaustion), and this betrayal leads to a feeling of contempt towards ourselves because we have ignored and disrespected our true self for so many years.

We buried it under layers of masking and years of playing a character. Our true self stopped getting any attention at the point at which we learned that we got higher rates of social approval when we played make-believe.

In the same way as we feel anger towards a romantic partner when they disrespect the relationship, we feel angrier about ourselves the more we hide and disrespect our true self. The disrespect builds up so gradually over the years that we don't even notice it until it becomes obvious when we realize how sensitive we are to criticism. A person who truly respects themselves is resilient to criticism because they know their worth; they don't hold the opinions of others over that sense of self-belief.

Many people who begin to mask from a young age develop a false self, a version of them that operates well in the world but is miles away from a true representation of who they really are underneath. This leads to them feeling like a child trapped in an adult's body. Regardless of how old the person with RSD is, they feel like a naive little child, bumbling through life, finding unique ways to do things that appear silly to others, and feeling like they are always about to be corrected by a proper adult. These corrections feel crushing because they instantly confirm the internal whispering voice that tells us we're useless and someone is going to find out soon enough.

With each little extra criticism we received when we were younger comes a little extra confirmation that we need to change who we are in order to be accepted by the world. People with RSD have 20,000 more reasons to mask than your average person, which means we have 20,000 more reasons not to respect ourselves. A lack of self-respect is the ultimate price we pay to not feel the sting of social disapproval. This lack of self-respect leaves us wide open to RSD, and RSD is the great equalizer, something that turns the effort of masking into a false economy. The soothing experience of being able to navigate life without feeling different is immediately overridden by the crushing experience of someone seeing through our mask.

RSD dies when self-respect grows

All this has led me to spend endless hours reflecting on the obvious question: 'How can someone who has disrespected themselves for such a long time start to rebuild their self-respect?'

The answer has become clear over time. It is a five-step process.

Step 1: Reconnect with your true self

We can't begin to build our self-respect if we continue to disrespect our true self on a daily basis. Our true self needs to be reassured that it is enough and that it is wanted by the world. However, many people with ADHD don't recognize which parts of their behaviour are a mask and which are authentic.

In order to understand which parts of our behaviour are authentically us, we can do basic self-awareness exercises. As often as you can, use your journal to ask yourself some simple questions such as:

- 'What did I do today that made me truly happy?'
- 'What did I do today that felt like I was exhaustingly dragging myself through it?'
- 'What conversations did I walk away from feeling energized?'
- 'What conversations did I walk away from feeling drained?'

You can refer back to Chapter 5 where I explored building self-awareness.

Step 2: Celebrate your achievements

People with ADHD rarely give themselves credit for their own accomplishments. We move on to the next thing really quickly and don't give ourselves time to store the memory of our achievements.

Intentionally celebrating our accomplishments gives us the self-worth we desperately need and makes us feel like less of a fraud when someone points them out.

I use the weekends to reflect on the week that has just passed and journal on what went well. Did I successfully complete a piece of work? Did I have a social encounter that felt amazing? Did I exercise, or make something really nice to eat? All these are accomplishments that, if not intentionally celebrated, will fade into our past and out of our memory. When we write down our accomplishments, we etch the memory of them into our subconscious – we are storing evidence of them in our mind, ready to use the evidence in moments when we need to be reminded of how amazing we are.

As our pile of evidence increases, so will our self-respect. And as this happens, we will stop looking at everyone else as superior to us and simply as another human being, on their own journey. Building your self-respect will simultaneously bring others down from the pedestals that they are placed on by so many people with ADHD, and this in turn will make their words or criticism less meaningful to you, and make you more resilient to RSD.

Step 3: 'This is not who I am'

Having self-respect comes with a level of self-awareness that accepts that you have both negative and positive traits. No one is perfect, and a well-rounded and self-aware person will know what their strengths and weaknesses are.

RSD is made weaker when we embrace our weaknesses and recognize that they are part of us but not the whole of us. Every human on this planet struggles with some things, and you are no different from everyone else in that regard. The difference between you and neurotypical people is that you were exposed to so many more negative messages, which have led you to believe that your weaknesses are all of you. You associate your challenges with your entire worth as a human being, so when someone criticizes or comments on one of your challenges it feels like your entire existence is being attacked.

You can remind yourself that your weaknesses are only a part of you by keeping a mantra written down in a place where you can see it daily. It could be a poster next to your bed that says 'Your challenges do not define you', or a bracelet that says, 'I am not my challenges'. This will give you a daily prompt to be nicer to yourself. It also acts as a visual reminder in moments when someone has criticized you.

I like to think of these visual reminders as similar to a hook that a rock climber hammers into a mountain face at various points of their climb. As the climber progresses up the mountain, they insert a safety hook into the wall, something to keep them safe should they lose their grip and fall. It ensures they fall only a certain distance, no farther than the height of their closest hook.

When we experience moments of doubt, or in moments where someone has criticized us, rather than spiral all the way down to a point of crisis, we can use our visual reminders in the same way that a climber uses safety hooks, to stop us falling too far and allow us to restabilize ourselves before we carry on with our day.

Step 4: Treat yourself like a child

So much of the shame associated with RSD comes from the fact that we never learned how to be compassionate towards ourselves; we were never given the opportunity to be truly vulnerable with ourselves because we were trying so hard to hide who we are. When someone with RSD becomes an adult, the mask is very thick. It has been built up slowly over many years. It is rarely obvious how much of a disconnect there is between the outward version of us and the true version of us within.

Our body grew into an adult but our psyche remained that of a child. Those 20,000 negative comments we received made us feel truly broken, so we abandoned ourselves in favour of a version that felt less painful and less susceptible to criticism. However, because this process happened over many years, we sometimes fail to recognize how much we have abandoned our inner child. We believe we are the adult we pretend to be and we hold ourselves to the standard that we think adults should be held to.

People with RSD need to treat themselves in the same way they feel children should be treated because, deep down, the part of us that is hurting is a human being who was never allowed to grow up. Many people with RSD are incredibly mean to themselves, they hold themselves to impossibly high standards, and they forget to give themselves the compassion that a neglected child is entitled to.

We can be very good at this when a friend comes to us with a problem. In fact, we can shower them with empathy and kindness. But we don't find it easy to treat ourselves with the same level of support. Our internal bias towards negativity can mean we talk to ourselves harshly, which can be very damaging when we are already at a low point.

Try to catch your negative self-talk and imagine how you would speak to a child under similar circumstances. For example, you might have burned dinner and your mind is torturing you with thoughts like 'You're useless at everything!' or 'Why can't I do anything right?' Picture a child making a similar mistake – can you see a difference in the way you would talk to them? Suddenly, 'You're useless at everything!' turns into 'What can you learn from this experience?' or 'What went wrong and how can you improve?'

Reframing our negative self-talk in this way enables us to replace it with a different narrative, one that reminds us that we are a work in progress. A narrative that says we need more compassion than most because we grew up in a world that made us believe we were not enough.

A narrative that belongs to someone who is only just discovering that they have been pretending to be an adult, holding themselves to a high standard, when in fact their subconscious, the part of them that's so easily triggered by the tiniest criticism, is just a child who was hidden by years of masking.

Step 5: Honour your self-promises

Many people with RSD have moments of great ambition. We can think of an amazing business idea, or we can plan a great trip. We can set incredible goals, or personal development milestones. For example, you could jump out of bed at 3 a.m. with an amazing idea for a new business venture, or you could suddenly buy a canoe with a plan to paddle down a local river.

These random acts often come out of nowhere and society has labelled them as 'impulsive', yet we need to listen to these urges, as they are often our true self, our inner child, giving us a reminder of what truly excites us. In these moments, it's easy to promise ourselves that we will follow through with the business idea, or that we definitely won't get distracted and forget about the canoe adventure. However, as time passes and our internal critic has time to creep in and convince us that we're not good enough to start that business or that going on a random canoe adventure would be too risky, we give up on our ambitious plans and place them in a graveyard of abandoned ideas. RSD tells us we are not prepared enough to start, nowhere near perfect enough for people not to laugh at us, so we throw in the towel and break the promise we made to that inner child. We quit.

However, every time we break our self-promises, every time we abandon an idea that we were excited for not long before, we invalidate the ambitions of our true self and give our inner child a signal that confirms their belief that they are inadequate.

That isn't to say that every new idea that pops into your mind needs to be considered as a signal towards your true passions. However, any new idea that makes you genuinely excited should be examined as a potential clue to your unmasked identity, something you need to paint a clear picture of as you journey towards building your self-respect and disarming RSD.

Final thoughts

RSD is by far the worst part of living with ADHD. It feels like my whole life has been unknowingly steered by a crippling fear of being triggered by a real or a perceived rejection. This fear has dominated my relationships, my career and my ability to maintain friendships.

It's easy to assume that the pain of rejection is the hardest part of RSD, that the physical and visceral reaction to criticism is the most damaging aspect of this horrible phenomenon.

I don't think that's true.

I think the most damaging aspect of RSD is the coping strategies we subconsciously put in place in our efforts to protect ourselves from the pain of being rejected: the people-pleasing, the perfectionism, the overworking, the saying 'yes' to advances when you really mean 'no', the putting everyone else's needs and wants ahead of your own, and the missed opportunities that pass us by because we don't feel we're good enough or worthy enough to give them a go.

So many of the diagnostic criteria for ADHD are focused on executive function challenges such as disorganization and forgetfulness. But I believe ADHD to be a condition with significant emotional consequences, caused by the many corrective messages we receive because of our differences.

It's so important for people who live with RSD to understand that they are living with RSD. Without this awareness, and armed only with a barrage of comments like 'You're so sensitive', we are doomed to live under the impression that we are hugely dysfunctional and not equipped for everyday life.

We are not dysfunctional; neither are we 'too sensitive'. We are human beings who were born with a set of differences that are misaligned with society's expectations of normality and because of this we came under attack 20,000 times from horrible reminders of those differences.

Our differences do not mean we are broken – there is nothing wrong or defective about us. We are brilliant in so many ways – we offer value, innovation and joy to the world. However, those 20,000 criticisms have left us feeling full of shame, that we are not good enough or deserving of any praise.

It broke my heart when I realized how many people are living their life without an understanding of RSD, how many people are going about their day to day with a fear of criticism, not knowing where it comes from. Their single explanation is that it must mean there is a moral defect in them somehow, as if they are in some way less than others.

I truly hope that picking up this book will be the first of many steps in your understanding of RSD, how it has impacted your life and where it comes from. My wish is that you will walk away from these pages with a new understanding of yourself, one that encompasses the entire ADHD experience and not one that simply defines ADHD as a disorder of attention.

My ambition is that this knowledge is shared with children as early as possible so that they do not grow up believing they are broken. I sometimes fantasize about having a time machine in which I could travel back in time, put my arms around the younger version of me, and reassure him that he's not broken, simply different, and that he has always been enough. But of course I can't do that. What I can do, however, is hope the words on these pages reach far and wide and end up in the hands or ears of a younger version of me, and that these words enable them to start their journey of self-awareness and self-compassion earlier than many of us did.

For all the late-diagnosed ADHD people reading this book, I can only apologize for not bringing an understanding of RSD to you sooner. I feel genuinely tearful when I imagine all the pain you have experienced, so much of which could have been reduced if there had been an understanding of RSD in your life sooner. I rage at the medical community for not giving us answers more quickly.

I applaud every single late-diagnosed ADHD adult who has battled tooth and nail through relationships, jobs and friendships, confused as yet another relationship, job or friendship succumbed to an inexplicable rage or a sadness-induced distance.

I cheer every single person who has ever looked confused as yet another person referred to them as Dr Jekyll and Mr Hyde, or said they have extreme mood swings.

I extend a hug to each and every person who has laid in bed all night crying, frustrated at their inability to understand why they can't maintain friendships like a 'normal' person.

I send my love to all the late-diagnosed ADHD adults who have burst into tears while looking at a picture (without them in it) of a group of friends on holiday together, shopping together or on a night out together.

I wish comfort for everyone who's ever been crippled with anxiety after someone has asked them for a 'quick chat', and to anyone who's ever been made to feel crazy by their doctor.

RSD is very real and it explains so much of our behaviour. The sooner we are able to understand and embrace RSD as a part of us, the sooner we are able to acknowledge that it is not all of us. As soon as we're able to see it for what it is – a consequence of our differences – we're able to hold it out in front of us, disarm it and take away its power.

And in those moments when RSD wins and we feel those intense feelings, we're able to remind ourselves that this is not us, this is RSD, and RSD is not our fault. We can accept the strong feelings into our life with a new viewpoint, one that now understands where these feelings are coming from, one that now understands that they will pass very soon.

RSD is a part of the ADHD experience; it's like wearing your heart outside of your chest. You feel the world very intensely, you're able to love more passionately and give more of yourself to others, but it also leaves you exposed to really feeling every single bump in the road. However, when we understand RSD, we're able to add an extra layer of armour to our toolkit, one that protects us from our own fragility and allows us to lean into the many qualities that come from having a brain that experiences the world a little bit differently.

I hope this book has helped you on your journey of self-awareness and enabled you to recognize that you are not broken, not too sensitive, not too much, not too energetic, not too anything. You are different, you live with RSD, and you have always been enough.

Acknowledgements

A special thank you to my family for your continued support, my partner for your encouragement, and my editor for your hard work.

Index

achievements, celebrating, 218–219
alcohol, 124–128, 213
amygdala, 5–6
anger, 108–109, 119
apologizing, 142, 149–150
asking for help, 182–183 *see also* support
assumptions, 119–120
attention deficit hyperactivity disorder (ADHD)
    diagnosis, 20–21
    early childhood memories, 2–3
    pain of rejection, xvi, 5
authentic self
    finding, 171–172, 218
    with friends, 14–15
    in relationships, 66–72, 80–81
    see also masking
avoiding RSD, 22–23

balloon analogy, 115–122
behaviour patterns, 16–18
blame, accepting, 39–40
body image, 52–53
boundaries, 32–34, 46–49
brain functions, 5–6, 131
burnout, 53–55

caring for someone with RSD, 185–192
childhood
    criticism in, xvi–xvii, 3–4
    early memories of ADHD, 2–3
children with RSD, 185–187, 188–190
    triggering your own RSD, 187–188
communication, 209–214
    with children, 186
    among friends, 111–114
    being seen as rude, 166
    talking to your partner about RSD, 86–87
    unambiguous, 92, 98–99
    in the workplace, 151–155

community, 128–129
compatibility, 194–198
confrontation, 12, 41–42, 77
constructive feedback, 12–13 *see also* criticism
criticism
    about minor things, 120–122
    based on subjective opinions, 193–198
    in childhood, xvi–xvii, 3–4
    extreme response to, 25–26
    responding to, 198–201
    towards someone with RSD, 191–192
    in the workplace, 156–157, 179–181

dangerous situations, 36
deference, 34–36
diagnosis of ADHD, 20–21
disproportionate responses to rejection, 8–9, 138–139
dopamine, 5–6

emails, 145–147
emotions
    all or nothing, 24–26
    high emotional states, 95–96
    hypersensitivity to, 82–86
    intensity, 9, 101–102
    managing at work, 175–179
equality, 167
evolutionary reason for rejection sensitivity, 6–7
exhaustion, 53–55, 167–168

false narratives, 117–118
feedback, 12–13
    seeing the negative in, 201–202
    see also criticism
fight or flight mode, 76
forgetfulness, 8

friends
    also with RSD, 116–117
    becoming authentic around, 14–15
    communication strategies,
        111–114
    drifting away from, 13
    taking advantage of your
        people-pleasing, 36–39

Gupwell, Matt, 132–134

Hallowell, Dr Ned, 15–16
helplines, 203–204
hyper-focus, 56, 177
hypersensitivity, 82–86, 95

impulsiveness, 104–109, 126–128, 139,
    175–179
injustice, 165–166
inner child, 42–45, 221–223
innovative ideas, 180–181, 223–224
intentions of others, 119–120
intrusive thoughts, 83–86

jealousy, 204–209
journalling, 144, 181, 192–193

letter to yourself, 130
long-lasting responses to rejection, 9
love languages, 64–66

mantra, 162–163
masking, 42–43, 50–51, 213, 216–217
    in relationships, 69, 79–80
    in the workplace, 180
    see also authentic self
meetings, trepidation about, 202
men, importance of talking about RSD,
    209–213
misunderstandings, 12, 78, 110–112
    in relationships, 66, 88–89

negative thinking, 64–66, 74–76,
    83, 136
    about feedback, 201–202
    about yourself, 222

nervous system, 143
neurotransmitters, 5–6
norepinephrine, 5–6

object permanence, 88–89
oversharing, 11
overwhelm, 58, 176
overworking, 164–165

pain of rejection, xvi, 5
    physical, 10
parenting a child with RSD, 185–187,
    188–190
    when you have RSD too, 187–188
people-pleasing, xviii, 30–32
    accepting blame, 39–40
    avoiding confrontation, 41–42
    deference to others' perspectives,
        34–36
    establishing boundaries, 32–34,
        46–49
    suppressing values, 42–45
    taking on emotional burdens, 36–39
    in the workplace, 159–165, 169–173
perfectionism, 38, 49–52
    all or nothing, 58–60
    breaking down tasks, 57–58
    in children, 189–190
    in the workplace, 159–160
physical appearance, 52–53
physical pain, 10
prefrontal cortex, 5–6
problem-solving, 179–180, 181

recharging, 114–115
rejection
    brain's response to, 5–6
    pain of *see* pain of rejection
rejection sensitivity dysphoria (RSD)
    avoidance, 77–81
    awareness of, 46
    community, 128–129, 151–152
    definition, xiv–xv, 23–24
relationships
    being your authentic self, 66–72,
        80–81

escalation of arguments, 93–94
feeling they are undeserved, 97–98
hypersensitivity in, 82–86
intensity of feelings, 62–64
jealousy, 204–209
love languages, 64–66
masking, 69, 79–80
misinterpretation of partner's mood, 66, 88–89
need for validation in, 72–74
negative thinking, 64–66, 74–76, 83
parent–child dynamic, 82–83
from the past, 14
RSD avoidance, 77–81
talking about RSD, 86–87
triggers for RSD, 89–96
removing yourself
from impulsive reactions, 107–109
for time out, 137–138
from triggers, 102–103
roller coaster analogy, 123–124
rudeness, 166

self-awareness, 60, 219–220
self-confidence, 79
self-criticism, 52–53
self-esteem, 97–98
self-respect, 215–224
self-sabotage, 19–20
self-validation, 103–104
shame, 19, 21–22, 211
  caused by perfectionism, 57
silence, 149
snapping at people, 175–179, 188
social exclusion, 12–13
social hierarchies, 166–167
social interaction, 139–140
  compatibility, 194–198
  indifference of other people, 134–137
  rehearsing, 78–79
  rudeness, 166
social media, 11
spirals, 83–86
  avoiding, 122–123, 141–144

state change strategy, 93–94, 96
STOP strategy, 132–134
subjective feedback, 193–198
suicidal thoughts, 205, 209
support, 128–129
  helplines, 203–204
  offers of help, 198–201
  seeking, 209–214
  in the workplace, 151–155, 182–183
suppression
  of self in relationships, 78
  of values, 42–45
survival instinct, 6–7

talking about RSD, 209–214
  to children, 186
  in relationships, 86–87
'them' problems, 173–174
time, perception of, 88–89
tone of voice, 14
triggers, 11, 89–96
  removing yourself from, 102–103

validation, 72–74, 103–104, 116
values, suppressing, 42–45
visual reminders, 220–221

withdrawal, emotional, 95–96
workplace
  asking for help, 182–183
  authentic self in, 171–172
  behaviour of others, 173–174
  criticism, 156–157, 179–181
  emails, 145–147
  getting fired, 165–167
  interaction with colleagues, 147–150
  managing emotions, 175–179
  masking, 180
  overcompensation, 169
  people-pleasing, 159–165, 169–173
  quitting your job, 155, 157–158
  talking about RSD, 151–155
writing down thoughts, 122–123
  journalling, 144, 181, 192–193
  letter to yourself, 130

Join the Sheldon Press community today, sign up for our newsletter!

- Select a **FREE eBook** or extract to read upon joining

- Keep up with our latest publishing and exciting author news

- Be the first to hear about book prize draws, free extracts, and upcoming author events

Simply scan the QR code below or head to www.sheldonpress.co.uk/newsletter to sign up.

RAISING READERS
Books Build Bright Futures

Dear Reader,

We'd love your attention for one more page to tell you about the crisis in children's reading, and what we can all do.

Studies have shown that reading for fun is the **single biggest predictor of a child's future life chances** – more than family circumstance, parents' educational background or income. It improves academic results, mental health, wealth, communication skills, ambition and happiness.[1]

The number of children reading for fun is in rapid decline. Young people have a lot of competition for their time. In 2024, 1 in 10 children and young people in the UK aged 5 to 18 did not own a single book at home.[2]

Hachette works extensively with schools, libraries and literacy charities, but here are some ways we can all raise more readers:

- Reading to children for just 10 minutes a day makes a difference
- Don't give up if children aren't regular readers – there will be books for them!
- Visit bookshops and libraries to get recommendations
- Encourage them to listen to audiobooks
- Support school libraries
- Give books as gifts

There's a lot more information about how to encourage children to read on our website: **www.RaisingReaders.co.uk**

Thank you for reading.

hachette UK

---

[1] OECD, '21st-Century Readers: Developing Literacy Skills in a Digital World', 2021, https://www.oecd.org/en/publications/21st-century-readers_a83d84cb-en.html

[2] National Literacy Trust, 'Book Ownership in 2024', November 2024, https://literacytrust.org.uk/research-services/research-reports/book-ownership-in-2024